Petroleum Fluid Phase Behavior

Taylor and Francis Series Emerging Trends & Technologies in Petroleum Engineering

Abhijit Y. Dandekar
University of Alaska, Fairbanks, AK, USA

Wax Deposition: Experimental Characterizations, Theoretical Modeling, and Field Practices
Zhenyu Huang, Sheng Zheng, and H. Scott Fogler

Hydraulic Fracturing
Michael Berry Smith and Carl Montgomery

Unconventional Oil and Gas Resources Exploitation and Development
Usman Ahmed and D. Nathan Meehan

The Legal Aspects of Industrial Hygiene and Safety
Kurt W. Dreger

Petroleum Fluid Phase Behavior: Characterization, Processes, and Applications
Raj Deo Tewari, Abhijit Y. Dandekar, and Jaime Moreno Ortiz

For further information about this series, please visit their website: www.crcpress.com

Petroleum Fluid Phase Behavior

Characterization, Processes, and Applications

Raj Deo Tewari, Abhijit Y. Dandekar, and Jaime Moreno Ortiz

CRC Press
Taylor & Francis Group
Boca Raton London New York

CRC Press is an imprint of the
Taylor & Francis Group, an **informa** business

CRC Press
Taylor & Francis Group
6000 Broken Sound Parkway NW, Suite 300
Boca Raton, FL 33487-2742

Printed on acid-free paper

International Standard Book Number-13: 978-1-138-62638-6 (Hardback)

Library of Congress Cataloging-in-Publication Data

Names: Tewari, Raj Deo, author. | Dandekar, Abhijit Y., author. | Moreno Ortiz, Jaime, author.
Title: Petroleum fluid phase behavior : characterization, processes, and applications / Raj Deo Tewari, Abhijit Y. Dandekar, Jaime Moreno Ortiz.
Description: First edition. | Boca Raton, FL : CRC Press/Taylor & Francis Group, 2019. | Series: Emerging trends & technologies in petroleum engineering | Includes bibliographical references.
Identifiers: LCCN 2018038801 | ISBN 9781138626386 (hardback : acid-free paper)
Subjects: LCSH: Petroleum—Migration. | Petroleum reserves. | Petroleum—Fluid dynamics. | Phase rule and equilibrium. | Oil reservoir engineering. | Gas reservoirs.
Classification: LCC TN871 .T386 2018 | DDC 622/.3382—dc23
LC record available at https://lccn.loc.gov/2018038801

Visit the Taylor & Francis Web site at
http://www.taylorandfrancis.com

and the CRC Press Web site at
http://www.crcpress.com

I would like to dedicate this book to my father, the late Paras Nath Tewari, and my mother, the late Bela Tewari, who were instrumental in inculcating the habit of reading, studying, and importance of education since childhood. My appreciation of the support and encouragement I have received from my wife, Sumitra Tewari, in the course of working on this book cannot be expressed adequately in words. She was very supportive despite her Parkinson's disease. My children and grandchildren were also very supportive and taking this effort as a source inspiration for them. A word of thanks also goes to my colleagues in the Petronas discussion, who have helped shape this book and to whom I owe much gratitude.

Raj Deo Tewari

I would like to dedicate this book to my grandfather, the late Purushottam V. Dandekar. A word of thanks also goes to my parents, Yashwant and Sunanda Dandekar, for their blessings, my wife Mrudula, my son Shamal, and my daughter Rama for their support during the course of writing of this book.

Abhijit Y. Dandekar

I would like to dedicate this book to my late father, Luis Alberto Moreno, and my mother Nelly Ortiz for their support and encouragement through my career. I would like to acknowledge the work, publications, and discussions of numerous engineers with whom I have had the privilege of working. I also owe a great intellectual debt to the subject matter of this book, to the researchers, and developers for their valuable contributions on the field. Finally, I would like to thank Schlumberger for its continuous support.

Jaime Moreno Ortiz

Contents

Preface

Chapter 1 covers the introduction of petroleum fluids, their types and phase behaviors, and characterization. It also discusses about various contaminants such as CO_2, H_2S, and mercury, which occur along with oil and gas resources. The chapter explains briefly the properties of hydrocarbons that are used in various reservoir engineering and production engineering calculations. It also explains how phase diagram can be used to describe the fluid types and equation of states for defining the fluid behavior at different pressure and temperature.

Chapter 2 covers various recovery mechanisms operative in oil and gas fields and fluid data required for reservoir engineering calculations. The chapter discusses about the importance of fluid characterization in field development planning, water and gas injection, enhanced oil recovery techniques such as gaseous, chemical, and thermal enhanced oil recovery. CO_2 sequestration and gas cycling are also briefly introduced in this chapter.

Chapter 3 covers the challenges and available technology for reservoir fluid sampling. The chapter discusses the main considerations for a proper reservoir coverage including fluid and reservoir heterogeneity, as well as the importance of the sampling operation. It describes the importance of well and reservoir conditioning prior and during the sampling, tailored to the type of operation, surface of downhole sampling, to ensure sample integrity and thus robust laboratory results. It also describes the newly available wire technology that is being used to identify and characterize fluid heterogeneity and hydrodynamic disconnections as well as improvements on downhole sampling. This chapter will enable the reader to recognize the importance and challenges of fluid sampling, particularly in heterogeneous reservoirs and familiarize with latest sampling technology and its applications for complex systems

Chapter 4 covers the planning of laboratory studies. The chapter discusses the main considerations for a fit-for-purpose fluid characterization accounting for the existing reservoir conditions as well as planned development, covering a range of miscible, near-miscible, and immiscible processes. It describes laboratory experiments, along with advantages and drawbacks, including quality checks to ensure consistency in the results, particularly for volatile and retrograde gas samples. This chapter will enable the reader to plan and quality check PVT laboratory experiments tailored for specific reservoir and expected recovery mechanisms, inclusive of samples with asphaltene precipitation challenges

Chapter 5 examines various phase behavior aspects of pure components such as CO_2 and typical reservoir fluids in the dense or supercritical region. This particular region may be achieved by manipulating the T&P conditions on a given fluid or this may be a natural occurrence in reservoir fluids such as gas condensates. When fluids exist at these particular T&P conditions, they possess some unique properties such as relatively high densities, low viscosities, and extractive or solvating capacity. These fluids are commonly referred to as "dense phase" or "supercritical." These unique properties are of great significance in a wide variety of applications ranging from long-distance transport in pipelines to CO_2-based enhanced oil recovery. Also included is an in-depth discussion on hydrate phase equilibria as it pertains to the dense phase behavior. The chapter concludes with practical applications of dense phase.

Chapter 6 covers the special fluid characterization for EOR processes, starting with equation of state (EOS) calibration requirements and consistency checks, followed by special calibration strategies for miscible and near-miscible PVT laboratory experiments in preparation for inclusion onto finite-difference numerical models. It describes and highlights the importance of fit-for-purpose dynamic testing of the EOS prior full-field deployment, including real-life project experience suggestions to ensure smooth transition and consistency from the laboratory to the numerical models. This chapter will enable the reader to consistently translate the laboratory results to finite-difference numerical models, including empirical calibration strategies and finite-difference full model considerations and requirements

Chapter 7 begins with the differentiation between the conventional and unconventional reservoir fluids based on reservoir rock and fluid properties and phase behavior together with prevailing reservoir temperature and pressures. Using published literature data, basic characteristics such as the molar compositions, properties of the single carbon number (SCN) fractions, and the C_{7+} fraction of conventional and unconventional gases and oils or condensates are compared. A large part of this chapter also is dedicated to a different class of unconventional reservoirs, i.e., those that contain excessive amounts of non-hydrocarbon gases such as CO_2, H_2S and even elemental Hg in some cases. In-depth discussion and illustration of phase behavior (including the phase behavior type) and properties of CO_2–H_2S systems, CO_2-hydrocarbon systems as well as Hg are covered. As a prelude to Chapter 8, this chapter ends with a discussion on the confinement effects (nano pore size) on fluid properties and phase behavior given the significance in shale reservoirs.

Chapter 8 is entirely dedicated to porous media or confinement effects on phase behavior of unconventional petroleum reservoir fluids and is basically an in-depth extension of the confinement aspects introduced in Chapter 7. First the fundamental general functional form of the capillary pressure (P_c) equation that is of significance for inclusion in the flash calculations is introduced that includes the prediction of gas–oil interfacial tension from parachor-type models. Various equations and relationships such as the fundamental equilibrium ratio and the Rachford–Rice flash functions that are typically used in EOS-based flash calculations for the determination of saturation pressures and phase split that incorporate the confinement effects via capillary pressure are established. A step-wise methodology for carrying out these types of calculations and numerical examples for bubble point, dew point, and VLE are provided, and the differences between the "bulk" and "confined" phase behaviors are highlighted. Results obtained from the tested systems in bulk and under confinement are compared with literature reported predictions and limited experimental data.

Chapter 9 mostly focuses on the exotic phase behavior of CO_2 and (heavy) oils that is pertinent to miscible type EOR processes. Using a pseudo-binary concept, a guideline of the phase behavior "type" that uses the oil molecular weight as a cut-off, produced by the CO_2 and various oil types, is introduced. Experimental observations from the selected literature and EOS modeling that also includes some aspects of Gibbs phase stability analysis for the CO_2–alkane and CO_2–oil phase equilibria supported by numerical examples are covered. Various equations and relationships such as the fundamental equilibrium ratio and the Rachford–Rice flash functions for use in EOS-based three phase split flash calculations are derived. An empirical Wilson-type correlation for the estimation of component partitioning between the CO_2 rich and oil rich phase is developed and proposed to fill the gap that currently exists. Using selected experimental literature data, the practical significance of the CO_2-rich liquid phase in EOR processes is highlighted. Finally, the "upgrading" of the heavy oil in microbial and in situ combustion EOR methods based on literature reported measured compositions and fluid properties are shown.

Chapter 10 discusses the importance of flow assurance in oil and gas production system. It describes about most likely flow assurance problems, such as formation of wax, asphaltenes, hydrate, emulsion, reservoir souring, corrosion, and aquathermolysis. It also covers characterization of flow assurance issues and their mitigation.

Chapter 11 covers the use of equation of state in numerical simulations, describes the selection process between black oil and compositional models, and highlights the necessary steps to validate the PVT data (experimental or otherwise) with field observations in preparation to inclusion onto finite-difference models. It provides alternatives to supply a complete set of PVT data when samples and/or experimental information is scarce as well as discussing challenges and key considerations when using models to describe complex processes, such as miscible to near-miscible displacements. This chapter will enable the reader to effectively validate and incorporate PVT data onto numerical simulation models.

Chapter 12 covers most popular industry correlations that can be used in the absence of laboratory data for fluid characterization and description. It also covers prediction of hydrate formation temperature and estimation of recovery factors which uses the fluid data.

Acknowledgments

The authors would like to gratefully acknowledge the considerable aid provided by the works of numerous scientists and engineers and their publications in the literature pertaining to the subject covered in this book. In particular, the authors would like to thank Professors Late Ali Danesh, William D. McCain Jr., Curtis Whitson, and Dr. Karen Schou Pedersen for laying the foundation and their fundamental contributions in the area of petroleum reservoir fluid properties and phase behavior.

Authors

Raj Deo Tewari has worked as a practicing reservoir engineer in various companies, namely: ONGC, GNPOC, Sudapet, and Petronas. He started his career in ONGC's Institute of Reservoir Studies in the 1980s, where he was involved with reservoir characterization and numerical simulation, oil and gas field development planning (FDP), and EOR laboratory studies. After 12 years, he moved to Mumbai as a reservoir manager and contributed to the redevelopment of the giant carbonate field, Mumbai High. His responsibilities included water flood management and identification of IOR/EOR initiatives for the field. He moved to North Africa in early 2000 to lead the Reservoir Engineering department at GNPOC, a joint venture company. During this period, he was responsible for FDP studies, and reservoir management, and surveillance of GNPOC fields, producing at plateau rate of 300 kbopd. Dr. Tewari joined Petronas in 2007 and since then has been actively involved in redevelopment studies of number of major fields incorporating EOR. In 2010–2011, he joined the national oil company of Sudan as Custodian PE and was involved with cyclic steam stimulation and steam flood EOR projects. He has extensive experience in light and viscous oil fields development and management, reservoir characterization and simulation, EOR, and reservoir management. He is actively involved in EOR studies and R&D projects in Petronas. He has conducted training courses on fluid characterization, rock properties, advance EOR and reservoir engineering. He was a visiting professor to Khartoum University and also teaches in Universiti Teknologi PETRONAS (UTP) on specialized topics. Dr. Tewari holds a Ph.D. degree in physics from Banaras Hindu University, India. He has published nearly 50 papers in journals and conferences. He has been a chairperson and technical committee member for a number of SPE technical workshops and conferences. Currently, he is industry advisor to UTP, Malaysia. He has been awarded SPE Asia Pacific Technical Award for Production and Operation in 2015.

Abhijit Y. Dandekar is professor and chair of the Petroleum Engineering program at the University of Alaska Fairbanks (UAF), where he has taught since January 2001. Before joining UAF, he was an assistant research professor at the Technical University of Denmark. In the summer of 2002, he also served as a visiting faculty at the University of Petroleum Beijing, P.R.C. He has also been a visiting professor at the African University of Science and Technology (AUST) in Abuja, Nigeria, and a visiting professor at the University of Witwatersrand, Johannesburg, South Africa (as Fulbright specialist). He holds a B.Tech degree in chemical engineering from Nagpur University, India, and a Ph.D. degree in petroleum engineering from Heriot-Watt University, Edinburgh, UK. Prof. Dandekar has authored more than 80 papers and a widely adopted textbook on *Petroleum Reservoir Rock and Fluid Properties* and he is also the editor for *Emerging Trends and Technologies in Petroleum Engineering* book series (Taylor & Francis). He has supervised more than 30 graduate students and conducted several funded research projects primarily in the area of PVT and SCAL. In 2014, he received the SPE Distinguished Membership and Fulbright Specialist awards. In 2015, he was the recipient of the SPE Western North America Region Distinguished Achievement Award for Petroleum Engineering Faculty. Over the last 16 years, he has contributed to numerous SPE activities. Currently, he is a petroleum engineering program evaluator (PEV) for ABET and has served on nine visits.

Jaime Moreno Ortiz is a reservoir engineer with focus on EOR R&D, instructor, and a principal engineer for Schlumberger. His strong expertise in field development and reservoir characterization with an emphasis in EOR was developed through a 20-year career, which started with Schlumberger's Holditch Reservoir Technologies in Denver as a simulation engineer working with NFR triple porosity reservoirs in the Gulf of Mexico, South America and unconventional reservoirs in continental USA. Throughout his career, he has worked as lead technical engineer in several field projects, including field development planning, in the USA, Malaysia, Australia, Middle East, and North Africa. Starting in 1998, he held a combined role in R&D for EOR, concentrating on screening (patented technology), pilot planning, monitoring, and surveillance. He has taught internal and external courses on EOR and compositional modeling. Mr. Moreno holds an M.Sc. degree from the Colorado School of Mines, Golden, CO. He has published nearly 20 papers in journals and conferences and has been a technical committee member of several SPE applied technical workshops and conferences in the Asia Pacific region.

1

Reservoir Fluid Properties

1.1 Introduction

Energy has played key role in the development of human civilization (Figure 1.1). In fact, consumption of energy has become index of development status of a country. There are a number of differences between developed and developing countries. Out of all energy consumed by mankind, fossil fuels dominate the energy mix and oil playing the pivotal role (Figure 1.2). Its existence has been known since ancient times because of seepage at the surface of the earth (Figure 1.3). The ancient people in America and the Middle East used it for variety of purposes such as medicinal, military, and other purposes, like waterproofing and controlling leakages in boats and ships. Oil-soaked arrows were used by the Persians in war with Athens in 480 BCE. The ancient Egyptians used to preserve their dead as mummies by soaking them in a brew of chemicals such as salt, beeswax, cedar tree resin, and bitumen. In fact, the word *mummy* was derived from Arabic word *Mumya*, after Mumya Mountain, where bitumen was found. Oil was needed and used very little until the nineteenth century. Growth of urban centers in the nineteenth century made it necessary to search for a better source of illumination. This requirement was fulfilled with whale oil, and whale hunting became an important industry. When whales became scarce due to excess hunting, kerosene derived from coal was used, but a better substitute for whale oil was needed as a source of fuel for illumination.

1.2 Gas and Oil

Oil and natural gas together make up petroleum, which is the mixture of naturally occurring hydrocarbons. It is a dark liquid substance, but exists in gaseous and solid forms as well. The liquid form of petroleum is called

FIGURE 1.1
Increase in energy consumption with progress of civilization.

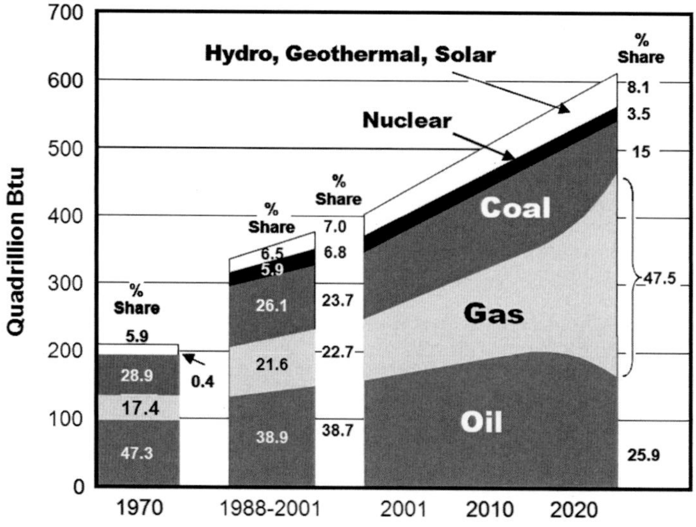

FIGURE 1.2
Pivotal of oil and gas in energy mix.

FIGURE 1.3
Seepage of hydrocarbon oil and gas to surface.

crude oil, its solid form is called asphalt, and its semi-solid form is called bitumen. Gas can occur together with oil or independently. It mainly contains carbon and hydrogen. Few impurities like nitrogen, sulfur, and oxygen are found in oil and gas fluids. An average composition of petroleum in all three states (liquid, gas, and solid) is shown in Table 1.1.

TABLE 1.1

Average Chemical Compositions of Natural Gas, Crude Oil, and Asphalt (From Levorsen[1] 1979)

Element	Crude Oil (wt%)	Asphalt (wt%)	Natural Gas (wt%)
Carbon	82.2–87.1	80–85	65–80
Hydrogen	11.7–14.7	8.5–11	1–25
Sulfur	0.1–5.5	2–8	Trace–0.2
Nitrogen	0.1–1.5	0–2	1–15
Oxygen	0.1–4.5	—	—

1.2.1 Gas Reservoirs: Hydrocarbon Gases Are Defined Based on Their Occurrences

Dissolved gas is defined as natural gas in a solution of in crude oil in a reservoir.[1–11] The reduction in pressure when oil is produced from a reservoir often results in dissolved gas being emitted from oil as free gas.

Associated gas occurs as a gas cap overlying in contact with crude oil within a reservoir. Nonassociated gas (NAG) is only found in natural gas reservoirs without the presence of crude oil.

Gas reservoirs produce gas. It can be dry gas or wet gas as described below.

1.2.1.1 Dry Gas Reservoirs

These types of reservoirs contain hydrocarbon in a single gas phase at initial pressure and temperature condition in subsurface, and remain in the gaseous phase on production when pressure and temperature decline. Very little or no condensation takes place at the surface. Dry gas reservoirs occur in both conventional and unconventional resources. Major composition is methane.

1.2.1.2 Wet Gas Reservoirs

These types of reservoirs contain hydrocarbon in single gas phase at initial pressure and temperature condition in subsurface. Some part of gas condenses at surface condition when pressure and temperature decline. Wet gas reservoirs occur both in conventional and unconventional resources and contain certain hydrocarbon that condenses at surface condition. Methane component is lower than dry gas and additionally contains ethane, propane, and butane in gaseous phase.

1.2.1.3 Gas Condensate Reservoirs

These types of gas reservoirs contain gas in single phases at reservoir pressure and temperature condition. Liquid dropout or condensation takes place in reservoirs, wellbores, and at the surface when pressure and temperature reduce or decline. The methane component is lower than dry and wet gas reservoirs with increased components of ethane, propane, and butane. These heavier gases, like propane and butane, facilitate condensation. The colors of condensate liquids range from clear to straw yellow. Liquid dropout increases with pressure drops to certain pressure levels, and starts vaporizing with further drops in pressure. Therefore, it is called retrograde condensate reservoirs.

1.2.2 Non-hydrocarbon Gases

Non-hydrocarbon gases are also found along with hydrocarbon gases. These are mainly carbon dioxide and hydrogen sulfide (H_2S). Both of these gases may be formed either inorganically or organically. Inorganic formation is usually associated with volcanic and/or geothermal processes.

1.2.2.1 Carbon Dioxide (CO_2)

CO_2 is a colorless, noninflammable gas that is odorless at low concentration and is 1.67 times heavier than air. It gives a sharp acidic odor at a higher concentration. Its molecular weight is 44. It is soluble in water. It is inert at atmospheric pressure and temperature, and becomes toxic if the concentration is more than 8%. This becomes very important when producing natural gas with higher percentages of CO_2 contained in it. This needs to be addressed in production, separation, and management to avoid the adverse impact on the environment. CO_2 is generated in nature by action of acids on carbonate and bicarbonate in igneous, sedimentary, and metamorphic rocks. Moreover, it is generated by the oxidation of hydrocarbons with contact with mineralized water. Heating of carbonates and bicarbonates and action of anaerobic bacteria reacting to hydrocarbon also generates the CO_2. There is a large number of reservoirs where the varying percentage of CO_2 is found both in the clastic and carbonate fields. Some fields in Borneo and Thailand contain CO_2 as more than 70% of total gas in reservoirs.

It is believed that the CO_2 in these fields has been generated by the heating action of igneous rocks coming in contact with limestone. The CO_2 is then driven off as it is in a lime kiln. It has also been observed that the depth of sediments in these fields is very high. In some cases, the depth is more than 10 km and the temperature is higher than 300°–500°. When CO_2 gas mixes with water, it forms carbonic acid and becomes corrosive.

1.2.2.2 Hydrogen Sulfide (H_2S) Gas

H_2S is an undesirable component of natural gas. The presence of H_2S not only can critically affect the economic value of hydrocarbon gas in the reservoir but it is also highly toxic and corrosive for production equipment. H_2S

is colorless with a characteristic foul odor of rotten eggs. This is soluble in water and hydrocarbon. Its solubility is higher in hydrocarbon than water. It is highly toxic and corrosive in nature. Therefore, it is a safety hazard in operation. It can be fatal even in small percentages. It has to be removed from the oil and gas stream and managed properly to avoid the safety risks, corrosion control of equipment's, and marketing of hydrocarbon gases free of it. H_2S exposure is fatal in petroleum industry operation. The clinical effects of H_2S depend on its concentration and the duration of exposure. H_2S is immediately fatal when concentrations are over 500–1,000 parts per million (ppm). Therefore, it has been referred as the "knock-down gas" because inhalation of high concentrations can cause immediate loss of consciousness and death. However, prolonged exposure to lower concentrations, such as 10–500 ppm, can cause various respiratory symptoms. The worst thing about H_2S exposure is that human senses stop working after initial exposure, and long-term exposure is always fatal.

H_2S originates through volcanic emanations, but bacteria in the absence of oxygen starts working on salts, thus generating H_2S. These are called sulfate-reducing bacteria (SRB). This typically happens in oil fields that are under water flooding for pressure maintenance. This process is called reservoir souring. The most favorable temperature for these bacteria to grow and generate H_2S is 25°C–50°C. Bacteria does not survive at higher temperatures. The calculated heat value of this gas is 956 BTU per 1,000 cubic feet. Its MW and SG are 34 and 0.973, respectively, and it yields 15 tons of sulfur per million cubic feet of H_2S.

1.2.2.3 Mercury (Hg)

Mercury occurs naturally in certain gas formation and is inevitably produced with the gas. The occurrences of mercury in traces originate from volcanic rocks, often underlying the gas reservoirs. It is environmentally hazardous element and has to be removed from the gas stream, preferably on the production site. This removal causes various problems for the operator. Laboratory analysis of reservoir fluid to quantify the quantity of mercury and identify the scope of the problem is important and depending upon the mercury concentration appropriate removal techniques have to be selected. The quality of reservoir fluid may also be affected due to the presence of mercury in gas. Removal of mercury in liquid condensate is more challenging than gas phase.

1.2.2.4 Coalbed Methane Gas

Methane gas is adsorbed in the microspores of coals. Gas is generated during the maturation of organic matter into coal and by microbes residing in the coal. Not all coals contain methane gas. Bituminous and subbituminous coals contain methane. Anthracite highly matured coal does not contain methane, and lignite and peat are highly immature and also do not contain methane gas. Coal seams have fractures that are called butt and face cleats. In total, 2%–5% methane gas occurs in these fractures. Coal seams are associated with water. To economically retrieve reserves of methane, wells are drilled into the coal seam, the seam is dewatered, and then the methane is extracted from the seam, compressed, and piped to market. The goal of dewatering is to decrease the pressure in the coal seam by pumping water from the well. The decrease in pressure allows methane to desorb from the coal and flow as a gas to the surface of the well. After desorption, methane gas diffuses to microspores and then follows a Darcy flow pattern from fractures to wellbore. The typical gas contents of methane coal seams are 100–400 Scf/ton. The San Juan Basin in the USA is one of the most prolific CBM producers, with an average gas content 300–600 Scf/ton. Jharia Coalfield in India has been burning for over 100 years due to the presence of methane gas. Coalbed wells production capacity ranges between 0.10 and 10 MMscf/d. CBM wells are low-pressure and low-rate, but their coal production lifespan is quite long, up to 30–40 years. Coal types and the typical performance of CBM wells are shown in following figures.

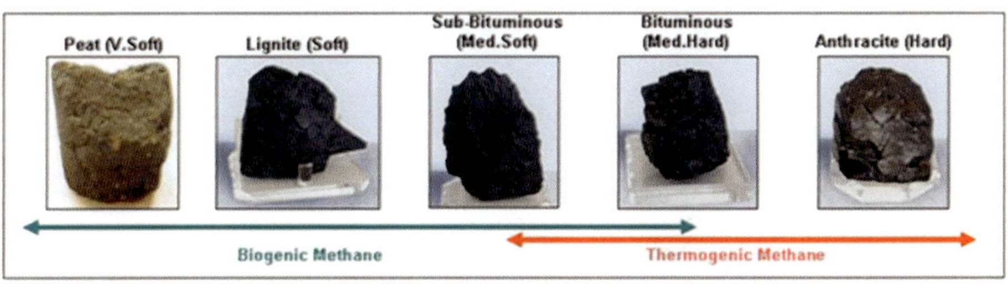

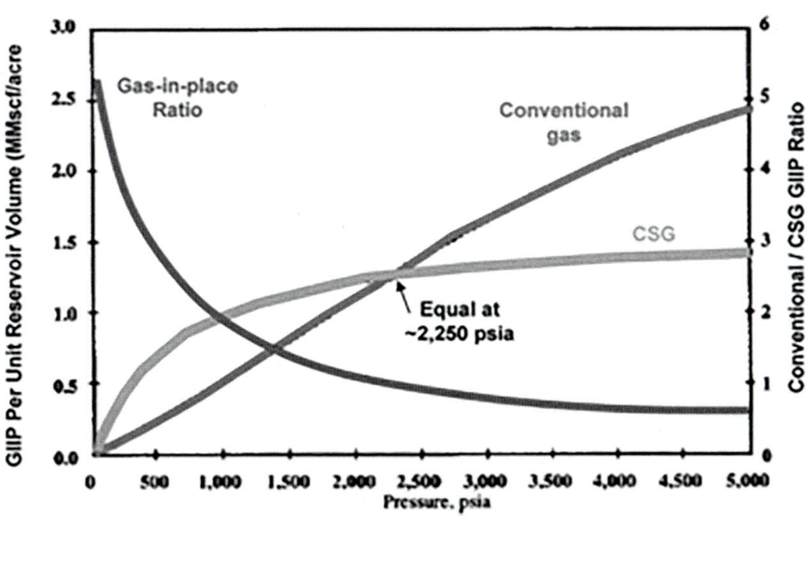

1.2.2.5 Shale Gas

Gas shale is organic-rich, fine-grained sedimentary rocks (shale to siltstone) containing a minimum of 0.5 wt% TOC. Gas shale may be thermally marginally mature (0.4%–0.6% Ro) to mature (0.6%–2.0% Ro) and contains biogenic to thermogenic methane. Gas is generated and stored in situ in shale as both sorbed (in organic matter) and free gas (in fractures and pores), similar to natural gas in coals. Shale gases are self-sourced reservoirs. Extremely low permeable in the order of nano-milidarcy and require extensive hydrofracturing to produce commercial quantity of gas. The typical shale containing gas and performance of well is shown in following figures.

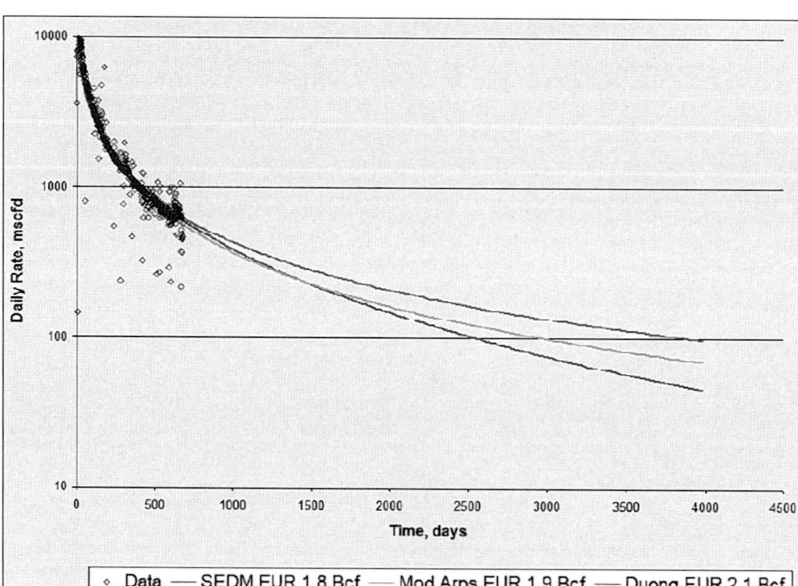

1.2.2.6 Gas Hydrates

Gas hydrates are an unusual form of subsurface hydrocarbon gas in which ice lattices, called clathrates, physically trap gas molecules in cage-like structures without the aid of direct chemical bonds. Hydrates look like wet snow.

There are two gas sites within the hydrate structure: a smaller one that can only contain methane and a larger one that can contain H_2S, carbon dioxide, and the larger hydrocarbon molecules of ethane, propane, and butane.

The composition of the gases in gas hydrates indicates that their origin is probably biogenic and shallow, rather than deep, much like the origin of marsh gas. Gas hydrates are formed as subsurface pressure increases and temperature decreases. They are stable only within a specific temperature range, which depends greatly on both the pressure and the composition of the hydrate.

Methane hydrates, which form at low temperature and high pressure, are found in sea floor sediment and the arctic permafrost. Currently, there is enormous interest in gas hydrates and much research ongoing to commercially exploit this massive resource.

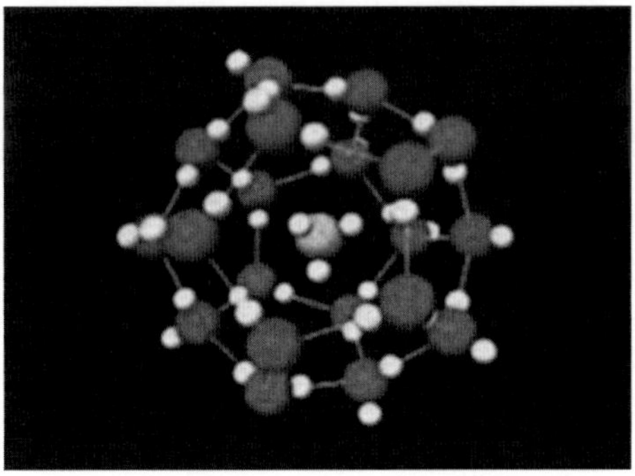

In order to produce the methane gas from hydrate, it has to be dissociated. The following dissociation methods are in development:

- Thermal methods
- Depressurization method

Gas recovery. Recovery factor of dry, wet, and gas condensate reservoirs are not same. Dry gas reservoirs have maximum recovery, even more than 90%. Wet gas has slightly lower recovery than dry gas reservoirs, but gas condensate has even lower recovery due to multiphase flow in the reservoir due to liquid dropout. The recovery factor is also impacted by other reservoir characteristics like permeability and aquifer support. Volumetric dry and wet gas reservoirs with reasonably good permeability deliver very high recovery. Tight gas condensate reservoir recovery is moderate around 60%. Aquifer-supported gas reservoirs are not good candidates for better recovery. A retrograde condensate reservoir is an excellent candidate for "gas cycling." Dry gas is injected into the reservoir to keep the pressure high and thus no liquids will drop out in the reservoir and prevention of unproducible residual oil saturation.

Sulfur and nitrogen are both undesirable elements within petroleum. Sulfur is most abundant in the heavier crude oils and in asphalt. It can also occur in natural gas mixtures such as the poisonous corrosive gas H_2S. Such natural gas is called *sour gas*. Nitrogen content is generally higher in asphalts as it occurs mostly in high molecular weight hydrocarbon compounds called nitrogen, sulfur and oxygen (NSO) compounds. CO_2 and the inert gases are also found in hydrocarbon. In some reservoirs of Borneo, and Papua New Guiney and Thailand offshore, CO_2 percentage is very high. However, in some areas, this percentage goes up to 70%–90%.

1.2.3 Physical Properties of Gas

1.2.3.1 Gas Density ρ_g

Density is defined as the ratio of mass and volume.[1–11] Gas density in pressure and temperature is calculated by

$$\rho_g = m/V$$

where for 1 mole m = molecular weight and $V = RT/P$ from gas law.

Therefore, we have

$$\rho_g = MW \cdot P/RT$$

1.2.3.2 Gas Gravity, γ_g

This is defined as the ratio of the molecular weight of the gas to the molecular weight of air (28.97). Gas is measured in moles, which is the weight of the gas divided by its molecular weight. The molecular weight of a gas of density ρ_g relative to air is 28.97 ρ_g.

1.2.3.3 Gas Viscosity, μ_g

Gas viscosities increase with pressure and temperature. The typical gas viscosities are found to be 0.01–0.05 cP. For a given pressure gradient, gas will flow 100,000 times faster than a heavy oil with a viscosity of 1,000 cP. Gas viscosities are often measured with a vibrating wire viscometer.

1.2.3.4 Gas Deviation Factor Z

Real gases do not follow the ideal gas equation and are defined as follows: $PV = ZnRT$, where Z is gas deviation or correction factor. This is also known as gas compressibility factor:

$$Z = V_{actual}/V_{ideal}$$

1.2.3.5 Gas Formation Volume Factor, B_g

Gas formation volume factor B_g is defined as the volume of gas under reservoir conditions divided by the volume of gas at surface under standard conditions. This can be estimated in a laboratory and can be calculated using the gas law, which is defined as $PV = nRT$:

$$B_g = \frac{V_{reservoir}}{V_{sc}} = \frac{(znRT/p)}{(z_{sc}nRT_{sc}/p_{sc})} = 0.3699Tz/p\,\text{m}^3/\text{m}^3 \rightarrow \text{or} \rightarrow 0.0283Tz/p\,\text{ft}^3/\text{ft}^3$$

where sc is standard condition,
$T_{sc} = 273$ K or 460 R,
$p_{sc} = 101$ kPa or 14.6 psi, and
$z_{sc} = 1.0$.

To convert B_g from ft³/ft³ to bbl/Scf, divide the above B_g by 5.615 to give

$$B_g = 0.005034Tz/p \text{ bbl/scf}$$

1.2.4 Oil Reservoir Types

Oil reservoirs contain hydrocarbon in the liquid form. Hydrocarbon in the form of liquid occurs associated with overlying gas and without gascap. Occurrence of oil in reservoirs could be light and heavy in terms of viscosity. Hydrocarbon contains mainly carbon and hydrogen, but constituent atoms can be arranged in numerous ways. Compounds with similar physical and chemical properties can be grouped into hydrocarbon series, of which four are particularly important in petroleum chemistry—the Paraffins, Naphthenes, Aromatics, and Resins and Asphaltenes.[1–11] The WPC categorization of crude oil is shown in following figure.

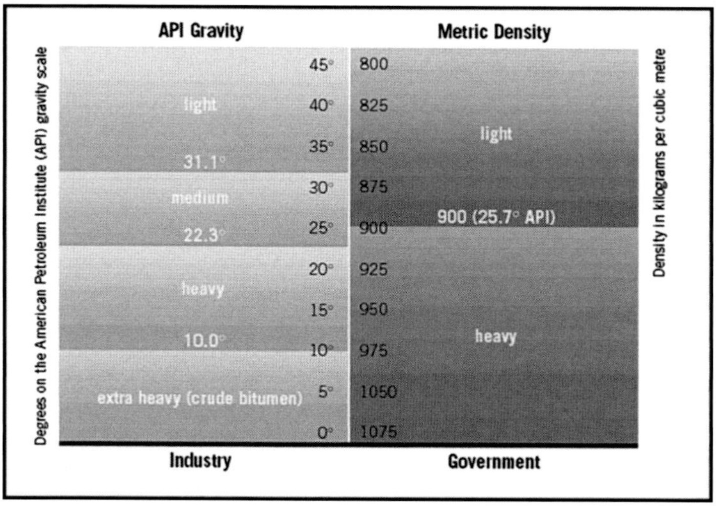

1.2.4.1 Black Oil Reservoirs

These reservoirs are called low-shrinkage oil. The color of these types of oils is generally lighter than that of the bitumen or heavy oils; it still tends to be rather dark. Color of black oil is green to black. Some of the typical characteristics of black oil are as follows:

Gravity: 20 < °API < low 30's

$10 < R_{si} < 90\,\text{m}^3/\text{m}^3$

$1.1 < B_o < 1.5\,\text{m}^3/\text{m}^3$

$100 > \mu_o > 2\text{–}3$ cP

1.2.4.2 Volatile Oil Reservoirs

These types of oil reservoirs contain lighter crude than black oil reservoirs. They are of high shrinkage. Some of the typical characteristics of volatile oil are as follows:

Gravity: low 30's < °API < low 50's

$1.5 < B_o < 2.5–3.5$ (m³/m³)

$90 < R_{si} < 1{,}000$ (m³/m³)

$2–3 > \mu_o > 0.25$ cP

The color of volatile oil is much lighter and sometimes as light as a bright gold.

1.2.4.3 Heavy and Extra-Heavy Oil

Heavy and extra-heavy crude oils and bitumen are petroleum or petroleum-like liquids or semisolids occurring naturally in porous and fractured media. Bitumen deposits are also called tar sand, oil sand, oil-impregnated rock, and bituminous sand. Heavy and extra-heavy oil reservoirs contain predominantly heavier components of crude. They are biodegraded oil and their typical characteristics are as follows:

Heavy oil: 10 < °API < 20

Negligible $< R_{si} <$ 10 m³/m³ (50 Scf/bbl)

Heavy oils from the Sparky Sandstone (Lloydminster, Alberta, Canada) have sufficient gas in solution to fuel engines on the lift equipment:

Extra-heavy oil: 4 < °API < 10

Some of the main properties of bitumen (at initial reservoir conditions) are as follows:

$R_{s,initial} = R_{si} \sim$ negligible

$B_o \sim 1.0$ Res. m³/m³ (bbl/STB)

$10{,}000 < \mu_o < 5{,}000{,}000$ cP

The color is usually dark or even jet-black; however, some are dark chocolate brown. Bitumen is essentially free of gas.

1.2.5 Physical Properties of Crude Oil

Crude oil is a natural mixture of hydrocarbons, which is liquid in subsurface reservoir and remains as liquid at surface condition even after passing through the processing facilities.

Typical appearance of crude oil is straw yellow, green, and brown to dark brown or black. It occurs in varying viscosity from very low less than 1 cP to very high million cP. On the surface, oils tend to be more viscous than when heated in the subsurface. Viscosity also increases with increasing the density of the crude oil.

Density of oil is measured using the American Petroleum Institute (API) definition:

$$\text{API gravity}(\text{degrees}) = \frac{141.5}{\gamma_o} - 131.5$$

where

γ_o = oil specific gravity at 15°C and 101 kPa, fraction.

The specific gravity (density of fluid divided by density of water) is measured relative to water whose specific gravity is 1.0.

Oil density, ρ_o, is then given by

$$\rho_o = 1000\gamma_o \text{ kg/m}^3$$
$$= \gamma_o \text{ g/cm}^3$$

Thus, the API of oil is inversely related to its density. It can also be seen that the API of fresh water is 10° API.

An API degree represents the quality of crude. The higher the value, the lighter the crude oil quality. The lighter crude oils have API gravities of 40°, which is equivalent to 0.83 relative densities, while the heavier crude oils have low API gravities.

According to WPC, the crude oil with API less than 23° is known as heavy oil.

1.2.5.1 Bubble-Point Pressure, P_b

This is the first pressure where, at constant temperature, the oil system releases a small bubble of gas. This gas can affect the oil recovery. Bubble-point pressure or saturation pressure varies with depth. The lower values of P_b are observed with an increase in depth. If bubble-point pressure is same as initial reservoir pressure, then reservoir is saturated oil reservoir. However, when bubble-point pressure is lower than initial reservoir pressure, it is undersaturated reservoir.

1.2.5.2 Oil Formation Factor, B_o

Oil formation volume factor B_o relates a reservoir volume of oil to a surface volume. The reservoir volume includes dissolved gas, while the surface volume is essentially dead oil and does not include the released gas. Initial formation volume factor B_{oi} is slightly lower than formation volume factor B_o at bubble-point pressure. Formation volume factor of oil will always be more than 1.0. Shrinkage of oil is expressed as inverse of formation volume factor $1/B_o$.

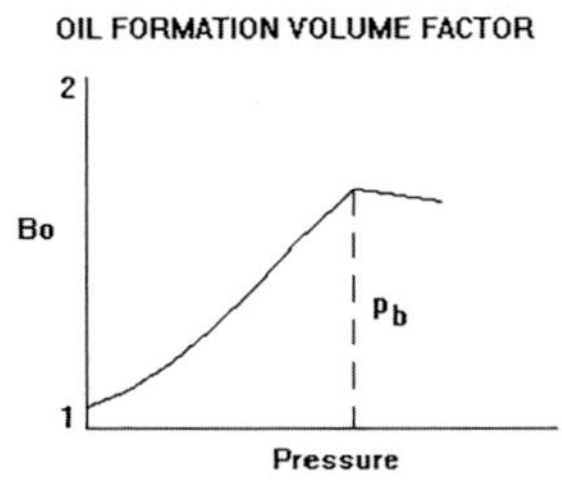

1.2.5.3 Gas Formation Volume Factor, B_g

B_g is defined as the volume of gas under reservoir conditions divided by the volume of gas at surface under standard conditions.

Let *sc* represent standard conditions. Then, we have

$$B_g = \frac{V_{reservoir}}{V_{sc}} = \frac{(znRT/p)}{(z_{sc}nRT_{sc}/p_{sc})} = 0.3699Tz/p \text{ m}^3/\text{m}^3 \rightarrow \text{or} \rightarrow 0.0283Tz/p \text{ ft}^3/\text{ft}^3$$

where

$T_{sc} = 273$ K or 460°R,
$P_{sc} = 101$ kPa or 14.6 psi, and
$Z_{sc} = 1.0$.

To convert B_g from ft³/ft³ to bbl/Scf, divide the above B_g by 5.615 to give

$$B_g = 0.005034Tz/p \text{ bbl/Scf}.$$

1.2.5.4 Solution Gas–Oil Ratio, R_s

R_s is a measure of the gas dissolved in the oil at any given condition. Its units are standard cubic meters of gas per stock tank cubic meter of oil, sm³/sm³ or standard cubic feet per stock tank barrel (Scf/STB) or thousands of standard cubic feet per stock tank barrel oil (Mcf/STB).

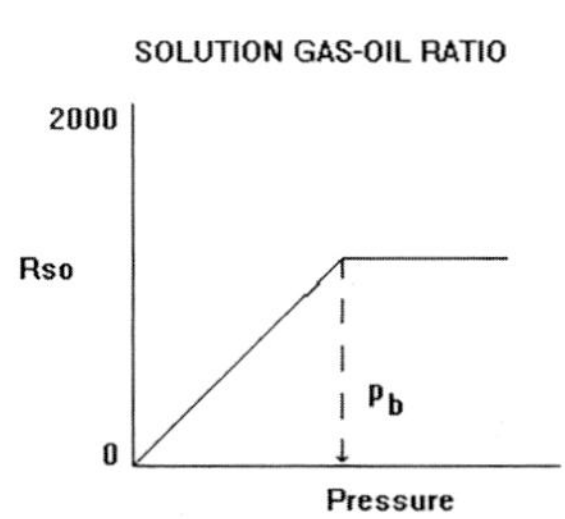

As a function of pressure, R_s is constant above the bubble point, and decreases with decreasing pressure below the bubble point as gas is released to become free gas.

Two types of gas liberation processes may occur:

a. differential
b. flash

In a differential liberation process, the evolved gas is allowed to escape from contact with the oil. In a flash liberation process, gas that is released from solution remains in contact with the oil. Flows in reservoirs with any appreciable vertical permeability will approximate a differentiable process; flow through tubing, surface equipment, and in reservoirs having continuous shales will approach a flash process. In general, R_s calculated from a differential process is slightly higher than that determined from a flash process.

1.2.5.5 Oil Viscosity, μ_o

Viscosity is the measure of resistance to flow. As the pressure is decreased below the bubble point, the oil viscosity increases due to release of solution gas. Above the bubble point, the liquid molecules are forced more closed together and oil viscosity increases as the pressure increases. At the bubble point, the oil viscosity is minimum. The oil viscosity is strongly dependent on temperature, dropping exponentially with increasing temperature.

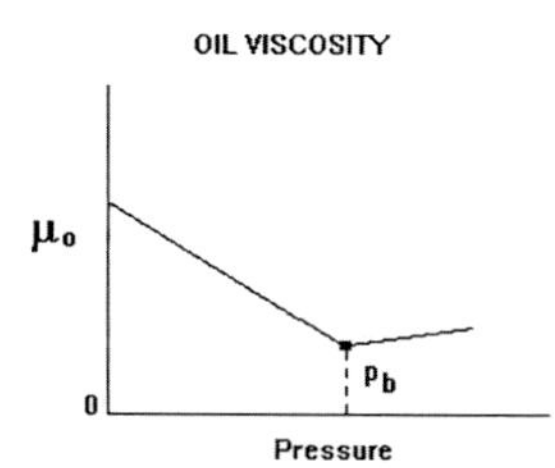

Often, correlations are used to determine oil viscosity. Two popular ones are Andrade's equation:

$$\mu_o = a\exp(b/T)$$

where a and b are constants and T is the absolute temperature. The other is Walther's equation, a log (log) relationship of viscosity to temperature:

$$\log(\log(\nu + 0.8)) = -n\log(T/T_1) + \log(\log(\nu_1 + 0.8))$$

where ν is the kinematic viscosity and ν_1 is the kinematic viscosity measured at absolute temperature T_1.

There are two common units of viscosity:

a. dynamic viscosity (μ)
b. kinematic viscosity (ν)

The dynamic viscosity, μ, has units of centipoise (cP) or Pascal-seconds (Pa-s). Multiply cP by 0.001 to obtain Pa-s. A poise is gm/cm/s. The kinematic viscosity, ν, has units of m^2/s. A Stoke is defined as cm^2/s, and thus kinematic viscosity is often expressed as centistokes (cs). (Water kinematic viscosity is about 0.01 cm^2/s or 1 cs.) To convert dynamic viscosity into kinematic viscosity, divide it by density, ρ, and thus we have

$$\nu = \mu/\rho$$

Viscosity can be measured by a rolling ball viscometer. Here, a steel ball rolls down a barrel which contains the oil sample. Viscosity is then calculated from measuring the time taken to fall through the oil. A typical barrel inside diameter is 6.60 mm; the ball diameter is 6.35 mm. The upper limit of oil viscosity that can be measured is 3,000 cP.

The pressure of gas bubbles in heavy oil can cause the ball to hang up for a long time. Viscometers have been redesigned for 12.7 mm balls and corresponding larger barrels. Electromechanical ball releases permit the use of tungsten carbide balls, which are denser than steel and will fall at faster rates to overcome the long hang-up time.

1.2.5.6 Oil Density, ρ_o

The oil density increases as the pressure is dropped. Oil density has units of gm/cm^3 or kg/m^3 and is commonly expressed in API degrees. (Specific gravity is defined as the weight of fluid/solid divided by the weight of an equal volume of a standard substance. For fluids/solids, the standard substance is distilled water at 4°C; for gases, the standard substance is air at 0°C and 1 atmosphere. The density of air at these conditions is 1.29 kg/m^3.) The density is measured by using a pycnometer or an oscillating tube densitometer.

1.2.5.7 Oil Compressibility, c_o

The oil compressibility, c_o, is defined as the rate of change of volume with pressure divided by volume at constant temperature. It thus has units of inverse pressure. (For an ideal gas, c_g is just the inverse pressure.) The oil compressibility is not strongly dependent on pressure. The typical c_o values are between 7.0 and 72.0×10^{-4}/MPa.

1.2.6 Chemical Properties

1.2.6.1 Paraffins

Paraffins have a general formula of C_nH_{2n+2}. The simplest and lightest molecule of the paraffin series is the gas methane, with a formula of CH_4.[1–11] Paraffins with less than five carbon atoms are gaseous at normal temperatures and pressures. In addition to methane, ethane, propane, and butane are also gases with carbon numbers 2, 3, and 4. C_5 through C_{15} paraffins are liquid at normal temperatures and pressures, while paraffins with carbon values greater than C_{15} are extremely viscous and may be solid waxes.

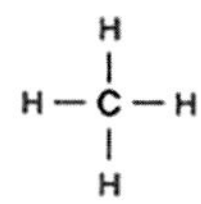	Methane	CH_4
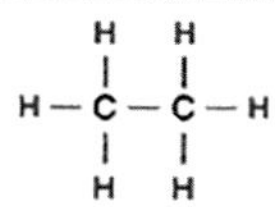	Ethane	C_2H_6
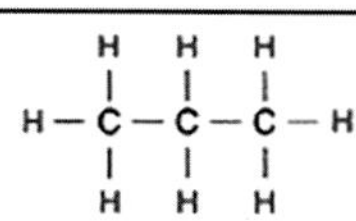	Propane	C_3H_8
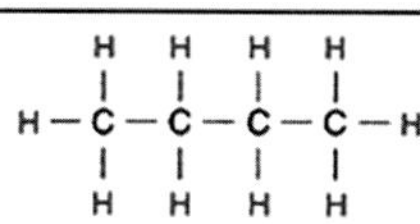	Butane	C_4H_{10}
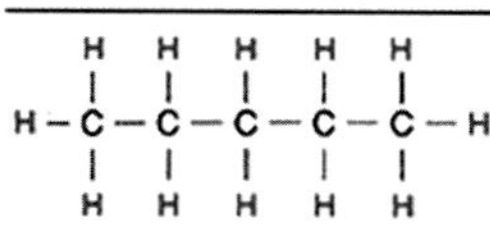	Pentane*	C_5H_{12}

1.2.6.2 Naphthenes

Naphthenes form as closed ring structures with the basic formula C_nH_{2n}. Compounds of the naphthenes series have chemical and physical properties similar to equivalent paraffins with the same carbon number. Together with the paraffins, naphthenes form the major components of most crude oils.

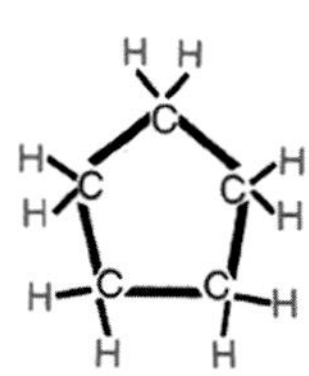

C_5H_{10}
Cyclopentane

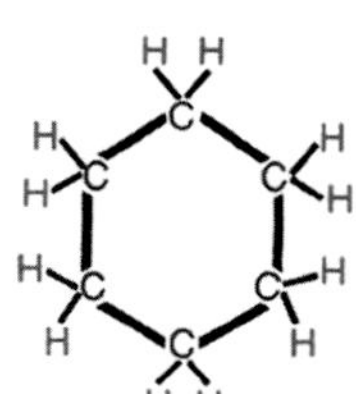

C_6H_{12}
Cyclohexane

1.2.6.3 Aromatics

Aromatics are the third group and have a structure based on a hexagonal ring of carbons, with alternate simple and double bonds. This basic unit is called the benzene ring, after the simplest and most abundant aromatic compound, benzene. Other aromatic compounds are made by substituting paraffinic chains or naphthenic rings at some of the hydrogen sites, or by fusing several benzene rings together.

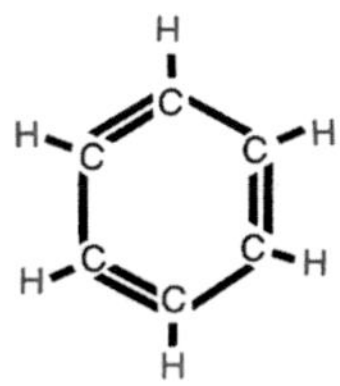

C_6H_6
Benzene

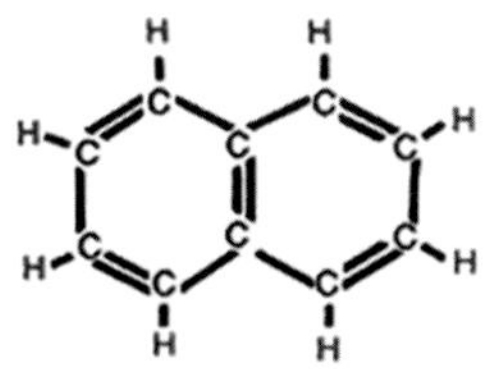

$C_{10}H_8$
Naphthalene

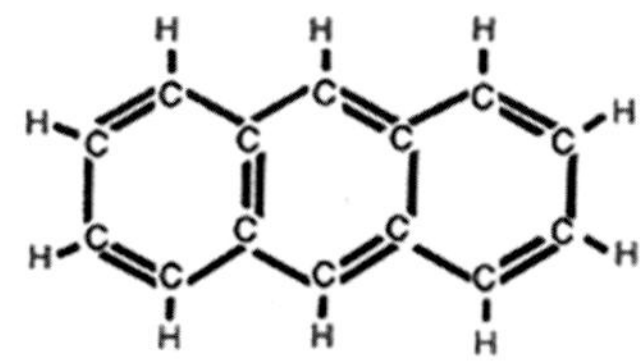

$C_{14}H_{10}$
Anthracene

1.2.6.4 Resins and Asphaltenes

These compounds are fused benzene-ring networks, but they contain other atoms and are not true hydrocarbons. These "impurities" are the high molecular weight NSO compounds. Resins and asphaltenes are the heaviest components of crude oil and the major components in many natural tars and asphalts.

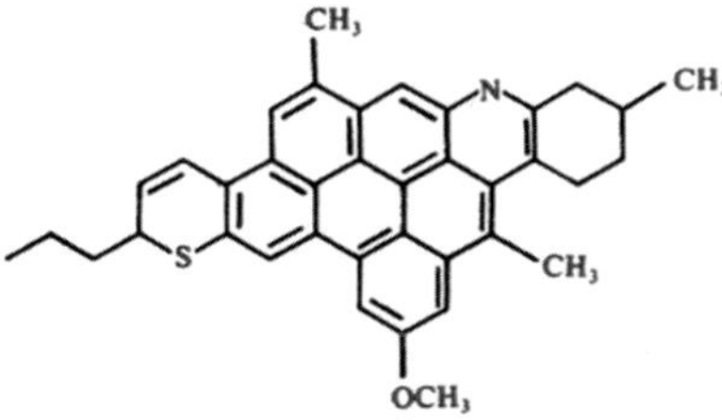

(from Waples 1981; reprinted with permission of Burgess Publishing Co.)

1.3 Non-hydrocarbon Crude Components

In addition to carbon and hydrogen, crude oil contains NSO. Sulfur is the third most abundant element in crude oils, after carbon and hydrogen, averaging 0.65% by weight. It can occur as free sulfur, H_2S, or as various organic sulfur compounds. Crude oils with less than 1% sulfur are considered as low sulfur oils, while crude oils with greater than 1% sulfur are considered as high sulfur oils. The oxygen content of most crude oils is on average 0.5% by weight. It occurs mostly in organic compounds that include acids and alcohols. The acids are especially common in young, immature oil. Some of these compounds are useful indicators that identify the kinds of organisms that gave rise to the crude. Nearly, all crude oils contain a small amount of nitrogen, between 0 and 0.9. Nitrogen is the most abundant compound in the NSO compounds in the heavy oils, but nitrogen can also occur in lighter compounds. Crude oils also contain small amounts of organo-metallic compounds, the most common being nickel and vanadium. Concentration of these metals ranges from less than 1 ppm up to 1,200 ppm.

1.4 Ternary Presentations of Crude Oil Classifications

Crude oils may be classified by their relative enrichment in the four primary hydrocarbon groups. Tissot and Welte (1978) proposed ternary plot paraffins, naphthenes, and the combination of aromatic and NSO compounds as three axes of a triangular graph and divides the graph into fields that represent six crude oil classes.[12]

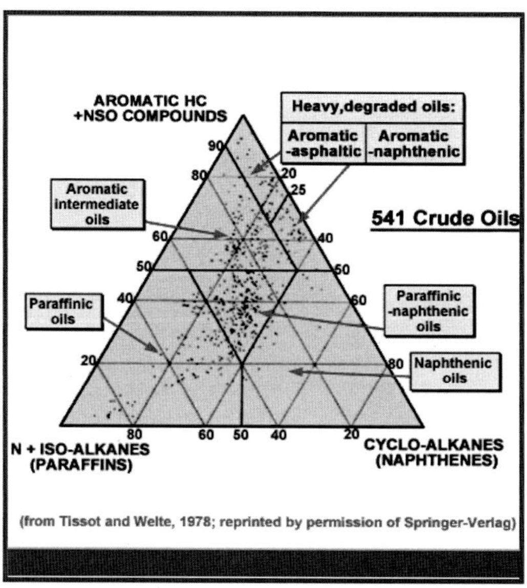

Most of the crude can be rich in paraffins (paraffinic oil), nearly equal amounts of paraffins and naphthenes which together make up more than 50% of the crude (paraffinic–naphthenic oil) and an have sub-equal amounts of paraffins and naphthenes, which total less than 50%, and the composition is dominated by the aromatics, resins, and asphaltene (aromatic intermediate oil).

Oil may degrade into heavy oil and tar as a result of bacterial action and of flushing by fresh meteoric waters of surface origin. This oil falls into one of two classes (aromatic–asphaltic or aromatic–naphthenic); both of which are enriched in aromatics. Some may contain naphthenes (aromatic–naphthenic oil) but the paraffin content is always very low. Deep burial, however, usually has the opposite effect in altering crude oil. It tends to make oil less dense and more paraffinic, through processes involving both thermal maturation and the precipitation and removal of asphaltic molecules.

Sometimes light, clear, high API gravity-liquids are produced with gas reservoirs. They are called condensate. Condensates begin as components of a heavier gaseous phase in the subsurface where they are highly compressed and at elevated temperatures. This gas phase contains some dissolved hydrocarbons which, when brought to lower surface temperatures and pressures, exolve. The subsurface phase then separates into distinct gas and liquid phases, the latter of which is called condensate.

1.4.1 Water Properties

Water is always associated with hydrocarbons in a petroleum accumulation, either as interstitial water in the hydrocarbon zone or aquifer water in the 100% saturated zone. The knowledge of chemical and physical water properties is necessary:

- to estimate the volume of water accumulation associated with the hydrocarbons, and locate its origin.
- to forecast the drive mechanisms originated by water, either water drive or water injection.
- to identify the source of the water produced and the production problems due to water production: scale, plugging, corrosion, hydrate formation, clay swelling, etc.

Water produced in surface is either

- aquifer water
- injected water
- water dissolved in gas

Interstitial and aquifer water properties are critical for planning the depletion strategies. Bottom hole sampling of reservoir water is collected for measurements in the laboratory. Water properties are mainly estimated by correlations from water samples collected in surface, separator, tank, or flow lines. The main water properties estimated for reservoir engineering are as follows:

- water composition: to identify the origin of water, its compatibility with water injected, and to determine its physical properties
- dissolved salts for scaling problem
- resistivity for log interpretation
- solubility, formation volume factor, compressibility, for material balance calculation
- density for estimation of water gradient and water oil contact determination—viscosity for mobility calculation
- compatibility of chemicals for EOR

1.4.1.1 Composition and Salinity

Water salinity is the amount of dissolved solids per unit quantity of solution (brine)

Salinity units are as follows:

Grams per liter (g/L)	g dissolved solids/liter brine
Parts per million (ppm)	g dissolved solids/10^6g brine
Weight per cent solids	g dissolved solids/100 g brine (= ppm 10^{-4})

g/L and ppm are strictly equivalent only if brine density is equal to 1.0 g/cc, but very often they are used indifferently.

Water salinity varies from as little as 200 ppm for fresh water to a maximum of approximately 400,000 ppm for saturated waters. Sea water contains approximately 35,000 ppm total solids. A water analysis is reported in terms of ionic analysis listing the amount of each ions in g/L or ppm. Main ions encountered in reservoir water are as follows:

Cation

- Sodium (Na^+)
- Potassium (K^+)
- Calcium (Ca^{++})
- Magnesium (Mg^{++})

Anions

- Chloride (Cl^-)
- Sulfate (SO_4^{2-})

- Carbonate (CO_3^{2-})
- Bicarbonate (HCO_3^-)

The presence of divalent ions is critical for chemical EOR projects and scaling during water injection. Reservoir water resistivity depends of temperature and nature and concentration of dissolved solids. Water resistivity decreases with temperature for a given salinity, and decreases with the amount of dissolved solids. Natural gas solubility in pure water is usually less than 30 Scf/STB (5.3 m^3/m^3) at reservoir conditions, and is a function of pressure and temperature.

1.4.1.2 Formation Volume Factor

Water FVF depends on pressure, temperature, solution gas, and salinity. Pure water FVF decreases with pressure and increases with temperature. Gas-saturated water has higher FVF than pure water at the same pressure and temperature conditions.

1.4.1.3 Density

Density of brine is calculated at surface condition as mass/volume and density at reservoir condition is calculated by density at surface condition by formation volume factor of water, neglecting the mass of solution gas.

1.4.1.4 Compressibility

Correlations of reservoir brine compressibility assumed the gas is dissolved in water, and applied to a monophasic liquid phase. In an oil/water system, the bubble point of the gas-saturated brine is equal to the oil bubble point. In a gas/water system, the water is considered to be at its bubble point at the initial reservoir pressure.

1.4.2 Water Solubility in Hydrocarbon System

Water and hydrocarbon have very low attraction. Solubility of water in hydrocarbon is low and has no practical application. Natural gas contains water in vapor phase. Presence of water vapor may form hydrate in certain pressure and temperature condition. Understanding of water content is important to predict for making design for dehydration for sale and transport.

1.4.3 Phase Behavior

Petroleum fluids occur subsurface and they have to pass through number of stages of processing in order to reach to usable form in industry, transport, and households. The primary responsibilities of a petroleum and natural gas engineer include developing, producing, separating, transporting, and storing petroleum fluids. At every stage of the petroleum exploration and production business, a good understanding of petroleum fluid characterization is critical.

Most of the fluid-handling protocols require the engineer to know *a priori* how the fluids will behave under a wide range of pressure and temperature conditions, particularly in terms of their volumetric and thermodynamic properties. Phase behavior has defining implications in petroleum and natural gas engineering processes. Pressure, volume, and temperature (PVT) relations are required in simulating reservoirs, evaluating reserves, forecasting production, designing production facilities, and designing gathering and transportation systems.[2–11]

Matter exists in three phases e.g. solid, liquid, and gases depending upon pressure and temperature conditions and they changes the form when these conditions are changes. Oil and gas occurs in liquid and gas phase in the reservoir. Fluid changes take place during production and processing at surface. Detail understanding would be required for making a suitable development plan. Therefore, it becomes critical to understand the variation in the properties.

A phase diagram is a concise graphical method of representing phase behavior of fluids. It provides an effective way of expressing the large amount of information about fluid's behavior at different conditions. Fluids are classified into two groups: pure-component system and mixtures.

Physical properties of interest are defined in terms of the pressure and temperature at which a hydrocarbon exists. Petroleum fluids as fluid in general exist under three different phases: liquid, solid, and gas. Of course conditions of pressure and temperature should be specified. It is usual also to classify fluids as liquids, gases, and vapors. Vapor being defined as any substance in the gaseous state which, under atmospheric conditions, is normally a liquid or a solid. An example is air saturated with water vapor, which gives water condensation at atmospheric pressure when temperature decreases. No distinction later in this text will be made between the gaseous state and vapor, the two words being synonymous and used indifferently. A phase is a portion of a system which is (1) homogeneous in composition, (2) bounded by a physical surface, and (3) mechanically separable from other phases which may be present.

The state of a system is defined through macroscopic variables as pressure, temperature, and specific volume. Each property should be defined based on dependent variables.

Gibb's law defined the variance of a system. The variance of a system is the number of independent thermodynamic variables necessary and sufficient to define the state of equilibrium of the system.

1.4.3.1 Gibb's Law

$$V = C + 2 - \Phi$$

V is the variance (number of parameters independent),
C is the number of constituents, and
Φ is the number of phases.

Example: for a pure constituent $C = 1$

Monophasic $\Phi = 1$; $V = 2$; volume $= f$(pressure, temperature) or $v = v(p,T)$ or $f(p,v,T) = 0$. This is the equation of state of a pure constituent
Biphasic $\Phi = 2$; $V = 1$; pressure $= f$(temperature)
triphasic $\Phi = 3$; $V = 0$; pressure, temperature, and volume are fixed: this is the triple point.

For multiconstituents $C = N$

$\Phi = 1$; $V = N + 1$,
$\Phi = 2$; $V = N$, and
$\Phi = 3$; $V = N–1$.

The phase diagram of a pure single component displays three phases: solid, liquid, and vapor which are separated by sublimation, liquefaction, and vaporization curves that join at the triple point denoted as "T." The vaporization curve, called the vapor pressure curve, terminates at the critical point denoted as "C." Beyond this point, any distinction between liquid and vapor is not meaningful.

1.4.4 Pure-Component Systems

In Figure 1.4, the P–T diagram indicates an increase in vapor pressure with an increase in temperature. Most of the liquid molecules escape into the vapor phase and increase the pressure that is the aggregate of all-vapor molecules exert on the system (i.e., vapor pressure). The curve in Figure 1.4 is called the vapor pressure curve or boiling point curve. The line also represents the dew-point curve and the bubble-point curve: one on top of the other. This curve represents the transition between the vapor and liquid states.

Line ACB represents isobaric heating. The phase change takes place at point C, where matter changes from liquid to vapor phase. Matter is liquid left to point C and vapor right to it on the line ACB. Up to this point, we saw that the heat added before the system reached the phase transition was used

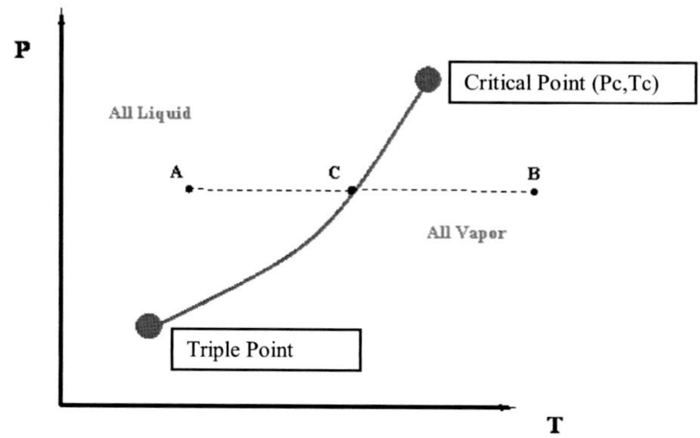

FIGURE 1.4
Vapor pressure curve or boiling point curve for a pure substance.

to raise the temperature of the substance. However, the heat added during the phase transition is not causing any temperature increase and is called latent heat. Therefore, these heats are termed as sensible heat and latent heat, respectively. Sensible heat causes an increase in temperature of the system and latent heat converts the liquid into vapor. Once all the liquid converted into vapor with latent heat and heat is further added to the system, the temperature will further increase and reach point B. This heat is sensible since it is causing the temperature of the system to rise. This process is reversible and termed as the isobaric cooling process when sensible heat is removed in order to cool the systems B to C and then latent heat of the vapor is removed to condense into liquid at transition point C. Further removal of sensible heat from C to A from the system brings cooling of the liquid.

Vapor pressure. The pressure that the vapor phase of a fluid exerts over its own liquid at equilibrium at a given temperature.

Dew point. The pressure and temperature condition at which an infinitesimal quantity of liquid (a droplet) exists in equilibrium with vapor. It represents the condition of incipient liquid formation in an initially gaseous system.

Bubble point. The pressure and temperature condition at which the system is all liquid, and in equilibrium with an infinitesimal quantity (a bubble) of gas. This situation is, in essence, the opposite of that of the dew point.

The P–T diagram of a pure compound is shown in Figure 1.5. This represents all three phases of solid, liquid, and vapor with a change in the temperature.

Two very important thermodynamic points bound the vapor pressure curve: the critical point at its upper end and the triple point at its lower end.

The triple point is the meeting point of the vapor pressure, solidification, and sublimation curves and represents the only condition at which all three phases of a pure substance (solid, liquid, and gas) can co-exist in equilibrium.

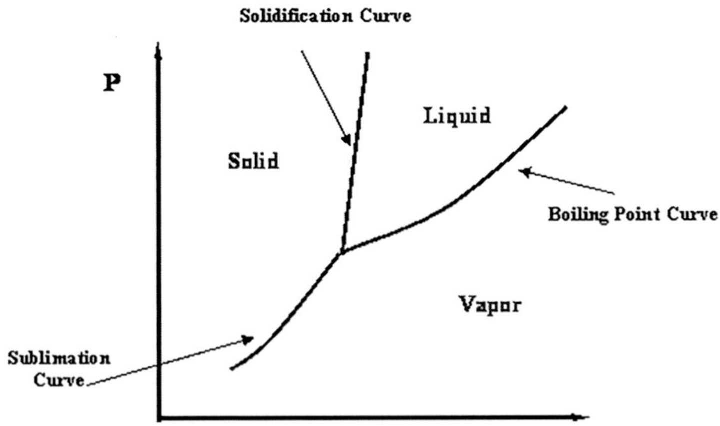

FIGURE 1.5
Complete P–T diagram for pure-component systems.

Critical point, gas, and liquid are in equilibrium without any interface to differentiate them; they are no longer distinguishable in terms of their properties. Critical point is clearly the maximum value of temperature and pressure at which liquid and vapor can be at equilibrium. This maximum temperature is called the critical temperature (T_c); the corresponding maximum pressure is called the critical pressure (P_c).

A combination of isobaric and isothermal change path ABCD is shown in Figure 1.6.

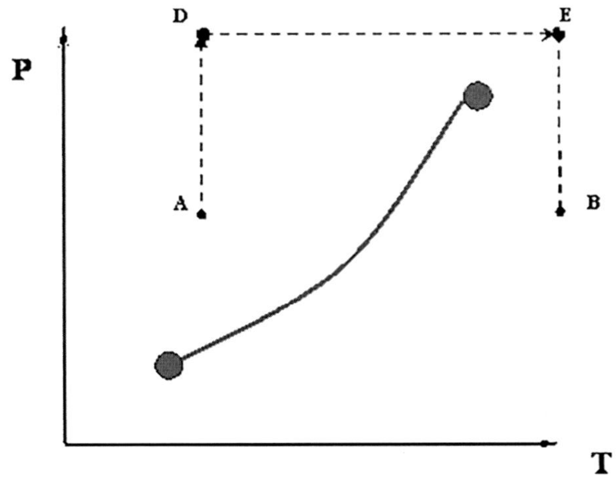

FIGURE 1.6
Vapor pressure curve and continuous path above critical point.

- Path AD: Isothermal compression
- Path DE: Isobaric heating
- Path EB: Isothermal expansion

1.5 PV Diagram for Pure Systems

P–T diagram explains the phase change at constant volume. Changes in phase behavior also take when either *T* or *P* is kept constant. In order to understand the volume changes in phase behavior, P–V or T–V diagrams are prepared keeping *T* or *P* constant, respectively (Figure 1.7).

In this case, temperature is being kept constant and substance goes under isothermal compression process. Starting at E (all-vapor condition), an increase in pressure will result in a rather significant reduction in volume since the gas phase is compressible. If isothermal compression is continued, then at point F the gas will be saturated and the first droplet of liquid will appear. Point F, where liquid (L) and vapor (V) co-exist in equilibrium, is termed as the dew point.

If the compression continued by decreasing the volume at constant pressure F- until the entire vapor has become liquid. Point G represents the last condition of liquid and vapor (L + V) coexistence, saturated liquid condition (liquid in equilibrium with an infinitesimal amount of vapor). If compressing (i.e., attempting to reduce liquid volume) is continued, then a rapid increase in pressure is observed. This is because liquid is virtually incompressible; hence, a great deal of pressure is needed to cause a small reduction in volume.

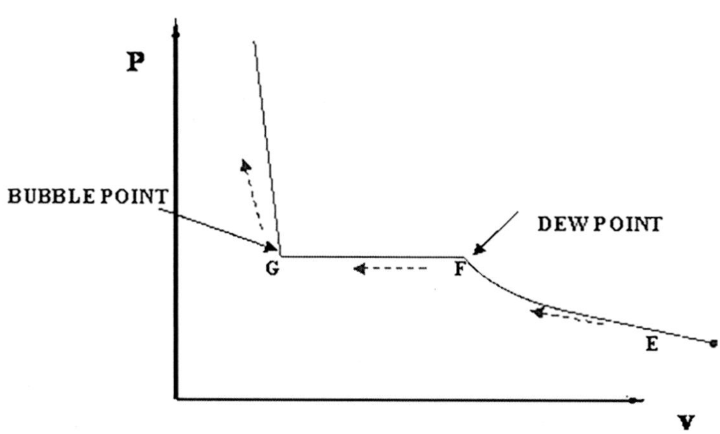

FIGURE 1.7
P–V Diagram of a pure component.

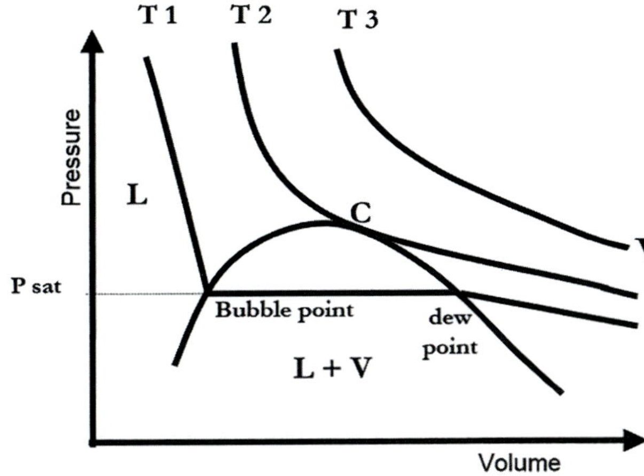

FIGURE 1.8
Family of P–V isotherms for a pure component.

It is important to recognize some points of this process. If we recall our previous definitions of basic concepts, we will recognize point F, where only a tiny quantity of liquid exists in an otherwise completely gaseous system (the dew point of the system at the given temperature). Similarly, point G is the bubble point; only an infinitesimally small bubble of vapor exists in an

A family of isotherms can be generated at different temperatures as shown in Figure 1.8. then compete vapor pressure are generated.

The zone where the isotherms become flat delineates the two-phase region. It is clearly seen that by plotting all the pairs in that zone (P_1, T_1), (P_2, T_2) … (P_c, T_c) vapor curve can be generated. Two curves meet at the critical point (P_c, T_c).

The critical point representing the maximum point (apex) of the P–V envelope is expressed as follows:

$$\left(\frac{\partial \rho}{\partial \tilde{V}}\right)_{P_c.T_c} = \left(\frac{\partial^2 P}{\partial \tilde{V}^2}\right)_{P_c.T_c} = 0$$

which are usually known critical conditions.

1.6 Binary Systems

P–V diagram of binary system is different than single-component system. In a binary system when volume is compressed isothermally and reached to dew-point pressure at B and further continuance of compression till point C,

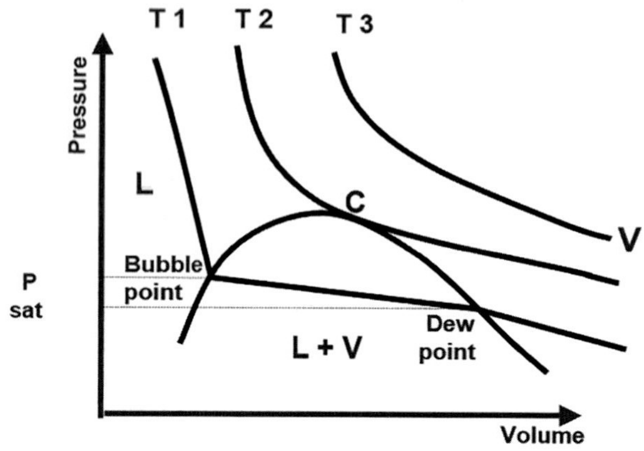

FIGURE 1.9
P–V diagram for a binary mixture.

bubble-point pressure will not keep constant pressure like single-component system, as shown in Figure 1.9.

It must be noted that bubble-point pressure and dew-point pressure do not coincide in the mixture or binary system like they do in the single-component pure system.

1.7 Effect of Composition on Phase Behavior

P–T diagrams of a binary systems A (methane, CH_4) and B (ethane, C_2H_6) will occur between two pure components depending upon their mole percentage (Figure 1.10).

Methane is more volatile than ethane, so it will have higher vapor pressure at lower temperature. Phase envelope of mixture of methane and ethane in different proportion can be *anywhere* within curves A and B (Figure 1.11).

The phase envelopes are bounded by the pure-component vapor pressure curve for component A (methane) on the left, that for component B (ethane) on the right, and the critical locus (i.e., the curve connecting the critical points for the individual phase envelopes) on the top. The range of temperature of the critical point locus is bounded by the critical temperature of the pure components for binary mixtures. Therefore, no binary mixture has a critical temperature either below the lightest component's critical temperature or above the heaviest component's critical temperature. However, this is true only for critical temperatures, but not for critical pressures. A mixture's critical pressure can be found to be higher than the critical pressures of both pure components—hence, we see a concave shape for the critical locus.

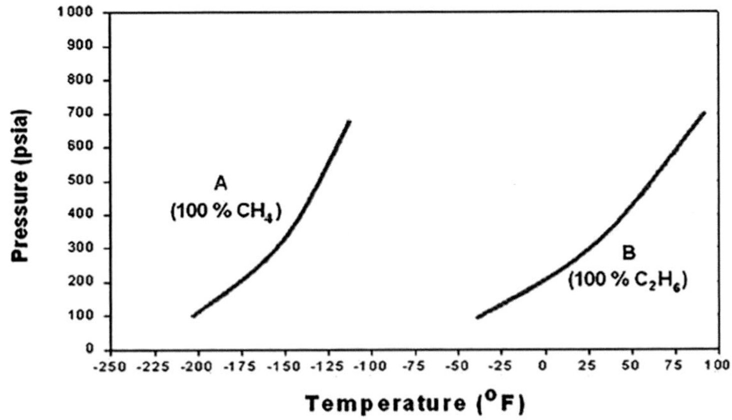

FIGURE 1.10
P–T diagram of two pure compounds.

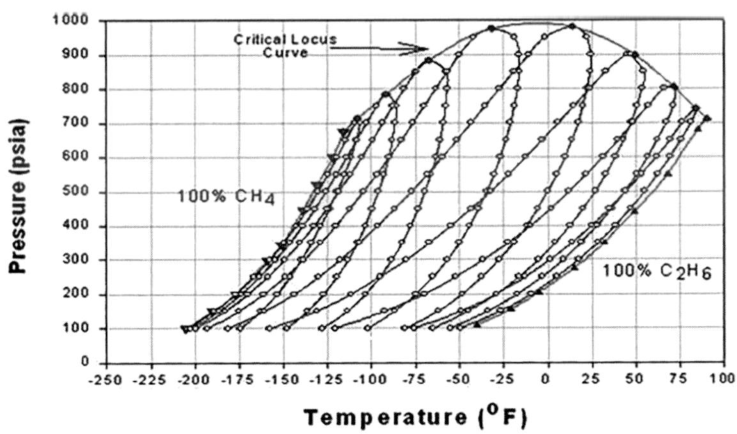

FIGURE 1.11
Effect of compositions on phase behavior.

1.8 P–x and T–x Diagrams

A P–x diagram for a binary system at constant temperature and a T–x diagram for a binary system at a constant pressure are displayed in Figures 1.12 and 1.13, respectively. The lines shown in the figures represent the bubble and dew point curves. End points represent the pure-component boiling points for substances A and B. In a P–x diagram, the bubble-point and dew-point curves bound the two-phase region. The single-phase liquid region is found at high pressures and the single-phase vapor region is

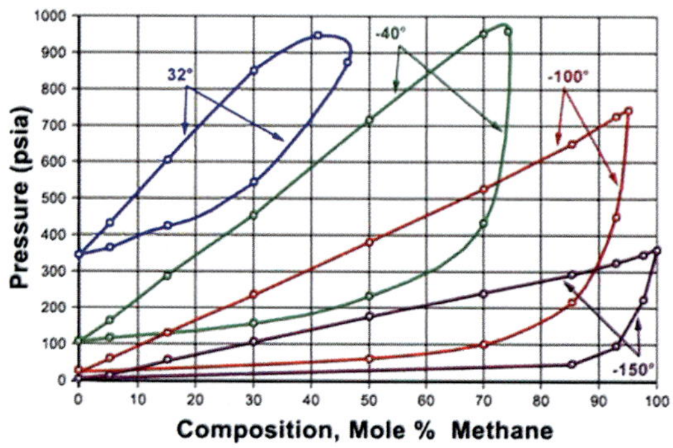

FIGURE 1.12
P–x diagram for binary system.

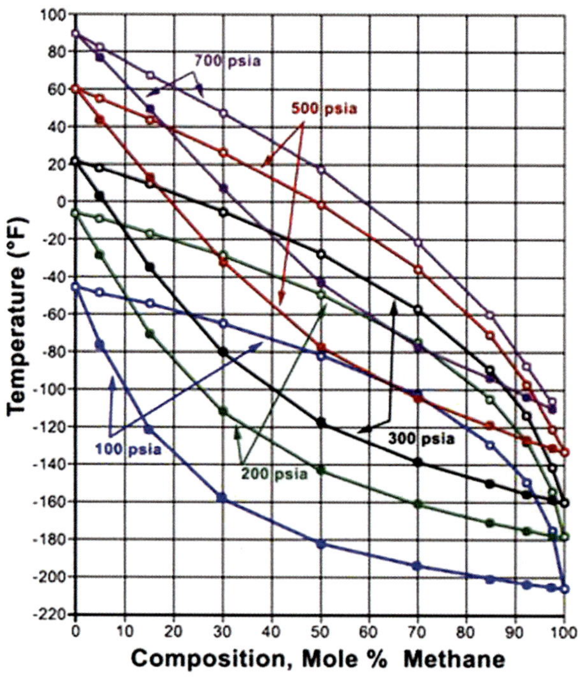

FIGURE 1.13
T–x diagram for a binary system.

found at low pressures. However, in T–x diagram the single-phase vapor region is found at high temperatures and the single-phase liquid region is found at low temperatures. Consequently, the bubble-point and dew-point curves occur at the bottom and the top of the two-phase region, respectively.

1.9 Retrograde Condensation

Critical point is the temperature and pressure at which liquid and vapor are same and indistinguishable. Within two-phase region, there can be temperature and pressure more than critical pressure and liquid. Thus, these can be defined as follows:

Cricondentherm (T_{cc}):

1. The highest temperature in the two-phase envelope.
2. For $T > T_{cc}$, liquid and vapor cannot co-exist at equilibrium, no matter what the pressure is.

Cricondenbar (P_{cc}):

1. The highest pressure in the two-phase envelope.
2. For $P > P_{cc}$, liquid and vapor cannot co-exist at equilibrium, no matter what the temperature is.

Retrograde phenomenon is explained in the below figure. When pressure of reservoir pressure (point 1) declines after production and reaches to dew-point pressure (point 2), the first drop of liquid is observed. Further reduction in reservoir pressure increases the liquid volume and reduces gas volume as depicted in the two-phase region. Liquid volume reaches to maximum at point 2 and further reduction in pressure causes reduction in liquid volume and increase in gas volume. Liquid starts vaporization and liquid volume become almost zero at low pressure. This phenomenon is called retrograde gas condensation.

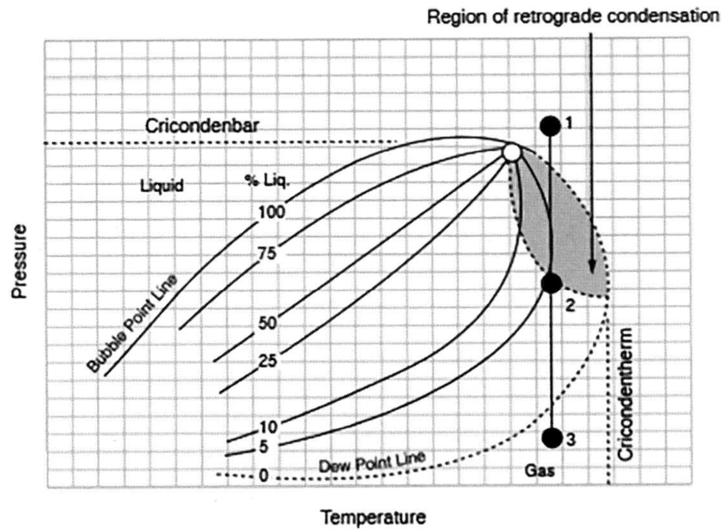

1.10 Multicomponent Phase Behavior of Hydrocarbon

The phase behavior of multicomponent hydrocarbon system is similar to binary system. However, mathematical and experimental analysis is complex. Hydrocarbons are defined using phase diagram.

Low-shrinkage oil (black and heavy oil)
High-shrinkage oil (volatile)
Retrograde gas condensate
Wet gas
Dry gas

1.10.1 Phase Behavior of Oil

1.10.1.1 Undersaturated Oil

This type of oil reservoir has bubble-point pressure lower than initial reservoir pressure. Difference in initial reservoir pressure and bubble pressure indicates whether reservoir fluid is moderately undersaturated or highly undersaturated. Separation from critical point is also noticeable. In heavy oil reservoirs, the difference between initial reservoir pressure and bubble-point is significant and initial GOR is very small.

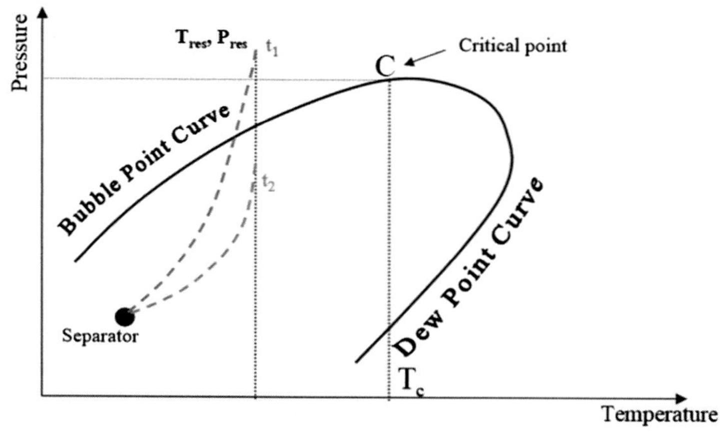

1.10.1.2 Saturated Oil

This type of oil reservoir has bubble-point pressure similar to initial reservoir pressure and initial pressure will lie on bubble-point curve.

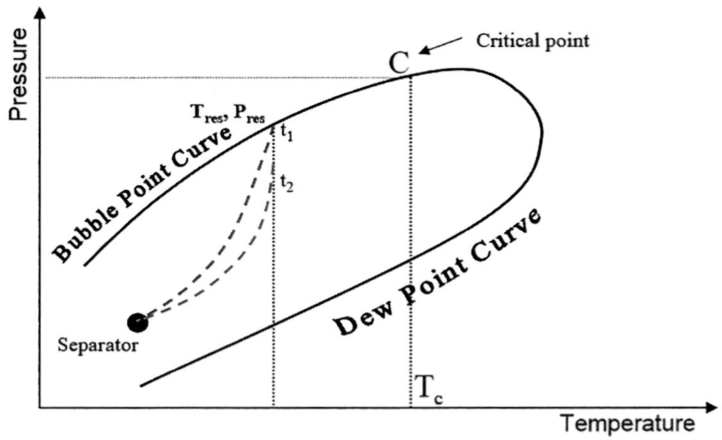

Some of the typical characteristics of black oil are as follows:

- Dark brown to black color
- GOR < 2,000 Scf/STB
- Formation volume factor < 2.0 Rb/Stb
- Oil gravity < 40° API
- Mainly liquid with significant heavier components C_{7+} > 30%

1.10.1.3 Volatile Oil

Initial condition of pressure and temperature of this type of reservoir fluid lies closer to critical point. The difference between initial reservoir pressure and bubble-point pressure is small, sometimes 100–200 psi. These are lighter compared to black oil. Some of the typical characteristics are as follows (Figure 1.14):

- Translucent orange to brown
- GOR < 2,000–3,000 Scf/STB
- Formation volume factor > 2.0 Rb/Stb
- Oil gravity > 40° API
- Mainly liquid with significant heavier components C_{7+} 12%–30%

1.10.2 Phase Behavior of Gas

1.10.2.1 Dry Gas

This type of fluid is presented in the phase diagram, at the extreme right of critical point and Cricondentherm line. Fluid is gas in the reservoir up

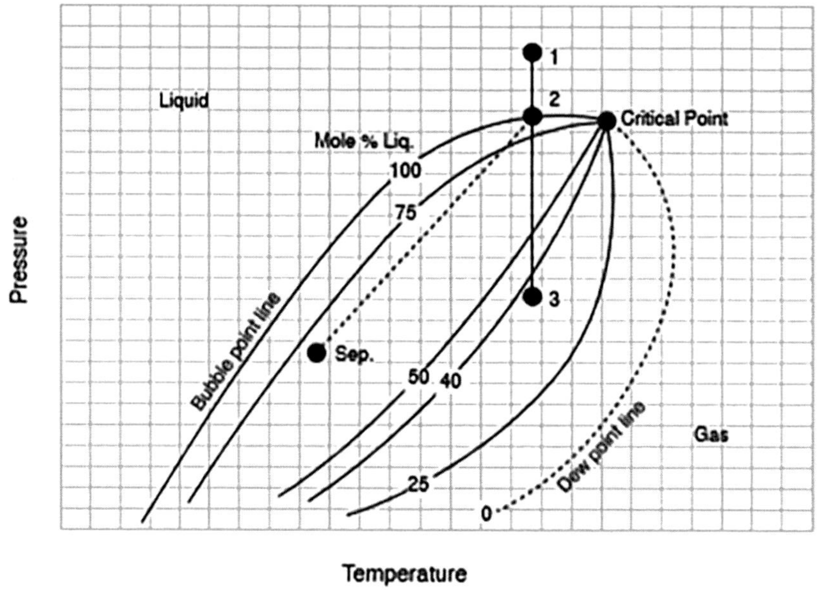

FIGURE 1.14
Phase diagram of volatile oil.

to separator conditions. No or very little liquid dropout takes place with pressure reduction. It always lies outside the two-phase envelop. Some of the typical characteristics are as follows:

- Mostly methane
- GOR > 100,000 Scf/bbl

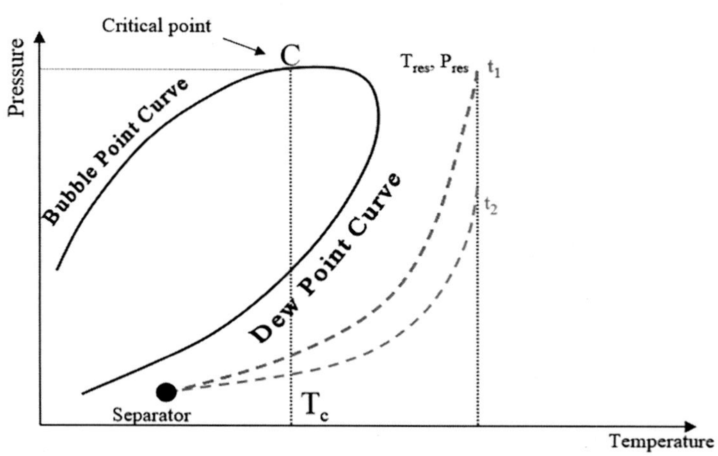

1.10.2.2 Wet Gas

Initial condition of reservoir fluid lies outside the two-phase envelop and right Cricondentherm line. Liquid dropout takes place and separator condition (*P*, *T*) lies within two-phase region. Some of the typical characteristics of wet gases are as follows:

- GOR < 100,000 Scf/bbl
- Condensate liquid > 50 degree API

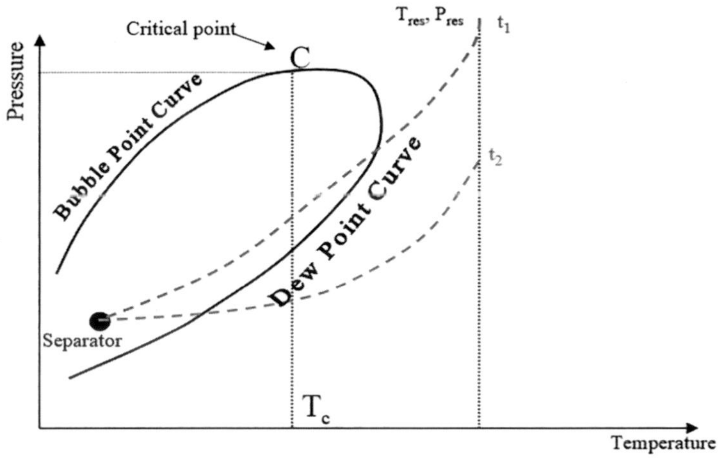

1.10.2.3 Gas Condensate

The initial condition of this type of reservoir fluid lies between the critical point and Cricondentherm line. The initial condition contains single phase. Reservoir pressure declines with production, and liquid dropout takes place at dew point. On further reduction in pressure, fluid enters in the two-phase region, and the liquid phase increases and gas phase decreases. In the later period at lower pressure, liquid starts vaporizing and enters the gas phase with insignificant liquid. Some of the typical characteristics of gas condensate reservoirs are as follows:

- GOR up to 60,000 Scf/bbl
- Condensate quality > 50 degree API
- Whitish color
- Higher percentage lighter hydrocarbon like propane and butane
- Liquid components C_{7+} < 1%–12%

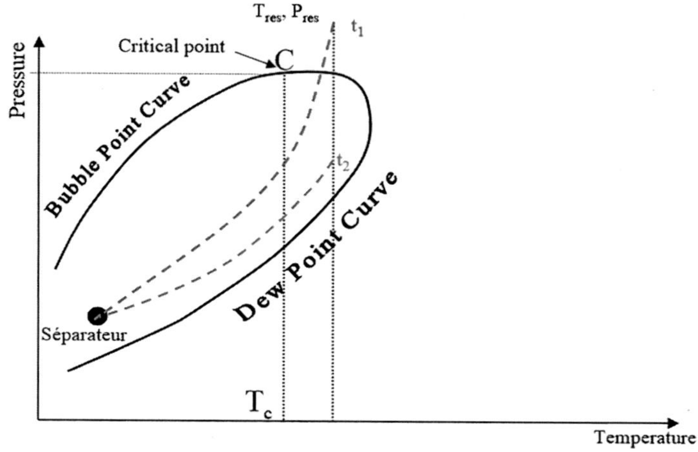

1.10.3 Comparison of Phase Diagram of Hydrocarbon Fluids

The following comparative table and graph gives an idea of all five fluid types. The black oil phase diagram has maximum two-phase region and critical point is in right most and dry gas phase diagram has the narrowest range of two phase and lies left side. The critical point is generally higher in volatile oil reservoir fluid and has the lowest of dry gas reservoir fluid.

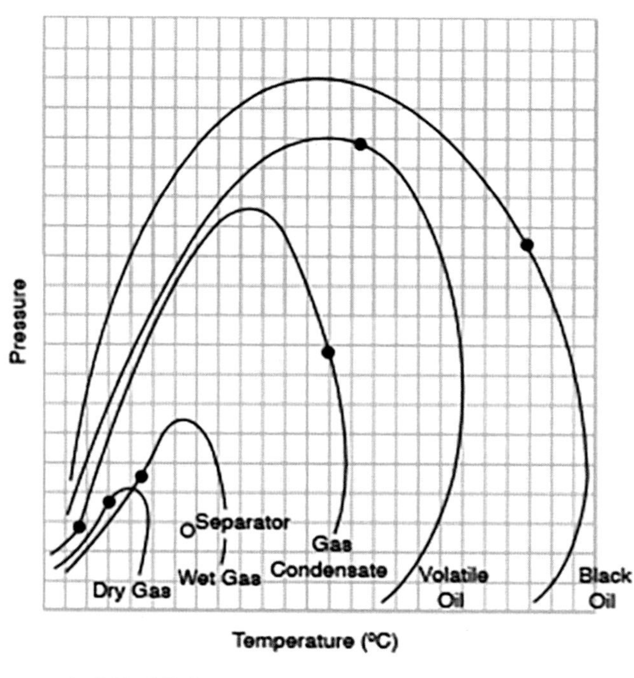

1.10.4 Phase Diagram of Reservoirs with Gas Cap

Oil reservoir with gas cap has two-phase envelops: one is for oil part and the other is for gas cap. Combined phase diagram of fluids lies in the middle. Interaction of vapor pressure line of gas cap and liquid line of oil part represents pressure at gas–oil contact. These points are depicted in following figure.

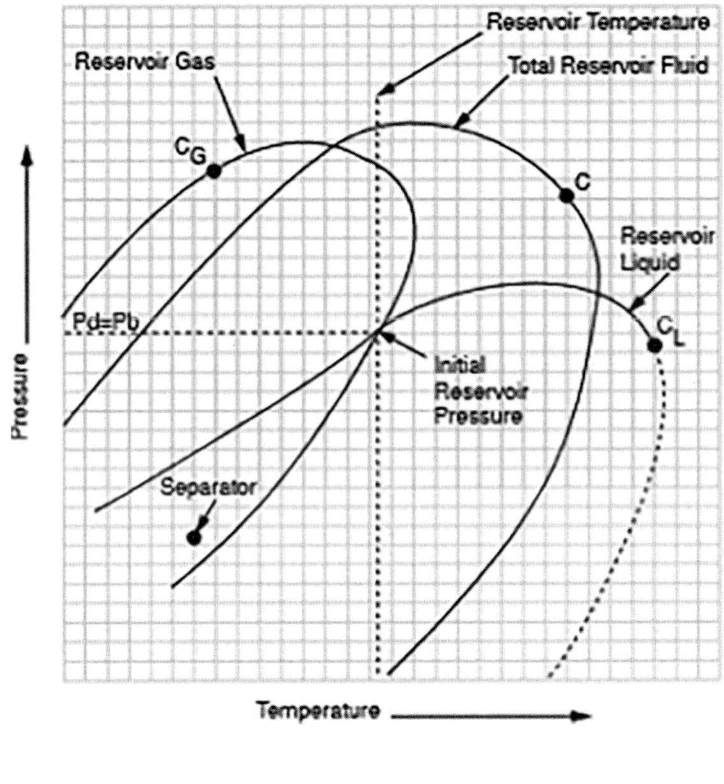

1.11 Ternary Diagram

Both phase plots and ternary diagrams show the shape and size of the two-phase region.

- A phase plot (or phase diagram) is at a fixed composition and is a function of pressure and temperature.
- A ternary plot is at fixed pressure and temperature and is a function of composition.

The ternary or triangular diagram was first introduced by Gibbs to represent phase relationships of three pure components. Multicomponent fluids can be displayed on the ternary diagram by grouping the *N*-component system down to three components:

- A light group, consisting of C1 and N_2
- An intermediate group, of CO_2, H_2S, and the hydrocarbons C2, C3...C6
- A heavy group, of all the heavy hydrocarbons, C_{7+}.

The detailed shape of the ternary plot will depend on exactly which components are grouped together, but the light group is always plotted at the top of the triangle, the intermediate at the bottom right, and the heavy group at the bottom left.

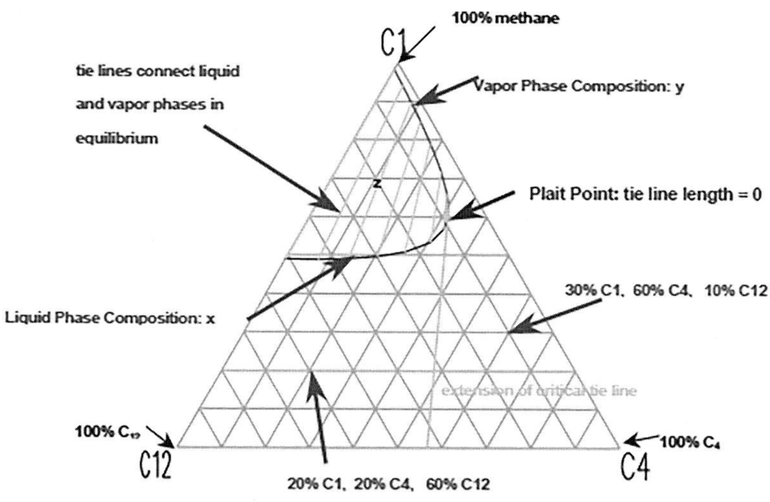

The phase envelope separates the single-phase and two-phase regions. The line at which a bubble of gas first appears is the equivalent of the bubble-point line in the phase plot, and the line at which a drop of liquid first appears is the equivalent of the dew-point line. The point at which these two lines meet is called the plait point, and is the equivalent of the critical point. At this point, the liquid and vapor compositions are identical. The ternary diagram is for all compositions of those three grouped components at a fixed temperature and pressure. A fluid with composition *z* at that temperature and pressure is shown to be in the two-phase region. It will therefore form an oil phase and a gas phase. The oil phase will have a composition *x* and the gas phase a composition *y*. Both of these are shown on the ternary diagram. The straight line joining *x*, *z*, and *y* is called the tie line. The tie lines get shorter as they get closer to the plait point. The tangent from the plait point to bottom axis effectively divides the fluid compositions into miscible and immiscible regions. More information on miscible/immiscible phase behavior is given in the chapter on miscibility.

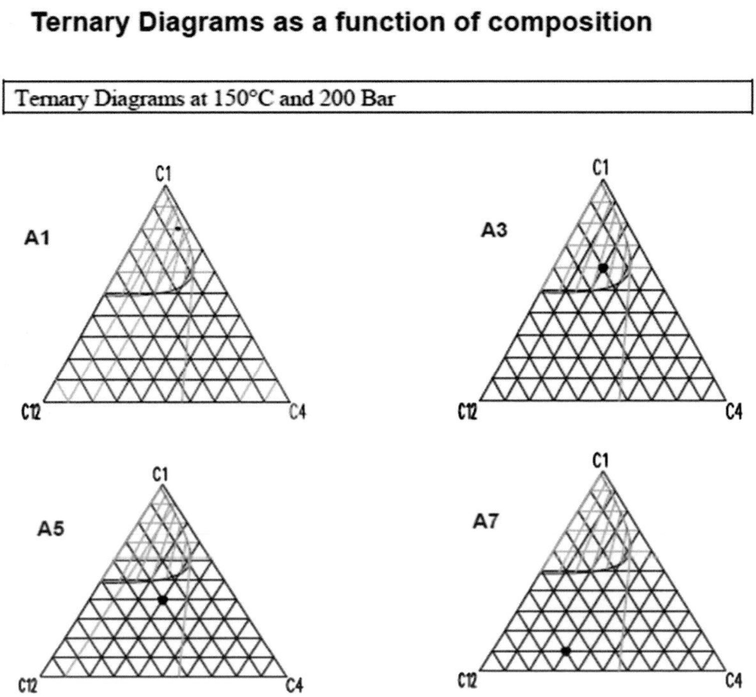

The four ternary diagrams above show four different fluids:

- Sample A1 is in the two-phase region and so will form two hydrocarbon phases. The sample is mainly composed of C1 and is close to the dew-point line, and therefore will be mainly gas with a little oil.
- Sample A3 is in also the two-phase region and so will form two hydrocarbon phases. The sample has less C1 than sample A1 and will be more oil than gas. Since A1 and A2 appear to lie on the same tie line, the composition of the oil from both samples will be the same, and the composition of the gas from both samples will be the same. The only difference will be the amounts of oil and gas.
- Sample A5 is in the single-phase region. We should avoid calling it either oil or a gas: the name of the phase will depend on the value of the temperature compared to the critical temperature. This can be determined by looking at a phase diagram but not in general at a ternary diagram.
- Sample A7 is also in the single-phase region.

1.11.1 Ternary Diagrams as a Function of Pressure

The following ternary diagrams show how the size and shape of the two-phase region changes with pressure at a fixed temperature of 500 K. The higher the pressure, the smaller the two-phase region. At high enough pressure, the two-phase region becomes so small that it cannot, in practice, be plotted.

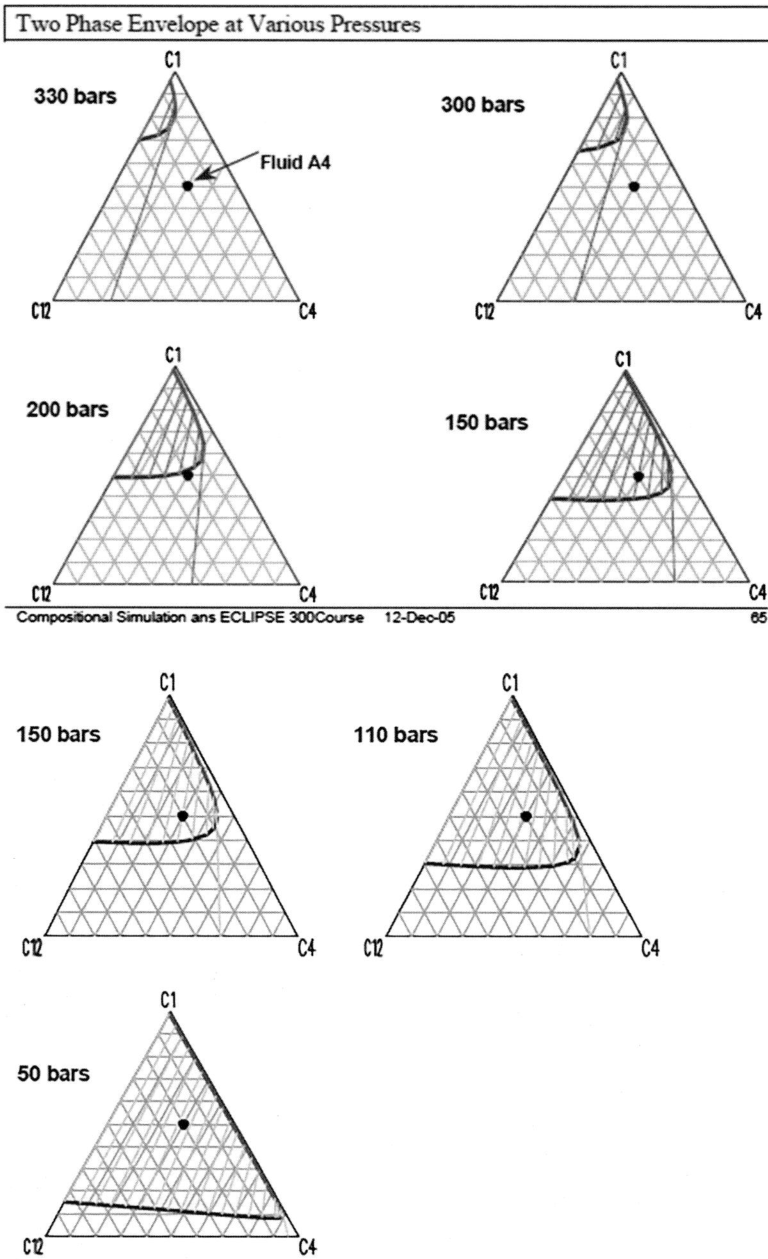

From these ternary diagrams, we can estimate the saturation pressure of the fluid sample to be just less than 200 bars. We can also call the sample an oil as we can see that the saturation pressure is a bubble-point pressure.

1.11.2 Equation of States

An equation of state (EOS) is a semiempirical functional relationship between pressure, volume, and temperature of a pure substance. This can be applied to a mixture by invoking mixing rules for various compositions. This has developed into an efficient and versatile means of expressing thermodynamic functions in terms of PVT data, and may be the best method for handling large amounts of PVT data. More than 300 years historical development of understanding the EOS is summarized as follows.

1.11.2.1 Period 1

- Before 1662, there was incomplete understanding of qualitative representation of the volumetric behavior of gases
- 1662: Boyle's law $PV = \text{constant}$
- 1787: Charles law $V \propto T$
- 1801: Dalton introduced the concept of partial pressure
- 1802: Gay–Lussac defined universal gas constant R
- 1822: Cagniard la Tour discovered the critical state (critical point)
- 1834: Clapeyron suggested $PV = R(T + 273)$

1.11.2.2 Period 2

- 1873: van der Walls provided most important contribution to EoS
- 1875: Gibbs defined thermodynamics of equilibrium
- 1902: Onnes theoretically confirmed critical states
- 1902: Lewis defined concept of fugacity
- 1927: Ursell proposed a series solution for EOS $PV = 1 + b/V + c/V^2 + d/V^3 + \ldots$. This virial EOS has better theoretical foundation.

1.11.2.3 Period 3

- 1940: Benedit Webb and Rubbin proposed "Cadillac" of EOS. It is complicated and easy to use
- 1949: Redlich and Kwong introduced temperature dependency in the attraction parameter "*a*" of vdW EOS

- 1955: Pizer introduced the idea of acentric factor to quantify the non-sphericity of molecules and was able to relate with vapor pressure data
- 1972: Soave modified the EOS by introducing the acentric factor
- 1976: Peng and Robinson proposed their EOS to have the best application for gas systems

The main equations of state are as follows:

Ideal gas law (Mariotte)	$PV = nRT$	
van der Waals	$P = RT/(V - b) - a/V^2$	(1873)
Redlich–Kwong	$P = RT/(V - b) - a/V(V - b)T½$	(1949)
Soave–Redlich–Kwong	$P = RT/(V - b) - a(T)/V(V + b)$	(1972)
Peng–Robinson	$P = RT/(V - b) - a(T)/(V^2 + 2bV - b^2)$	(1976)
Benedict–Webb–Rubin (pure components)		

$$Z = \{PV/RT\} = 1 + (1/V)\{BRT - A - C/T^2 + D/T^3 - E/T^4\}$$
$$+ (1/V^2)\{BRT - a - d/T\} + \alpha/V^5\{a + d/T\}c/VT^2\{1 + \gamma/V^2\}\exp\{-\gamma/V^2\}$$

Cubic equations of state. These equations are called "cubic" because volume determination at constant pressure and temperature needs to solve a third-order polynomial equation.

These can be presented as follows:

$$Z^3 + UZ^2 + VZ + W = 0 \text{ with}$$

$$Z = PV/RT \text{ and } U, V, W, \text{depending of two constants } A \text{ and } B$$

$$A = ap/R^2T^2$$

$$B = bp/RT$$

The number of solution of this equation gives the number of phases. The equations presented here are two-constant cubic equations.

1.11.3 van der Waals Equation

This equation differs from the ideal gas law by the introduction of two parameters:

- *a*: attraction parameter
- *b*: repulsion parameter

The term + a/V^2 represents an attempt to correct pressure for the forces of attraction between the molecules. The actual pressure exerted by the gas on the walls of the cell is reduced by the attraction between molecules. The volume occupied by the molecules reduces the total volume available; the constant *b* is regarded as the correction to the total volume and is often called covolume.

Modified the ideal gas law (becomes van der Walls equation $P + a/V^2$) $(V - b) = nRT$

The two constants *a* and *b* are defined by

$$\left(\delta_p/\delta_v\right)_{p_c,T_c,v_c} = 0$$

$$\left(\delta p^2/\delta v^2\right)_{p_c,T_c,v_c} = 0$$

The critical isotherm is tangent to the saturation curve at critical point in PV diagram and presents an inflexion point. *a* and *b* can then be expressed as follows:

$$b = 3/8\ RT_c/P_c$$

$$a = 27/64\ R^2T_c^2/P_c \quad \text{and}$$

$$V_c = 3/8\ RT_c/P_c \text{ resulting in } Z_c = 3/8 = 0.375$$

van der Waals equation has limited use, since it is accurate only at low pressure. The polynomial form of van der Waals equation is $Z^3 - (1+B)Z^2 + AZ - AB = 0$.

1.11.4 Redlich–Kwong Equation

The equation can also be written as follows:

$$P = RT/(V - b) - a/V(V + b)$$

The second term is made dependent also of temperature.

The constants *a* and *b* are determined in the same way as for van der Waals equation:

$$b = 0.08664\,RT_c/P_c$$

$$a = 0.42748R^2T_c^2/P_cT^{0.5}$$

RK EOS can also be written in polynomial form as

$$Z^3 - Z^2 + (A - B - B^2)Z - AB = 0$$

1.11.5 Soave–Redlich–Kwong Equation

Vapor–liquid equilibrium (VLE) of RK equation was improved by introducing a component-dependent correction α

$$\alpha = \left[1 + m\left(1 - T_r^{0.5}\right)\right]^2 \text{ and}$$

$$m = 0.480 + 1.574\omega - 0.176\omega^2$$

new a constant becomes (b being the same)

$$a = \alpha * 0.42748R^2T_c^2/P_c$$

The polynomial form of RK EOS remaining is the same. The SRK equation is the most widely used RK EOS. It offers an excellent predictive tool for accurate predictions of VLE and vapor properties. Volume translation is highly recommended when liquid densities are needed.

1.11.6 Peng–Robinson Equation

The PR EOS gives very similar results to SRK EOS. The values of a and b constants are as follows:

$$b = 0.07780\,RT_c/P_c$$

$$a = \alpha * 0.45724R^2T_c^2/P_c$$

$$\alpha = \left[1 + m\left(1 - T_r^{0.5}\right)\right]^2 \text{ and}$$

$$m = 0.37464 + 1.54226\omega - 0.26992\omega^2 \text{ for } \omega \le 0.49$$

$$m = 0.379642 + 1.48503\omega - 0.164423\omega^2 + 0.016666\omega^3 \text{ for } \omega > 0.49$$

PR EOS can also be written in the polynomial form as

$$Z^3 - (1-B)Z^2 + \left(A - 2B - 3B^2\right)Z + \left(AB - B^2 - B^3\right) = 0$$

1.11.7 Benedict–Webb–Rubin Equation

An equation with eight empirical constants was developed, which have been tested extensively for many hydrocarbons and also derived compounds. Modified BWR EOSs were proposed by various authors. As the cubic EOS, they should not be used outside the pressure and temperature ranges for which they have been tested.

Volume translation. Peneloux et al. used volume translation to improve volumetric calculations of the SRK EOS. It works also with PR EOS. The volume shift does not affect VLE calculations for pure components or mixtures.

A simple correction term is applied to the EOS calculated molar volume:

$$V_{corrected} = V - cV$$

EOS calculated volume

where c is the volume translation, component specific constant, and
c_i is determined for each component and
for a monophasic mixture

$$c = \sum x_i c_i$$

1.12 Comparative Assessment of RK, SRK, and PR EOS

Over the years, these EOSs have been tested, and some comparisons can be made. As an engineer, you have to be able to decide which EOS best fits your purposes.

1.12.1 Redlich–Kwong EOS

- Generally good for gas phase properties.
- Poor for liquid-phase properties.
- Better when used in conjunction with a correlation for liquid-phase behavior.

- Satisfactory for gas phase fugacity calculation @ $P_r < T_r/3$.
- Satisfactory for enthalpy departure and entropy departure calculations.

1.12.2 Soave–Redlich–Kwong and Peng–Robinson EOS

- Serve similar functions as the Redlich–Kwong EOS but require more parameters.
- PR obtains better liquid densities than SRK.
- Overall, PR does a better job (slightly) for gas and condensate systems than SRK. However, for polar systems, SRK always makes a better prediction, but in the petroleum engineering business we do not usually deal with those.

1.12.3 Z_c as a Measure of Goodness of an EOS

The experimental Z_c for various substances are as follows:

$CO_2 = 0.2744$
$CH_4 = 0.2862$
$C_2H_6 = 0.2793$
$nC_5 = 0.2693$
$nC_6 = 0.2659$

1.12.4 "Universal" Z_c Predicted by Different EOS

Ideal EOS = 1.000
vdW EOS = 0.375
RK EOS = 0.333
SRK EOS = 0.333
PR EOS = 0.301

Ideally difficult to get such average value of Z_c as 0.27 or so. None of the equations of state give this value.

1.12.5 Vapor–Liquid Equilibrium

VLE describes the distribution of a vapor phase and a liquid phase. Understanding of VLE is important in designing the following:

- Separators
- Reservoir

- Pipelines
- Wellbore
- LNG processing
- NGL processing
- Storage
- Oil and LNG tankers

The VLE data are represented in terms of *K*-values (vapor–liquid distribution ratios) and are defined by

$$K_i = y_i / x_i$$

where y_i and x_i are mole fraction of y and x phases.

K-values are also used for determining the numerical simulation.

1.12.6 Pressure and Temperature

Oil- and gas-bearing reservoirs occur at varying depth. Pressure and temperature increase with increasing depth due to overburden and fluid contained in the pore spaces. The pressure and temperature of the reservoir control the properties of fluids in reservoir. Hydrocarbon fluid composition, pressure, and temperature determine whether the fluid would initially exist in a single phase (oil or gas) or in two phases (oil with gas cap). It is a common practice to measure the pressure and temperature in exploratory and appraisal wells. These data are used to generate parameters that are used in various reservoir engineering calculations for volume estimation, filed development planning, and deciding the reservoir management plan.

Reservoir pressure plays an important role in determining numerical simulations where knowledge of pressure variation with time is required to understand the future filed performance and pressure maintenance strategies. Major methods for estimating the pressure are deploying downhole pressure gauges and through RFT/MDT at various depths. The estimation of pressure in reservoir is based on the following factors:

- Hydrostatic pressure gradient with fresh water
- Specific gravity of formation water
- Depth of reservoir

The hydrostatic pressure gradient of fresh water is 0.433 psi/ft. Here, psi stands for pound per square inch.

Formation water is not fresh. Number salts are dissolved in it and that increases the specific gravity of formation water. Interstitial water is fresh as it occupies the pore space during deposition of sedimentary rock whereas formation may enter the pore space post depositional period. Therefore, the hydrostatic gradient of formation water is expressed as follows:

$0.433\ \gamma_w$ psi/ft.

where γ_w is specific gravity of formation water. Typical range for hydrostatic pressure gradient varies between 0.433 and 0.50 psi/ft. If the gradient is more than normal hydrostatic gradient of fresh water, then reservoir has over pressure gradient and may have abnormal pressure. Certain gas reservoirs have been discovered with pressure gradient of 0.8 psi/ft. Reservoir would be sub-hydrostatic when the gradient is lower than normal pressure gradient of 0.433 psi/ft.

When depth of reservoir and specific gravity of formation water are known, the reservoir pressure can be estimated as follows:

$$P_r = P_{atm} + 0.433\gamma_w D \text{ psia}$$
$$= P_{atm} + \left[dP/dD\right]_w D \text{ psia}$$

where

P_r is reservoir pressure, psia,
D is depth of reservoir, ft,
γ_w is specific gravity of formation water, dimensionless,
$[dP/dD]$ formation pressure gradient, psi/ft.

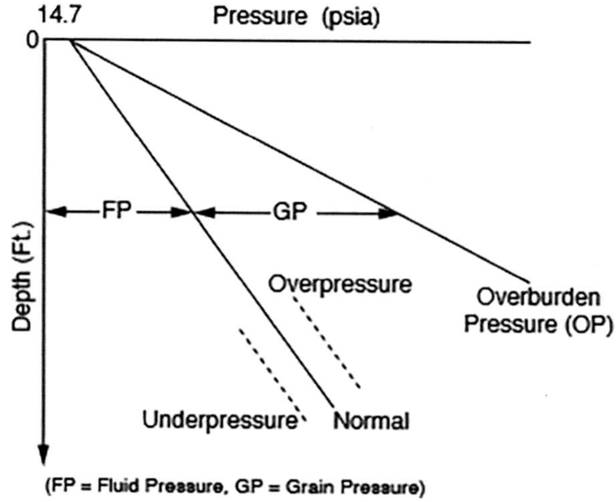

1.12.7 Fluid Contacts

RFT measurements provide fluid gradients and are used to estimate the gas–oil and oil–water contacts. This is depicted in Figure 1.15.

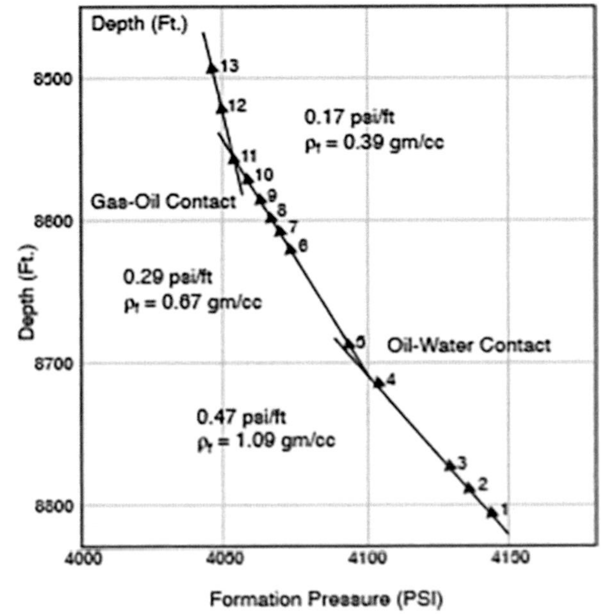

FIGURE 1.15
RFT pressure expressing the contact.

1.12.8 Reservoir Temperature

Estimation of accurate subsurface formation temperature is important and plays a key role in hydrocarbon resource estimations, well and surface facility designs. Accurate temperature assessment is required for the following reasons:

- Determining the fluid properties such as viscosity, density, formation volume factor, and solution GOR, etc.
- Initial volume-in-place calculation and performance analysis
- Applicability of sensitive EOR methods like chemical and miscible gas process
- Well head, tubular, and facility design

The temperature measured in a well is actually the temperature of the well fluid, not the formation. The drilling mud temperature is cooler than the formation and it invades the formation while drilling and cools it efficiently via heat convection. The temperature of the bore hole thus reaches equilibrium in between the drilling mud and formation temperature. Once the circulation is stopped, the borehole gradually regains the true formation temperature. This is achieved via a slower heat conduction process.

1.12.8.1 Impact of Temperature on Subsurface Parameters

Formation volume factor is crucial for volume calculations. Changes in temperature can affect the B_g value and thus the in-place numbers.

Gas formation volume factor, B_g. The formation volume factor relates a reservoir gas volume to a surface gas volume. It is directly proportional to temperature:

$$B_g = 5.035\frac{zT}{p} \text{ res. bbl/MSCF} \tag{1.1}$$

Gas initially in place. The GIIP has B_g in its denominator. Any increase in temperature increases the value of B_g. However, the increased value of B_g, being in the denominator, impacts on the GIIP by a decrease in value:

$$\text{GIIP} = \frac{Ah\phi(1-S_{wc})}{5.615B_g} \tag{1.2}$$

where GIIP is gas initially in place, Scf,

A is the area in ft^2,

h is the average pay thickness, ft,

ϕ is the porosity in fraction,

B_g is the gas formation volume factor, reservoir barrel/MScf,

P is the reservoir pressure, psia,

V is the volume of gas, ft^3,

z is the gas deviation factor and can be read from the chart of Standing and Katz,

R is the universal gas constant: 10.732 ft^3 psi/°R lb-mol,

T is the temperature in Rankine = T °F + 460,

S_{wc} is the connate water saturation in fraction.

1.12.8.2 Subsurface Gas Density

An increase in temperature can decrease the density of fluid in the given condition since, $pV = zRT$

$$\text{Subsurface gas density} = \rho = \frac{M}{V} = \frac{pM}{zRT} \tag{1.3}$$

Increasing temperature would decrease the subsurface density of gas in reservoir condition. Since the estimated pore volume remains the same, the total mass or moles of gas would reduce accordingly.

Subsurface gas pressure gradient (GPG) is the pressure exerted by one cubic ft of gas over an area of 1 square inch under reservoir condition. Once gas density is determined, GPG can be easily calculated as follows:

Gas Pressure Gradient, psi/ft

$$= \text{subsurface gas density}\left(\text{lbs/cub ft}\right)/144\left(\text{square inch/square feet}\right)$$

$$\text{GPG}, \frac{\text{psi}}{\text{ft}} = \frac{\rho}{144} = \frac{pM}{144\, zRT} \tag{1.4}$$

1.12.8.3 Subsurface Fluid Viscosity

With an increase in temperature, the viscosity of reservoir oil decreases. However for gases, gas viscosity increases are more visible at reservoir conditions (Figure 1.16).

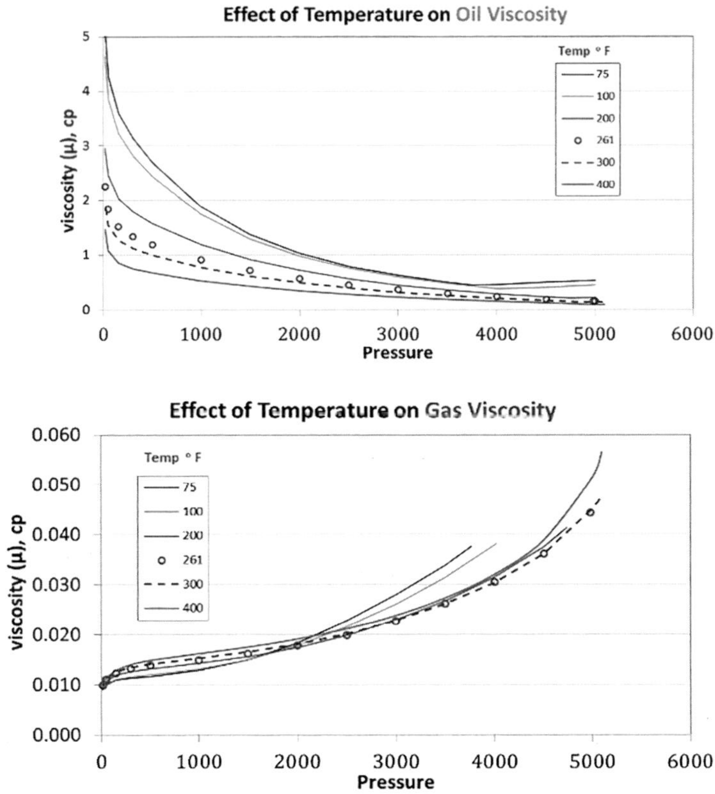

FIGURE 1.16
Impact of temperature on oil (same subsurface composition) and gas.

1.12.8.4 Geothermal Gradient

The geothermal gradient or geotherm is the changes in formation temperature with increasing depth. Its typical values are 20–35°C/km, although higher values (up to 85°C/km) can be observed in tectonically active area while lower values (0.05°C/km) in stable continental platforms have been seen. The geothermal gradient for a particular region, often assumed to be a constant, does show variation due to differences in thermal conductivities of rocks. A low thermal conductivity rock such as shale acts as a thermal insulator and has a large temperature gradient, while high thermal conductivity rocks, such as salt permits the conduction of heat efficiently, have a small temperature gradient across them. Temperature in Sarawak offshore, Malaysia at subsea 3,000 m depth is found to be approximately 150°C ± 20°C. However in Gulf of Mexico, USA at a reservoir depth of 7,000 m, the formation temperature is found to be approximately 90°C ± 10°C.

1.12.8.5 Sources of Temperature Measurement

- MDT: MDT measurements of pressure and temperature are made just after reaching the drilling target. Though MDT pressure data are the most accurate pressure of the reservoir at the drilled position, MDT temperatures are not advised to be used for temperature estimation as they are reportedly lower than the actual formation temperature, affected by drilling/mud circulation.
- DST: Temperatures recorded during pressure build-up (PBU) are treated as the representative static temperature. An important QC of static bottom hole temperature (SBHT) from PBU is to extrapolate it to perforation depth using appropriate thermal gradient and confirm that its value is higher than the FBHT recorded at gauge depth.
- Well temperature log: Each logging tool combination is equipped with a temperature sensor. Temperature logs are the only logs to be measured, while the sensor is being lowered into the well. Once the fluid circulation has stopped the temperature in the well gradually start to changes toward the actual formation temperature; thus, log measurements in successive wire line runs get closer and closer to the real formation temperature. Thus, the bottom hole temperature recorded needs to be corrected. The most common method is the Horner Plot.
- Analog: In the early stages of development and production of a field or reservoir, the measured bottom hole temperature data may be too sparse to provide a reliable estimate of initial conditions. In this case, a regional temperature data base is a handy reference for temperature estimation. Temperature of nearby oil and gas fields along with the available field data are required to be plotted against depth.

1.12.9 Reservoir Fluid Analysis

Reservoir fluid analysis is required to relate surface production of oil and gas to underground withdrawal at different pressures and temperatures. For most primary and secondary recovery mechanisms, isothermal conditions are assumed. The reservoir fluid analysis is required in the oil business to understand and address the following:

- The design of production facilities
- Compatibility with pipeline transport
- Product sales value
- Refinery maintenance costs
- Reservoir asset values in general

Thus, the analysis is carried out for the following:

1. PVT analysis for reservoir engineering analysis, the formation factors of oil, gas, and water B_g, B_o, B_w, gas/vapor oil ratios, R_s and R_v. These are important for hydrocarbon in place and reserves estimations.
2. Hydrocarbon composition to understand the fluid type and its economic value so as to decide what would be the most optimal production strategy. In total, 20% CO_2 in Kasawari gas has an impact on fuel value of produced gas. The shale gas has to be delivered, say, at a limiting CO_2 mol% $\leq$ 6.5%.
3. Knowledge of certain component contents for further planning and future drilling.
    a. Sulfur components
    b. CO_2
    c. Corrosiveness of fluid
4. Knowledge of the fluid's ability to flow through production tubing, pipelines, and other flow lines and possible problems that may arise because of viscosity changes, the precipitation of wax, and/or asphaltenes.
5. Determine the contaminants that can affect the plant design such as mercury content, sulfur contents, and radio active compounds.
6. If the saturation pressure is very close to the initial reservoir pressure, then a second phase may be present. This is particular relevant for gas reservoir where further drilling may discover an oil or condensate leg.

The use of reservoir fluid analysis for different PE disciplines is as follows.

- Reservoir engineering: Reservoir simulation, FDP, reserve estimation, MBAL, fluid flow in porous media, drive mechanisms, displacement efficiency, well test design, and interpretation.
- Production engineering: Artificial lift evaluation, completion design, material specification, and PLT interpretation.
- Facilities/process: Facilities design, flow assurance calculations, and pipelines design.
- Geology: Reservoir correlation, geochemical studies, hydrocarbon, and source studies.

1.12.9.1 Compositional Variations

General understanding and assumption is that fluid composition is constant with respect to depth in oil and gas reservoir. However, in light and gas condensate reservoirs with large thickness, the composition variation exists. The variation in composition exists due to following reasons:

- Gravity equilibrium
- The effect of curved surface
- Irreversible thermodynamics and steady state
- The effect of temperature gradient without convection
- The effect of temperature gradient with convection

1.12.10 Non-equilibrium Distribution of Hydrocarbons

Equilibration methods such as those described above are the normal method of initializing a simulation. The aim is to set up a static initial configuration: one in which phases present are in equilibrium and in which the inter-block flows are zero.

In some cases, however, it may be required to start a study from a point in existing production. An example might be initialization of a compositional study from the status of a black oil model at some stage in its history. As flow patterns have developed, this cannot be done by equilibration, and you must specify the initial pressures and saturations. In such a non-equilibrium initialization, a sufficient set of data is required to define the initial state.

References

1. Levorsen, A.I. 2006. *Geology of Petroleum*. New Delhi: CBS Publishers and Distributors.
2. Sattar, A. et al. 2008. *Practical Enhanced Reservoir Engineering*. Tulsa, OK: Penwell Press.
3. Bahdori, A. 2017. *Fluid Phase Behavior for Conventional and Unconventional Oil and Gas Reservoirs*. Cambridge, MA: Elsevier.
4. Pedersen, K. et al. 2015. *Phase Behavior of Petroleum Reservoir Fluids*. Boca Raton, FL: CRC Press.
5. Standing, M.B. 1977. *Volumetric and Phase Behavior of Oil Field Hydrocarbon System*. Richardson, TX: SPE.
6. Carlson, M. 2003. *Practical Reservoir Simulation*. Tulsa, OK: Penwell Press.
7. Kelkar, M. 2008. *Natural Gas Production Engineering*. Tulsa, OK: Penwell Press
8. *Reservoir Engineering*. UK: HWU.
9. Baker, R. et al. 2015. *Practical Reservoir Engineering and Characterization*. Oxford, MS: Elsevier.
10. Ahmed, T. 2016. *Equation of States and PVT Analysis*. Cambridge: Elsevier.
11. Lake, L.W. 2007. *Petroleum Engineering Handbook*. Richardson, TX: SPE.
12. Tissot, B.P. and Welte, D.H. 1984. *Petroleum Formation and Occurrence*. Berlin: Springer-Verlag.

2

Fluid Characterization and Recovery Mechanism

2.1 Introduction

Upon discovery, the next question is whether this field is worth developing. For proper and optimum development, an extensive knowledge of the geological environment of hydrocarbon accumulation has to be gained. The extent of reservoir has to be delineated to know the likely in-place oil and/or gas volumes and ascertain the presence of aquifer. The properties of reservoir fluids and their PVT relationships are required to be determined as accurately as possible to apply to hydrodynamic calculations especially in investigating the development of reservoirs, limiting conditions of well performances, level of recovery, and productivity from reservoirs.

To know, understand, and interpret the properties of reservoir fluid, and their movement within the reservoir during production phase, it is essential to study the in-depth reservoir rock-fluid systems. By combined knowledge of entire system, and the likely reservoir recovery processes, it is possible to forecast the recoverable reserves.

Before the development plan is evolved, it is essential to drill few delineation wells for collecting the basic information about the field such as

- Delineating the areal extent of reservoir/fields and develop understanding of the reservoir architecture
- Acquiring reservoir rock, fluid samples, and petrophysical properties (porosity, permeability, saturations, hydrocarbon-bearing thicknesses, etc.)
- Understanding depositional environment
- Drilling stem tests for well productivity
- Acquiring pressure data and their variations
- Understanding the OWC and GOC
- Understanding the reservoir energy and drive mechanisms

Understanding of reservoir fluid properties especially density and viscosity is very relevant for field development. Reservoir water is highly saline resulting in higher water density. The higher the water density, the greater is the density contrast between oil and water favoring oil recovery in vertical displacement. The higher temperature leads to lower oil and water viscosities. However, increasing the temperature on the viscosities of oil leads to a relative improvement in the relative mobility. Consequently, the risk of recovery losses through bypassing oil and channeling of water flow is normally less in high-temperature than in low-temperature reservoirs. Gas density and viscosity have no significant impact on the level of oil recovery.

The low-density contrast between oil and water has an adverse effect on oil recovery and high viscosity of heavy oil results in low productivity of wells, whereas the density difference between oil/water and low viscosities in medium and light oils causes a favorable mobility ratio and thus a chance of better recovery.

2.2 Reservoir Recovery Processes

Oil- and gas-bearing formations occur at varying depths and contain potential energy associated with them, which is termed as drive mechanism. These reservoirs may contain the following types of natural reservoir energy that controls the reservoir performance [1–4]:

- Rock-fluid expansion
- Solution gas or depletion drive
- Gas-cap drive
- Water drive
- Gravity segregation drive
- Combination drive

It is important to understand the drive mechanisms of the fields in order to formulate the appropriate field development plan (FDP). The objectives of FDP are as follows[5];

- FDP is an important document in upstream business of oil and gas industry.
- It comprises strategy and concept for development and management of oil and gas fields including secondary and tertiary recovery plans.
- A systematically prepared comprehensive FDP document incorporating all the influencing factors in evaluating, planning, and carrying out the oil and gas FDP is necessary for all the operating companies.

- This is a master document that dictates to know among others the most likely production rate, plateau period, producing life of the fields, facilities requirement, capital investment, and reserves.
- It also incorporates the appropriate reservoir management plan that is used to monitor the field performance and to carry out the corrective measures during production stage in case of deviation is observed with respect to prediction.
- Reservoir simulation is a widely used tool for making FDP decisions on the development of new fields, the location of infill wells, reservoir performance analysis, and the implementation of enhanced recovery projects.
- It is the focal point of an integrated effort of geosciences, petrophysics, reservoir, production and facilities engineering, computer science, and economics.

2.2.1 Rock-Fluid Expansion

An undersaturated black oil reservoir with initial reservoir pressure higher than bubble-point pressure away from critical point (cp) may produce oil at isothermal condition due to liquid and rock expansion. The expansion of rock occurs due to a decline in reservoir pressure and is accompanied by compaction of pore space. In this type of reservoir, pressure drops fast up to bubble point. The single-phase oil produced above the bubble-point pressure and evolution of gas takes place at surface condition when pressure is below the bubble point. This is weak drive mechanism and provides typical recovery factor of 1%–5%. Thus, compressibility of rock and oil plays an important role in calculation of in-place volume and reserves:

$$N = N_p B_o + \left(W_p - W_e\right) / B_{oi} C_t \left(P_i - P_{ws}\right)$$

where C_t is composite compressibility and is defined as follows:

$$C_t = C_o S_o + C_w S_w + C_g S_g + C_f$$

The other parameters are as follows:

N_p = cumulative oil production, bbl;

B_o and B_{oi} = formation volume factor at P_b and P_{ii};

W_e = water encroachment;

W_p = water production;

N = initial oil-in place;

P_i and P_{ws} = initial and average reservoir pressure, respectively.

2.2.2 Depletion or Solution Gas Drive Reservoirs

These types of reservoirs do not have gas cap or aquifer associated with them. Dissolved gas in oil provides the energy for flow. Stratigraphic trap reservoirs are often depletion drive reservoirs, due to the nature of trap, which precludes the presence of aquifer associated with them. Some of the typical characteristics of such reservoirs are as follows:

- Rapid pressure decline on production
- Water-free oil production
- Rapidly increasing gas–oil ratio (GOR) when reservoirs go below the bubble-point pressure and remain constant at R_{si} level up to P_b
- Above the bubble-point pressure—low compressibility of oil, connate water, and pore space—limited expansion and rapid decline in pressure.
- Below the bubble-point pressure—gas liberated from oil—its expansion provides the drive force and pressure declines less quickly
- Liberated gas can also migrate vertically and form secondary gas cap
- Due to rapid pressure drop artificial lift required in early years.
- Well locations low to encourage vertical gas migration
- Low ultimate recovery 5%–30%

Oil production from depletion drive is usually inefficient recovery methods. Arps and Roberts[6] presented statistical analysis of a large number of depletion drive reservoirs and suggested that recovery in such reservoirs depends on relative permeability ratio, oil gravity, and solution gas–oil ratio. Performance plots of depletion drive reservoirs are shown in Figure 2.1a, b. Generally, depletion drive reservoirs are low oil density, low oil viscosity, and high oil bubble-point pressure and require secondary recovery schemes to improve the recovery from the fields.

The material balance equation above the bubble point can be expressed as follows:

$$N_p/N = (B_t - B_{ti})/B_o + (R_p - R_{si})B_g$$

where

N_p = cumulative oil production, bbl;
N = STOIIP, bbl;
B_{oi}, B_o = formation volume factor of oil at P_i and P_b;
B_t = total formation volume factor;
B_{ti} = total Volume factor at initial condition;
R_p = gas oil ratio, Scf/bbl;
R_{si} = initial gas oil ratio, Scf/bbl;

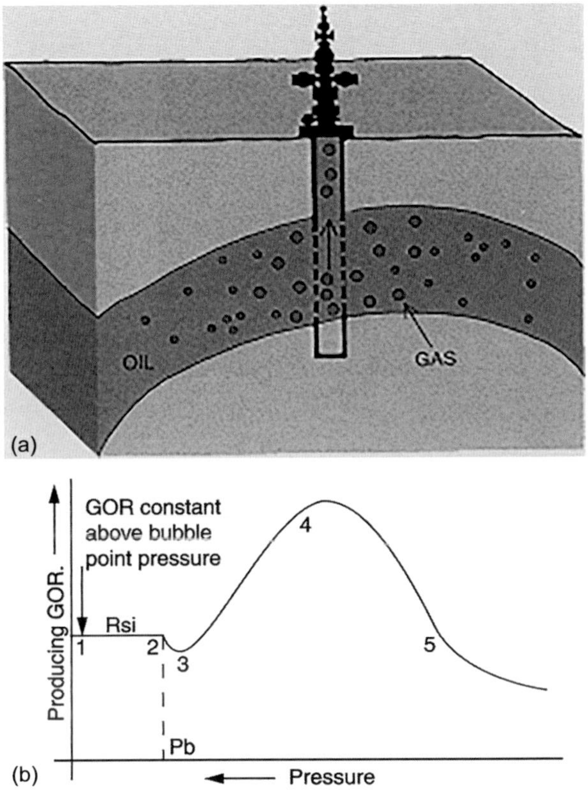

FIGURE 2.1
(a) Schematic of depletion drive reservoir. (b) Typical performance of GOR in depletion drive.

2.2.3 Gas-Cap Drive

Gas-cap reservoirs contain a free gas at the upper portion of reservoirs with little or no water at the lower portion of reservoirs. The size of gas varies and can be expressed as m, which is the ratio of gas volume to oil volume. Source of energy in gas-cap drive reservoirs is mainly due to highly compressible gas in gas cap and some energy from solution gas in oil. Oil expansion is low in such reservoirs. Some of the typical characteristics of gas-cap reservoirs are as follows:

- Production decline is less significant than solution gas drive
- Pressure declines slowly due to readily expansion of gas cap
- Initially, GOR is steady but it increases rapidly later as gas bypassing oil
- Negligible water production
- Well performance—longer, depends on gas-cap size

- Range of recovery factors 20%–40%
- Prone to viscous fingering due to higher mobility of gas. Therefore, in order to maintain the effectiveness of displacement oil by gas reservoirs should not be produced with very high rates (Figure 2.2a, b).

Well placements in gas-cap reservoirs are made away from GOC in a gas-cap drive oil reservoir. Some of the typical properties of gas-cap drive reservoirs are as follows:

- Low oil viscosity
- Relatively large ratio of gas cap to oil zone
- High reservoir dip angle
- Thick oil column

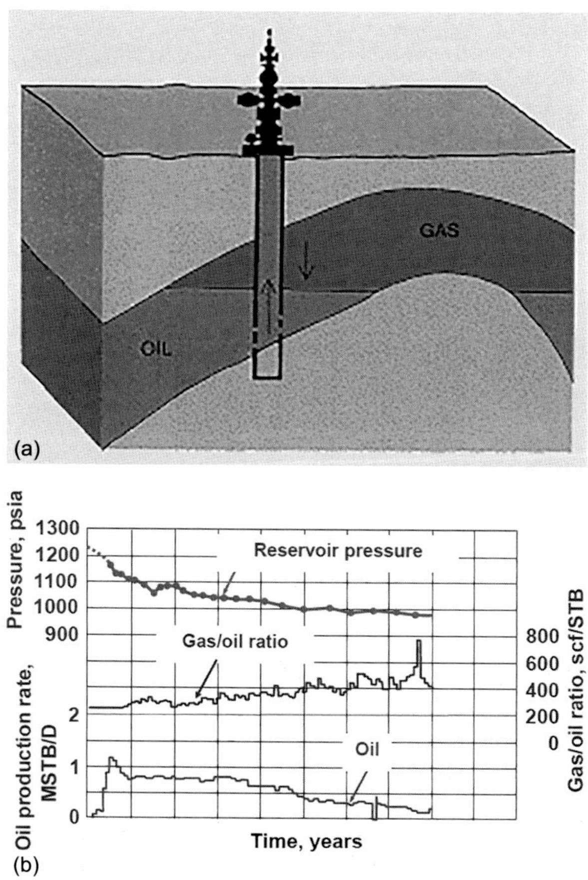

FIGURE 2.2
(a) Schematic of gas-cap drive reservoir. (b) Performance of gas-cap drive reservoirs.

2.2.4 Water-Drive Reservoirs

Water-drive reservoirs are more efficient in terms of recovery. They are associated with aquifer present in the lower part of reservoirs. The main source of energy in water-drive reservoirs is influx of aquifer water—edge-water or bottom water. The typical characteristics of water-drive reservoirs are as follows (Figure 2.3a, b):

- Pressure declines relatively slowly
- GOR is kept relatively low and stable
- Water production increases steadily
- Oil production declines when water breakthroughs
- Reasonable sweep—it depends on mobility ratio
- Bottom water-drive reservoirs are probe to coning

Some of the typical characteristics of water-drive reservoirs are as follows:

- Large aquifer
- Low oil viscosity
- High relative oil permeability
- Little reservoir heterogeneity and stratification

2.2.5 Combination Drive

Reservoirs generally contain combination of energy sources and drive mechanisms. A large aquifer and a large gas cap can occur in a reservoir simultaneously. Sometimes oil rim exists as sandwich between gas cap and aquifer. Dominant drive mechanism depends on the "strength" of each of

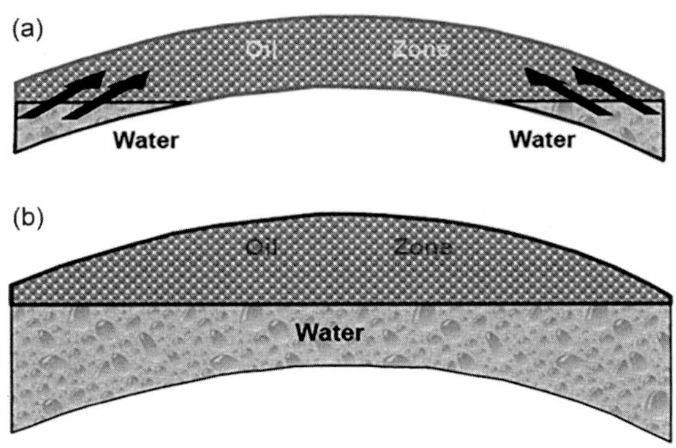

FIGURE 2.3
(a) Edge water-drive reservoir. (b) Bottom water-drive reservoir.

the source(s) of reservoir energy: size of original gas cap, "strength" of aquifer (size and permeability), and degree of pressure decline (Figure 2.4).

2.2.6 Segregation Drive

Gravity drainage may occur in any type of reservoir and can improve recovery when reservoir has sufficient time to act. This is very important in depletion drive and gas-cap reservoirs. A secondary gas cap can be formed in depletion drive reservoirs when liberated gas migrates to structurally up-dip of reservoir and displace the oil. In gas-cap drive reservoirs, oil in gas-invaded region drains down to rejoin oil column, reducing residual oil saturation, thus improving recovery efficiency (Figure 2.5).

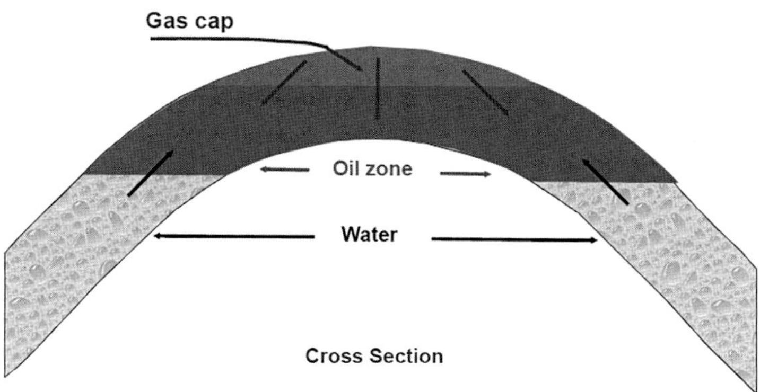

FIGURE 2.4
Schematic of combination drive reservoirs.

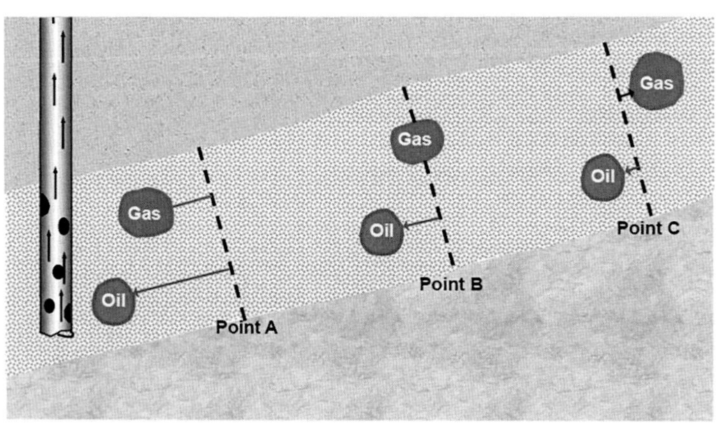

FIGURE 2.5
Schematic of segregation drive.

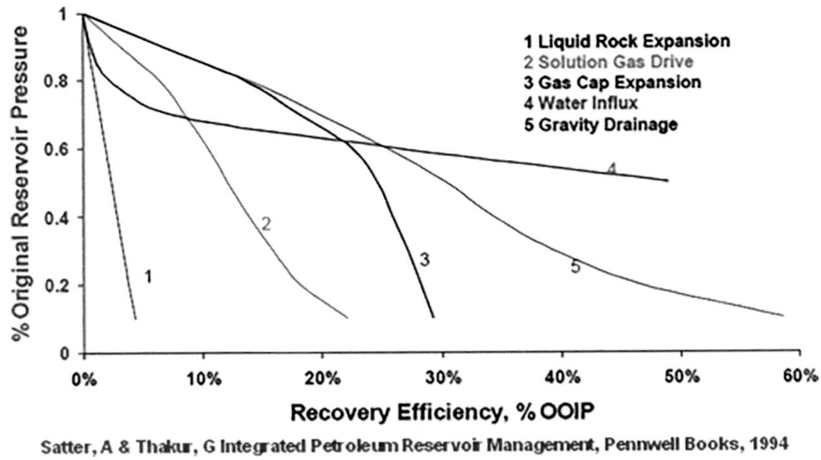

FIGURE 2.6
Comparative performance of oil fields with various drive mechanisms.[6]

Recovery efficiency for a gravity-drainage reservoir may lead to better than water-drive reservoirs (Figure 2.6).

2.2.6.1 General Material Balance Equation

$$N_p\left[B_o + B_g\left(R_p - R_s\right)\right] + W_pB_w - W_{inj}B_{winj} - G_{inj}B_{ginj}$$

$$= N\left\{\left[B_o - B_{oi} + B_g\left(R_{si} - R_s\right)\right] + \frac{B_{oi}}{B_{gi}}m\left(B_g - B_{gi}\right) + B_{oi}(1+M)\left(\frac{c_f + c_wS_w}{1-S_w}\Delta p\right)\right\}$$

$$+ W_gB_w$$

The above equation used fluid properties viz. formation volume factors for oil and gas at initial and different pressure conditions and compressibility values along with production and injection data.

2.3 Gas Reservoirs

Gas reservoirs have the following drive mechanisms[1–4]:

- Volumetric gas reservoirs
- Water-drive gas reservoirs

2.3.1 Volumetric Gas Reservoirs

These types of gas reservoirs are not associated with aquifer. Production from reservoir is pressure dependent. If the initial gas in place is known, then the ultimate gas recovery can be calculated as follows:

$$\text{Ultimate Recovery} = \frac{V \cdot \phi \cdot (1 - S_{wc})}{B_{gi}} \cdot \text{RF}$$

$$\text{HCPV} = V \cdot \phi \cdot (1 - S_{wc}) = G \cdot B_{gi}$$

Reservoir pressure drops with production. Corresponding to a drop in pressure $\Delta_p = P_i - P$, at the new pressure the gas FVF is B_g and cumulative gas production can be expressed as follows:

$$G_p = G \cdot \left(1 - \frac{B_{gi}}{B_g}\right)$$

Or

$$\frac{P}{z} = \left(\frac{P}{z}\right)_i \cdot \left(1 - \frac{G_p}{G}\right) \text{psia}$$

The ratio G_p/G is termed the fractional gas recovery at any pressure P and abandonment pressure as follows (Figure 2.7):

$$\frac{G_p}{G} = 1 - \frac{B_{gi}}{B_g}$$

$$\left.\frac{G_p}{G}\right|_{abandon} = 1 - \frac{B_{gi}}{B_{ga}}$$

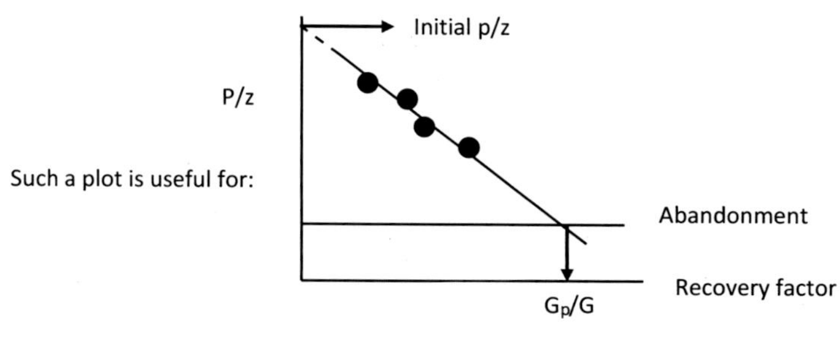

FIGURE 2.7
p/*z* plot of volumetric gas reservoirs.

The material balance equation indicates a straight line when p/z is plotted vs. G_p/G:

$$\frac{P}{z} = \left(\frac{P}{z}\right)_i \cdot \left(1 - \frac{G_p}{G}\right)$$

- Diagnostics—drive identification
- Determination of Initial conditions, original hydrocarbon in place, etc.

2.3.2 Water-Drive Gas Reservoir

There are gas reservoirs that have various degrees of aquifer support, and encroachment of water into the gas reservoir results in a decrease in pore volume after some production and a decline in pressure in gas column:

Production = Initial Gas-in-place Remaining gas

$$G_p = G - \frac{(\text{HCPV} - W_e)}{B_g}$$

$$= G - \frac{\left(G \cdot B_{gi} - W_e\right)}{B_g}$$

This can be expressed in terms of p/z as follows:

$$\frac{G_p}{G} = 1 - \frac{B_{gi}}{B_g} + \frac{W_e}{G \cdot B_g} = 1 - \frac{B_{gi}}{B_g}\left(1 - \frac{W_e}{G \cdot B_{gi}}\right)$$

$$\frac{p}{z} = \frac{p}{z}\bigg|_i \cdot \left(1 - \frac{G_p}{G}\right) \Bigg/ \left(1 - \frac{W_e}{G \cdot B_{gi}}\right)$$

The quantity $W_e/G{\cdot}B_{gi}$ is W_e/HCPV, the fraction of the HCPV that is invaded by the aquifer (≤1). The reservoir pressure is maintained at a higher level due to water encroachment.

2.3.3 Gas Condensate Reservoirs Production

Gas condensate is characterized as gas phase in reservoir and condensate dropout at surface condition. Assuming ideal gas law, the gas equivalent (GE) volume of N_p can be calculated as follows:

$$\text{GE} = V = \frac{nRT_{sc}}{p_{sc}} = \frac{350.5 \cdot RT_{sc}}{p_{sc}} = 133{,}000\frac{\gamma_o}{M_o}$$

Then the total gas produced is as follows:

$$G_p = G_{s1} + G_{s2} + G_{s3} + \text{GE} \cdot N_p$$

where G_{s1}, G_{s2}, and G_{s3} are separators. If the gas volumes from each separator are not measured, then the vapor equivalent is used from certain available nomograph:

$$G_p = G_{s1} + G_{s2} + G_{s3} + V_{\text{eq}} \cdot N_p$$

If V_{eq} is the equivalent volume of gas, then we have

$$V_{eq} = 133{,}000 \cdot \frac{\gamma_o}{M_o}$$

where $M_o = M$ wt of cond. liquid

This volume of gas will have a specific gravity of $M_o/28.97$.

Therefore, the sp. gravity of equivalent well fluid as a volume-weighted average of fluid sp. gr. is as follows:

$$\gamma_{well} = \frac{\gamma_1 \cdot R_{s1} + 133{,}000 \cdot \frac{\gamma_o}{M_o} \cdot \frac{M_o}{28.97} + \gamma_2 \cdot R_{s2}}{R_{s1} + 133{,}000 \cdot \frac{\gamma_o}{M_o} + R_{s2}}$$

If M_o is unknown, then it can be calculated from $M_o = \dfrac{42.43\gamma_o}{1.008 - \gamma_o}$

Detailed compositional analysis and liquid drop out with respect to pressure from, CVD experiment is required for condensate COMPOSITIONAL simulation modeling. Table 2.1 describes the fluid properties data for reservoir engineering calculations.

2.3.4 Fluid Characterization for Reservoir Simulations

Reservoirs under primary recovery stage utilize natural energy. Rock and fluid characterization along with energy systems in reservoir are required for understanding and predicting the performance of reservoirs. The following fluid properties are critical to have measured in the alb for engineering calculations[7]:

- Initial reservoir pressure
- Initial reservoir temperature
- Bubble-point pressure
- Dew-point pressure

TABLE 2.1

PVT Data Requirement for RE Calculations

Topic	Fluid Properties	Source of Data
Reservoir types	Pressure, temperature, composition of fluid	Laboratory analysis of fluid sample, well test, and log analysis
Volumetric estimates	Formation volume factor of oil and gas	Differential and flash lab experiment or correlations
Primary and secondary reservoir performance	Specific gravity, viscosity, Initial GOR and its variation with pressure, *Z* factor, compressibility of initial and varying reservoir pressure, formation volume factor of oil, gas, and water, bubble-point pressure and initial pressure, critical value of pressure and temperature, composition, MW	Laboratory analysis of fluid sample
Reservoir management and surveillance	Pressure and temperature, water and gas composition, hydrate, scale, WAX, and asphaltenes	Well test and log analysis, laboratory analysis of fluid
Design of facilities	Fluid composition, impurities, gas oil ratio, temperature, viscosity	Laboratory analysis
Enhanced oil recovery	Viscosity and its variation with temperature, IFT, miscibility, MMP, fluid compatibility with chemicals	Laboratory analysis

- Density of oil and gas
- Gravity of oil and gas
- Composition of gases
- *Z* factor for gas
- Formation volume factor of oil and gas
- Viscosity of oil, gas, and water
- Compressibility of oil, gas, and water
- Formation water salinity
- Condensate gas and oil ratio
- Size of aquifer and gas cap *m*

2.4 Secondary Recovery

Natural production depends on reservoir internal energy and arises due to the higher pressure in the rock pore space that is bottom of the well. This phase of recovery is called primary recovery. Reservoir energy diminishes with time, thus leading to lower primary recovery and leaving large of volume of oil inside the reservoir. Therefore, reservoir energy is required to be replenished to maintain the reservoir pressure and improve the oil recovery. The following are the two common methods of pressure maintenances[1–4]:

- Water injection
- Gas injection

2.4.1 Water Injection

Pressure maintenance by water injection is employed in solution gas reservoirs to enhance ultimate recovery. There are several advantages of initiating water injection at or near the bubble-point pressure prior to the development of a significant gas phase. One advantage is linked to the nature of favorable relative permeability. The oil and water relative permeabilities are higher when a gas phase is not present than when an appreciable gas saturation is established. Thus, for the same pressure drawdown at the wells, the oil production rate is higher. Another important advantage is that the rate of water injection required maintaining a certain level of production is lower. To maintain reservoir pressure, the injection rate must be equal to the total fluid withdrawal in reservoir volumes. The total fluid withdrawal associated with one surface volume is as follows:

$$B_t + \left(R_p - R_{si}\right)B_g$$

where
B_t = two-phase formation-volume factor, RB/STB;
R_p = cumulative produced GOR, Scf/STB;
R_{si} = initial gas in solution, Scf/STB; and
B_g = gas formation-volume factor, RB/Scf.

At or near to the bubble point, $R_p = R_{si}$ and the withdrawal is approximately $B_{ti} = B_{oi}$, Conversely, if pressure maintenance is started at a pressure level significantly below the bubble-point pressure, the value of $(R_p - R_{si})\ B_g$ can be significant and the total withdrawal associated with one surface volume is considerably larger than B_{oi}. The voidage replacement ratio (VRR) is expressed as follows:

$$\text{VRR} = \frac{W_I B_w + G_I B_g}{B_o N_p + B_g\left(G_p - R_s N_p\right) + B_w W_p - W_e}$$

where
W_I = water injected, m³/month;
B_w = water formation volume factor, m³/m³;
G_I = gas injected, m³/month;

B_g = gas formation volume factor, m^3/m^3;
B_o = oil formation volume factor, m^3/m^3;
N_p = oil produced, m^3/month;
G_p = gas produced, m^3/month;
R_s = solution GOR, m^3/m^3;
W_p = water produced, m^3/month; and
W_e = aquifer encroachment, m^3/month.

Here, B_o, B_g, B_w, and R_s are functions of that month's measured reservoir pressure. If G_p is less than $R_s{*}N_p$, then that term is set equal to zero.

Oil recovery can be determined at any time in the life of a waterflood if the following four factors are known:

- First, the amount of oil-in place (OIP) at the start of a waterfloods must be identified. This is a function of the oil saturation and floodable pore volume, which is highly dependent on parameters such as permeability and porosity cut-offs.
- Second, the areal sweep efficiency (the percentage of the reservoir area that the water will contact) must be known. It depends mainly on the relative flow properties of oil and water, the pattern and pressure distribution between injection and production wells, as well as directional permeability.
- Third, the vertical sweep efficiency (the fraction of a formation in a vertical plane that the water will contact) must be identified. It depends mainly on the degree of reservoir stratification.
- Fourth, the displacement sweep efficiency must be known. This represents the fraction of oil that will be displaced by water in the invaded zone.

By using Darcy's law, the oil and water fractions in the flowing stream can be calculated as follows:

$$f_o = \frac{q_o}{q_o + q_w}$$

and

$$f_w = \frac{q_w}{q_o + q_w}$$

By using Darcy's law, the oil and water rates can be expressed as follows:

$$q_o = \frac{AK\,K_{ro}}{B_o\mu_o}\frac{dp}{dx}$$

$$q_w = \frac{AK\,K_{rw}}{B_w\mu_w}\frac{dp}{dx}$$

Substituting the above equations provides the following fractional flow equations for oil and water:

$$f_o = \frac{\dfrac{K_m}{B_w\mu_o}}{\dfrac{K_m}{B_o\mu_o} + \dfrac{K_{rw}}{B_w\mu_w}}$$

and

$$f_w = \frac{\dfrac{K_{rw}}{B_w\mu_w}}{\dfrac{K_m}{B_o\mu_o} + \dfrac{K_{rw}}{B_w\mu_w}}$$

The fractional flows for water plots are extensively used in understanding the performance of water flood (Figure 2.8)

2.4.2 Gas Injection

Hydrocarbon gas injection is applied in oil reservoirs either in oil zone or in gas cap. When a gas cap originally exists in oil reservoir or has been formed due to segregation during primary recovery, gas injection helps to maintain the reservoir pressure while forcing gas into oil zones and driving the oil toward production well. When gas injection takes place in a reservoir

FIGURE 2.8
Fractional flow water.

without a gas cap, the injected gas flows radially from the injector wells, thus driving the oil toward production wells. The principal factor involved to decide the gas injection is the availability of gas in sufficient quantity.[1–4,8] Recycling of produced gas is a major source. Gas recycling or gas injection may give better recovery in a light oil reservoir with extension of gas cap or with high initial water saturation or into a reservoir with sufficiently high vertical permeability. Laboratory displacement studies are carried on core sample to measure the incremental oil recovery. Moreover, the PVT properties of injected gas are carried out for treating the gas for desulfurization, removal of CO_2, dehydration, etc.

2.4.2.1 Flue Gas Injection

Flue gas is used in miscible and immiscible flooding instead of hydrocarbon gas injection. It would be more economical than hydrocarbon gas, especially if it can be obtained easily from combustion sources such as power plants, gas turbines, gas engines, or heaters.

Flue gas can be quite corrosive, however, because it contains NO_2, SO_2, CO, CO_2, H_2O, and O_2. Other oil fields that have used flue gas for pressure maintenance have encountered corrosion problems, leading to, in some cases, the abandonment of the process. The flue gas should be dry. Corrosion problems in the production equipment can occurr where acid gases contact water in the reservoir. Flue gas is slightly more miscible than pure nitrogen and will therefore break through into the associated gas sooner than nitrogen. The cost of removing the CO_2 and N_2 from the produced sales gas would have to be added to the total cost at some point in the future.

2.4.2.2 Nitrogen Injection

Nitrogen has been used successfully in pressure maintenance and enhancing reservoir-drive mechanisms for both miscible and immiscible projects. It provides a higher reservoir displacement volume per standard volume of nitrogen than any of the other injection gases, which is the lowest volume requirement for pressure maintenance. The nitrogen sent to the field for injection has a purity of less than 10 ppmv oxygen. This is very low-concentration of oxygen to react with reservoir fluid and anaerobic activity is minimized. Nitrogen is also noncorrosive. Therefore, no special metallurgy is required for the injection equipment, no corrosion inhibitors need be added to protect the existing equipment, and no materials must be replaced on the existing production equipment. Nitrogen is environmentally acceptable and does not contribute to the greenhouse effect when it is rejected and released to the atmosphere. It is also readily available from air, and a plant can be located almost anywhere. Nitrogen production by cryogenic air separation is a proven technology. There are some limitations in terms of fluid characteristics: it has to be light, miscibility of nitrogen occurs at high pressure, and use of nitrogen as injection gas instead of associated natural gas releases additional gas for sale.[9]

2.4.2.3 Carbon Dioxide Injection

CO_2 is found to be a good potential factor for achieving the miscibility in the oil reservoir because of its lower cost and lower miscibility pressure than required for other miscible solvents such as hydrocarbon-based solvents (ethane, propane, etc.). A miscible displacement using CO_2 is achieved if we have high enough pressure and an oil degree API is generally greater than 30° API. The interfacial tension between the injected CO_2 and oil is very small. The oil swells and its viscosity is reduced as well. Residual oil saturation (SORG) decreases and moveable oil volume is increased. Incremental oil recoveries can be between 10% and 15% OOIP.

If the reservoir pressure and oil gravity is too low, the injected CO_2 will yield an immiscible displacement where recoveries will be lower. CO_2 is injected as a slug, repeatedly alternated with a slug of water (WAG). This helps to reduce the requirement of gas and controls the viscous fingering due to water injection alternatively.[10]

2.4.3 Gas Cycling

Recovery in gas condensate reservoir is adversely affected when reservoir pressure reaches below the dew-point pressure. Lean gas is injected in the reservoir in order to maintain the reservoir pressure above dew-point pressure. Then, condensate is collected at the surface and lean gas is injected back in the reservoir. This injection of lean gas and production from reservoir is continued until produced gas contains little condensate. Injected gas sweeps condensate to wellbore and also retards retrograde condensation. If there is water influx, such that reservoir pressure drops very little, then

$$\text{Recovery} = \frac{\left(1 - S_{wi} - S_{gr}\right) \cdot V \cdot \phi \cdot B_{gi} \cdot F}{\left(1 - S_{wi}\right) \cdot B_{gi}}$$

where F is the fraction of the PV swept by the water.

Phase behavior of gas condensate reservoir has to be studied to be modeled properly.[11]

2.4.4 CO_2 Sequestration

CO_2 sequestration in the form of injecting CO_2 for enhancing recovery of oil has been ongoing since the last 3–4 decades in US fields. Significant incremental oil (10%–15% of in-place volume) has been produced from these projects. One unknown in sequestering CO_2 in geologic formations is the length of time that the CO_2 remains in these formations. It is also known to cause asphaltenes precipitation in the reservoir, which could lead to formation plugging. Even though the CO_2 is dry, it will pick up moisture in the reservoir and increase the possibility of corrosion in the oil production equipment. In addition, as the CO_2 breaks through in the oil and then appears in the associated gas, it will increase the produced gas that must be treated to remove the CO_2 prior to sales.[12]

2.5 Tertiary Recovery

Primary recovery using only natural energy of the reservoirs typically recovers less than 50% of STOIIP, usually 30%–35% (Figure 2.9) and secondary recovery which is pressure maintenance by water and gas injection provides additional 10%–15%. Nearly, 70% of world's oil production comes from mature fields and majority of these fields are at various stages of declining phases of their producing life. The worldwide average recovery factor is poised to be around 35% with current development strategies and practices, thus leaving huge oil volumes underground in the reservoirs. The average recovery factors from offshore fields are lower compared to onshore fields predominantly due to larger well spacing, inadequate reservoir characterization, and shorter life cycles. The major reasons for low recovery include the following:

- Fields containing medium to high viscous crude with experiencing resistance to flow in porous medium

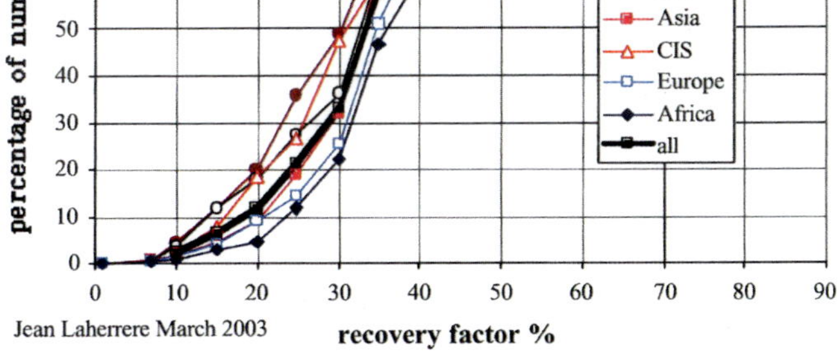

FIGURE 2.9
Trend of average recovery factor.

- Suboptimal location of producers especially in majority of small to medium-sized fields and less number of drainage points
- Viscous fingering in adverse mobility reservoirs and abrupt rise in water cut
- Decline in production due to pressure decline in some fields
- Reservoir pressure decline—inadequate WI and poor injection quality
- Higher residual oil saturation resulting in less-moveable oil volume recovery
- Sand production
- High pour point
- Wax and asphaltene problems

These are major challenges faced by upstream oil and gas industry. Some companies are making efforts to convert these challenges into opportunities by applying EOR techniques to maximize the recovery as remaining oil both in offshore and offshore fields are much larger. These efforts are gaining momentum which were not pursued earlier due to various constraints, such as higher CAPEX & OPEX, logistics, surface facilities constraint, and environmental regulations.

Current development strategy of the fields in primary mode with continuous optimization of artificial lift will lead to modest recovery. One of the challenges in offshore field is the limitation of the slots on the platforms, drilling limitations in reaching the suitable bypassed/undrained targets in the subsurface. Another major limitation in multi-stacked reservoirs is the limited applicability of horizontal wells, which can otherwise increase the recovery by controlling the drawdown for delaying the excess water and gas production along with oil. Sometimes these limitations compel to drill the suboptimal wells. At this stage of mature development, continuous drilling will help to improve the reserves, but this effort will be approaching limits for increasing recovery factor. Therefore, a new approach for maximizing recovery needs to be part of the redevelopment strategy with rejuvenation or enhancing the facility capacity.

One important point in improving resistance factor (RF) and making the EOR process successful is by not delaying the deployment in the field. By exhausting all primary and secondary development options and then embarking on tertiary EOR techniques applications may not be a prudent strategy here. Studies suggest that further delay may result in unviable EOR projects. Therefore, selection of appropriate EOR techniques, laboratory and modeling studies, staged implementation for better understanding and managing the uncertainties and risks, infusion of regular R&D efforts for improving the process and mitigation of challenges will make the process a success. Improving the well density and applying appropriate EOR in a phased manner with proper injection volumes along with good conformance control without delay in project initiation would result in an incremental RF. Tertiary recovery provides over 5%–20% primary and secondary recovery except heavy and extra-heavy oil where thermal EOR is the only technique to produce from these fields.

The selection of appropriate EOR techniques for a particular field depends mainly on the following factors:

- Rock and fluid characteristics
- Size of field
- Depth of reservoir
- Location of reservoirs
- Development stage of field
- Availability of identified injectants
- Available technology
- Mindset of company

The recovery efficiency is defined as follows:

$$\text{RE} = E_d * E_v$$

where

RE is recovery efficiency,

E_d is microscopic displacement efficiency, and

E_v is volumetric sweep efficiency which is product of vertical and areal sweep efficiencies.

Sweep efficiency depends on reservoir characteristics, mobility ratio (density and viscosity), and well spacing. Vertical sweep efficiency is strongly dependent on heterogeneity or stratification of reservoirs (Figure 2.10)

Mobility ratio is a function of viscosity and relative permeability, which depends mainly on saturations and can be expressed as follows:

$$M = \lambda_w/\lambda_o = \left[K_{rw}(S_w)/\mu_w\right]/\left[K_{ro}(S_o)/\mu_o\right]$$

where λ_w is the mobility of water K/μ and λ_o is the mobility of oil. Mobility ratio is a function of viscosity and relative permeability of water and oil.

The mobility ratio plays an important role in the displacement of oil in the reservoir. Ideal displacement takes place when mobility ratio is 1 or less than 1. Displacement in reservoir is not ideal or like piston when mobility ratio is more than 1. These types of reservoirs face problem of viscous fingering. To improve the overall recovery from the reservoirs, all three factors, microscopic displacement efficiency, vertical and areal sweep efficiencies,

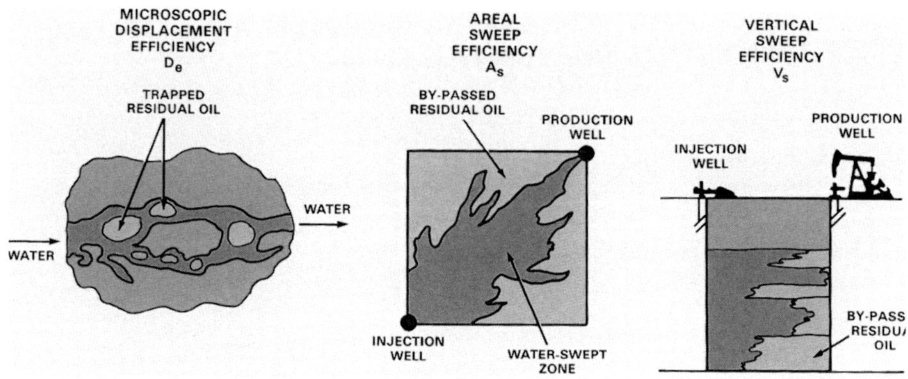

FIGURE 2.10
Schematic of various sweeps in reservoir.

have to be enhanced. If we take an example where microscopic displacement efficiency is 70%, areal and vertical sweep efficiency is also 70%, then RE becomes 35%:

$$RE = 0.7 * 0.7 * 0.7 = 0.35$$

If we apply economic efficiency of 70%, then over all economical recovery becomes 24.5% which is very low. This suggests that enhanced oil recovery requires careful planning for improving all the influencing parameters. If we improve displacement and volumetric sweep efficiency to 90% and E_c 90%, then RE becomes

$$RE = 0.9 * 0.9 * 0.9 = 73\%$$

The following three are widely-used EOR processes[8–10,13–20]:

- Chemical EOR
- Gaseous EOR
- Thermal EOR

There are three dimensionless numbers, capillary number, bond number, and gravity number, which are interplay of capillary, viscous, and gravity forces and govern the flow in porous media. Interplay of these forces becomes important in various EOR processes.

Capillary number describes the relationship between viscous and capillary forces and is expressed as follows:

$$N_c = v\mu/\sigma$$

where
v is Darcy velocity of injected fluid in m/s,
μ is viscosity in cP, and
σ is interfacial tension in mN/M.

Typical value of capillary number in water-injected case is 10^{-7} to 10^{-5}. This has to be reduced to level of 10^{-3} in the case of EOR.

Bond number is defined as the relationship between gravity and capillary forces and can be expressed as follows:

$$N_b = \Delta\rho \cdot g \cdot k/\sigma$$

where
$\Delta\rho$ is the density difference between oil reservoir and injected fluid, kg/m^3;
g is gravitational constant, m/s^2;
k is vertical permeability in m^2; and
σ is interfacial tension in N/m.

Gravity number is defined as the relationship between gravity and viscous forces and can be expressed as follows:

$$N_b = \Delta\rho \cdot g \cdot k/v \cdot \mu$$

where
Δ_ρ is the density difference between oil reservoir and injected fluid, kg/m^3;
g is gravitational constant, m/s^2;
k is vertical permeability in m^2;
μ is viscosity of injectant, cP; and
v is velocity of injectant in m/s.

2.5.1 Chemical EOR

The chemical EOR involves a variety of techniques. Some of these techniques are defined as follows:

- Polymer flooding
- Surfactant flooding
- Alkaline surfactant polymer flooding

The main objectives of the chemical EOR processes are to modify the microscopic and macroscopic sweep of the reservoir. Adding chemicals to increase the viscosity of displacing fluid improves the macroscopic sweep efficiency and surfactant in injected water reduces the IFT between water and oil in pores and increases the moveable oil by reducing the residual oil saturation, thus improving the microscopic efficiency. Alkali is also added in injected water which reacts with acidic component in the crude oil and generates in situ soap and reduces the IFT. Additional alkali reduces the requirement of surfactant and adsorption of expensive surfactant on rocks. Various combinations of alkali, surfactant, and polymers or standalone chemical such as alkali and surfactant (AS), surfactant and polymer (SP), alkali– surfactant and polymer (ASP) are applied in the fields depending upon rock and fluid characteristics and dominant flow mechanism in the reservoir.

Polymer is one of the successful CEOR applications. Two types of polymer are common in use.

2.5.1.1 Polysaccharides Biopolymer

- Polysaccharides or biopolymers are formed from a bacterial fermentation process. This process leaves substantial debris in the polymer product that must be removed before the polymer is injected.
- The polymer is also susceptible to bacterial attack after it has been introduced into the reservoir. These disadvantages are offset by the insensitivity of polysaccharide properties to brine salinity and hardness.
- Polysaccharides are more branched than are HPAM's and the oxygen-ringed carbon bond does not rotate fully; hence, the molecule increases brine viscosity by snagging and by adding a more rigid structure to the solution (Figure 2.11).
- Biopolymers are mainly water-thickening agents. They increase the viscosity of the polymer solution. They do not materially affect the water relative permeability—K_{rw}:
    - Xanthan gum
    - Guar gum

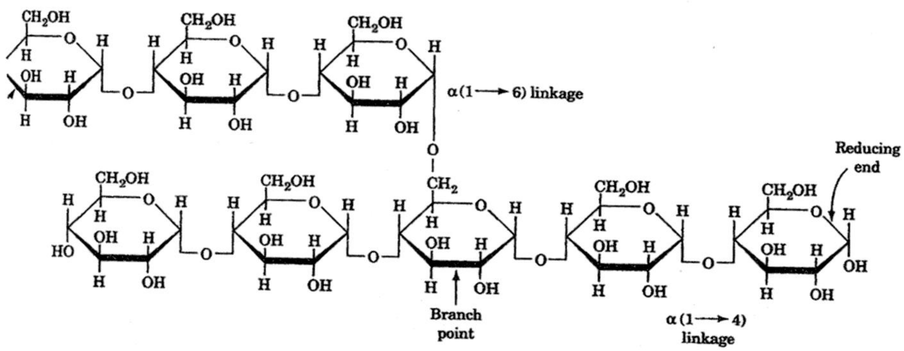

FIGURE 2.11
Molecule of polymer.

2.5.1.2 Synthetic Polymer

- This class of polymer is synthetically generated. Partially hydrolyzed polymers increase the viscosity of the polymer solution and decrease K_{rw}. PAMs of polymers are observed. Monomeric unit is the acrylamide molecule.
- Typical degrees of hydrolysis are 30% or more of the acrylamide monomers; hence, HPAM molecule is quite negatively charged.
- The viscosity-increasing feature of HPAM lies in its large molecular weight (MW), which is accentuated by the anionic repulsion between polymer molecules and segments on the same molecule:
  - Poly acrylic acid
  - Polyacrylamide (PAM)
  - Hydrolyzed polyacrylamide (HPAM)
  - HPAM/AMPS

The polymer added in the injected water increases the viscosity of displacing fluid and makes the mobility ratio more favorable for oil displacement. In order to make the polymer flood successful, it is important to characterize the rock and fluid properties extensively. The following parameters need to be studied for appropriateness of polymer flooding:

- Reservoir fluid viscosity and drive mechanism
- Reservoir temperature
- Properties of polymer at ambient and reservoir temperature
- Formation salinity
- Fluid–fluid interaction—the interaction with formation water and injection water with polymer
- Viscosity variation of injected water with polymer
- Variations of polymer properties in different shear conditions
- Rock properties in terms of permeability, clay contents, and mineralogy
- Optimization of concentration of polymer and displacement studies for incremental oil estimation
- Optimization of slug size in terms of pore volume for injection
- Change in mobility and mobility ratio

Success of polymer flooding depends on proper characterization of reservoir rock and fluid, polymer, and interaction of polymer with rock and fluid. Change in fluid properties like viscosity of injection water due to polymer is measured in the lab.

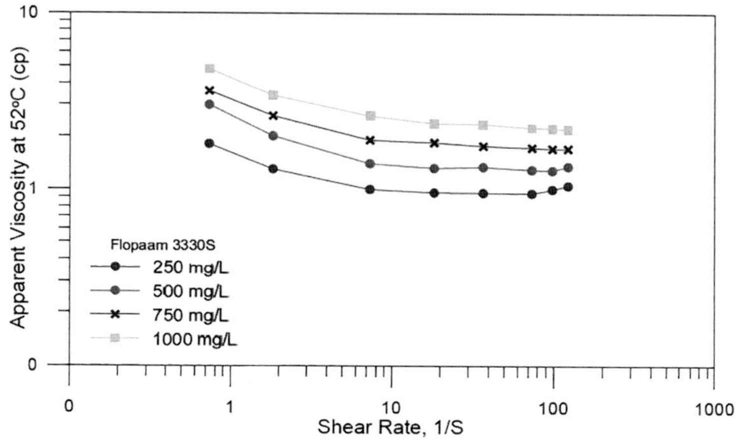

FIGURE 2.12
Shear rate, concentration, and apparent viscosity relationship.

2.5.2 Effect of Concentration and Shear Rate on Viscosity

Polymer molecules have very high MW. The higher the concentration of polymer solution, the higher will be viscosity. In addition, polymers with high MW will have longer chain, so viscosity is higher. Small chain polymers will have lower viscosity. Therefore, higher concentration has to be used for maintaining higher viscosity (Figure 2.12).

2.5.3 Rheology of Polymer and Shear Impact

Apparent viscosity of polymer changes with change in the shear rate. When a polymer solution is injected into a reservoir, the flow rate, which is related to shear rate, will change from wellbore to reservoir and that will cause the solution viscosity will also change from near wellbore to reservoir correspondingly.

2.5.4 Salinity Impact

Salinity of water has an adverse impact on polymer solution viscosity. Distilled water or less-saline water preserves the viscosity, whereas water with higher salinity adversely impacts the viscosity of polymer solution. Adsorption of polymer also increases with concentration (Figure 2.13).

2.5.5 Adsorption

One of the major problems in polymer injection is degradation of polymer molecule with time and its adsorption on pore surfaces. Adsorption increases with increasing the concentration of polymer in the solution. The schematic behavior is shown in Figure 2.14.

RF is defined as the ratio of mobility of brine and polymer and can be expressed as follows (Figure 2.15):

$$\text{Resistance Factor} = \text{brine mobility}/\text{polymer mobility}$$

$$= (k/\mu)_{brine} / (k/\mu)_{polymer}$$

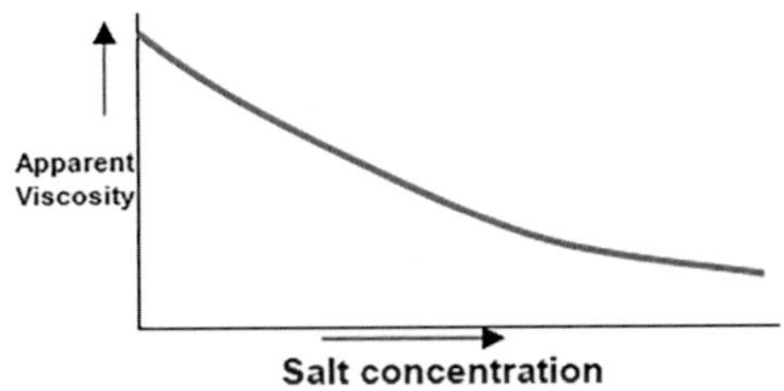

FIGURE 2.13
Impact of salinity on polymer solution viscosity.

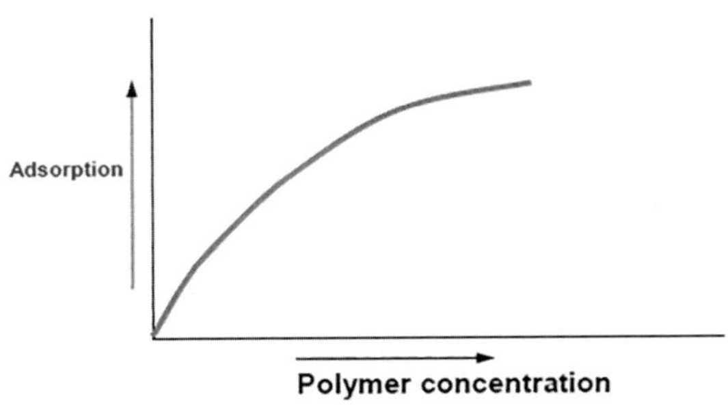

FIGURE 2.14
Polymer adsorption trend with concentration.

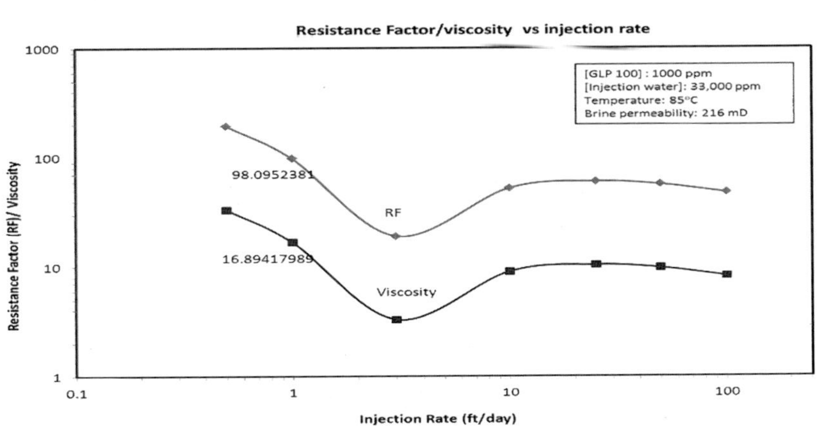

FIGURE 2.15
Residual factor and residual resistance factor.

Residual resistance factor (RRF) is used to indicate the absolute permeability reduction after polymer flow (Figure 2.15):

$$\text{Residual resistance factor} = \text{Initial brine permeability} / \text{Final}(\text{after polymer}) \text{brine permeability}$$

MW has a direct impact on viscosity of polymer. Higher MW polymers will provide the large viscosity, while the lower MW polymers will provide the small viscosity. In order to maintain the similar viscosity, low MW polymer concentration should be high but the advantage over high MW polymer would be less entrapment.

Polymer flood is adopted at an early stage in waterfloods because of its capability to control the breakthrough and improve areal sweep efficiency. Injection of polymer not only improves the area sweep but also reduces the water cut. The fractional flow behavior change is shown in Figures 2.16–2.18.

A schematic graph shows the incremental recovery due to polymer flooding.

The major challenges observed in polymer flooding are as follows:

- Lower injectivity than with water can adversely affect oil production rates in the early stages of the polymer flood.
- Acrylamide-type polymers loose viscosity due to shear degradation, salinity, and divalent ions.
- Xanthan gum polymers cost more, are subject to microbial degradation, and have a greater potential for wellbore plugging.

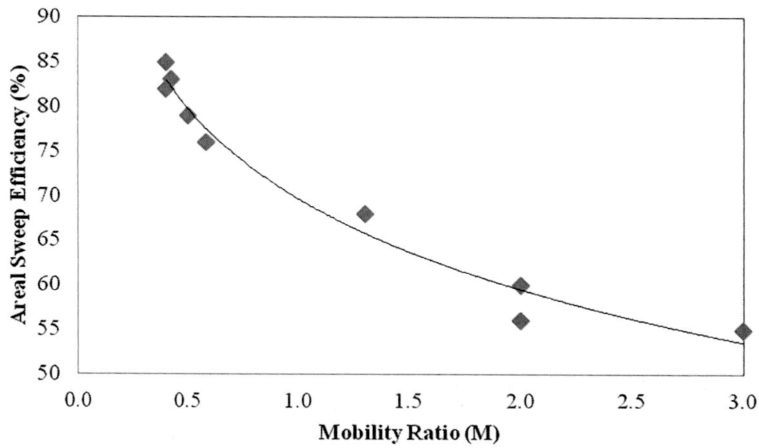

FIGURE 2.16
Mobility ratio vs. areal sweep efficiency.

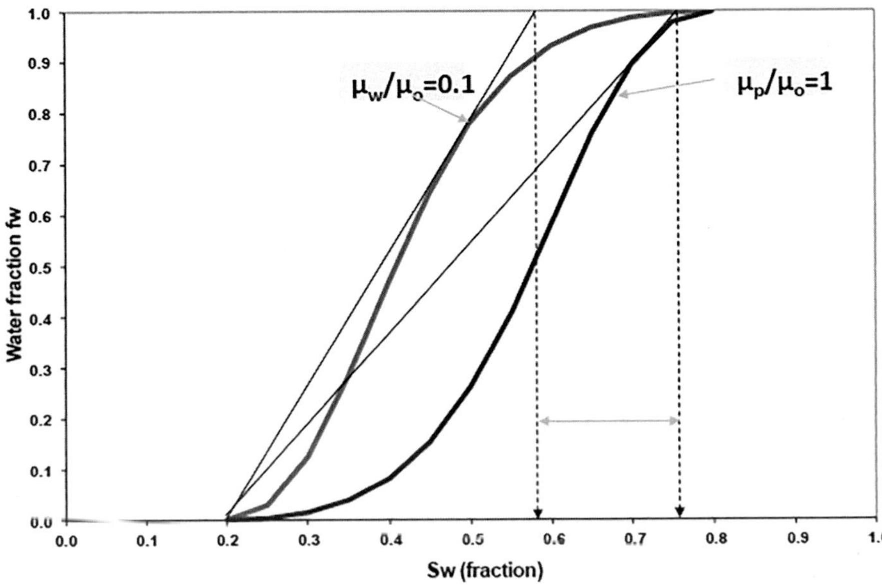

FIGURE 2.17
Modification in fractional flow due to polymer flooding.

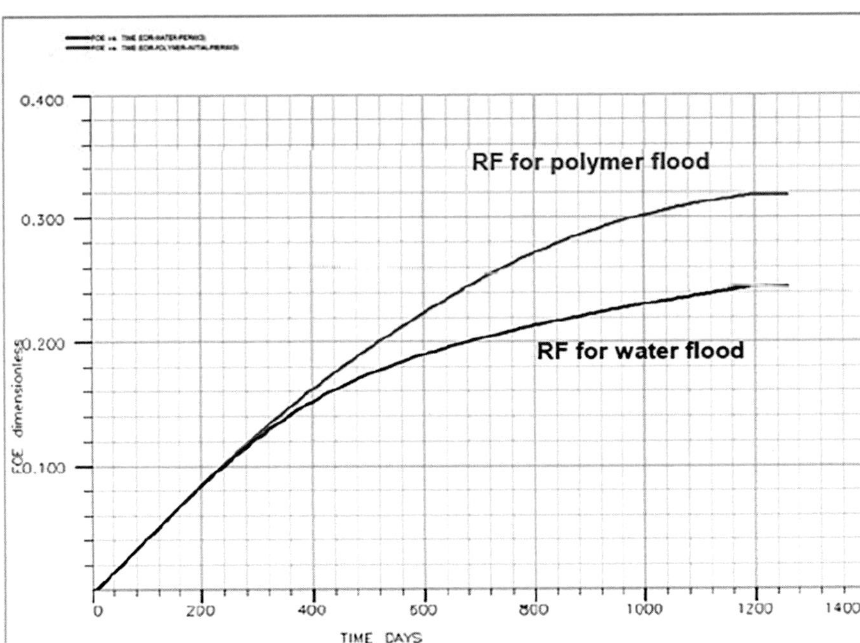

FIGURE 2.18
Incremental oil recovery due to polymer flooding.

2.5.6 Surfactant Flooding

Surfactants are surface active agents (Figure 2.19). It is used to reduce the IFT and in-turn reduction in the S_{ORW} and to increase the moveable oil volume. The suitability of surfactants is studied in the lab along with fluid properties:

- Compatibility with formation water
- Phase behavior changes due to surfactant
- Solubilization of surfactant
- Reduction in IFT
- Adsorption of surfactant
- Thermal stability of surfactant
- Capillary number
- Critical micellar concentration (CMC)
- Emulsion formation
- Optimization of concentration and displacement studies for estimation of residual oil saturation change due to surfactant
- Optimization of slug size and duration of injection

Surfactants can be classified into the following three categories:

Anionic (−). Negatively charges in aqueous solution and surfactant head has negative charge.

Cationic (+). Positively charges in aqueous solution and surfactant head has positive charge.

Non-ionic surfactant. Neutral and surfactant head is larger than tail group.

- Anionic and non-ionic surfactants are commonly used in CEOR
- Anionic surfactant is widely used because it is relatively stable, has lower adsorption on reservoir rock, and can be manufactured commercially

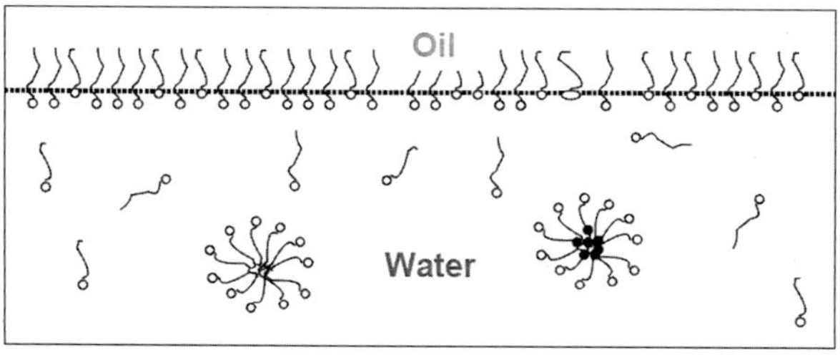

FIGURE 2.19
Surfactant in oil–water system.

- Non-ionic surfactant is widely used as co-surfactants to improve robustness of chemical formulation system
- Cationic are not commonly used—high adsorption to reservoir rock
- The CMC is defined as the concentration of surfactants above which micelles form and all additional surfactants added to the system go to micelles. The CMC is an important characteristic of a surfactant.
- IFT between surfactant solution and hydrocarbon phase is a function of salinity, temperature, surfactant concentration, surfactant type and purity, and the nature of hydrocarbon phase (Figure 2.20).
- IFT decreases sharply with increasing the surfactant concentration until CMC. However, a little change can be observed in IFT beyond CMC (Figure 2.21).

2.5.6.1 Fluid–Fluid Interactions

Water for dissolving ASP chemicals must be compatible with surfactant, polymer, and alkali. Precipitate formation is undesirable because it can cause formation damage, i.e., pore plugging and reduced injectivity. Also, alkaline agents such as Na_2CO_3 react with the acids present in the oil to form in situ surfactants that enhance the final ASP performance. It is also important to know the phase behavior between a surfactant solution and the target oil. Phase behavior or spontaneous emulsification testing is important because it helps to define those surfactants that reduce interfacial tension to ultra-low values. The synergistic effect of combining alkalis with surfactants and some polymers is another advantage of the ASP technology. To account for these

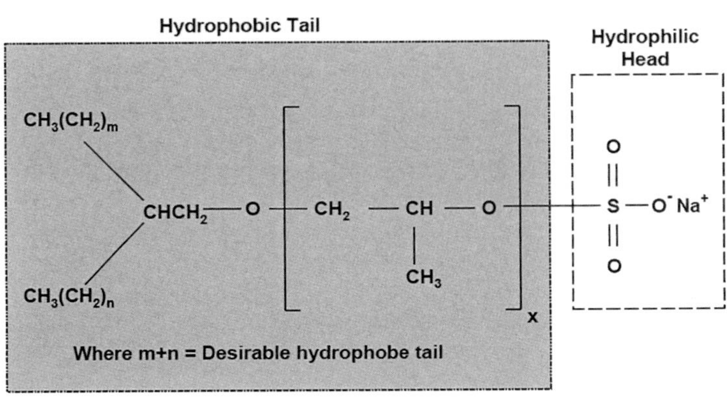

FIGURE 2.20
Surfactant molecule.

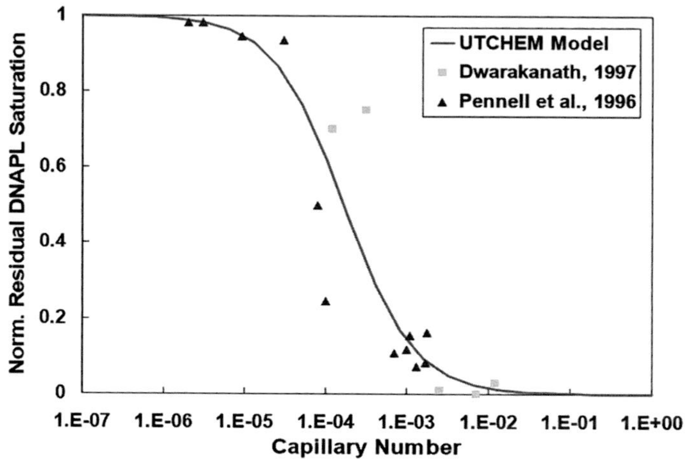

FIGURE 2.21
Capillary desaturation curve.

effects, compatibility tests must be performed with the solutions reformulated as necessary.

2.5.6.2 Phase Behavior

- Effect of electrolyte
- Oil solubilization and IFT reduction
- Microemulsion densities
- Surfactant and microemulsion viscosities
- Coalescence times
- Identify optimal surfactant-co-solvent formulations
- Identify optimal formulation for coreflood experiments

Ternary diagram is used to understand the phase diagram of surfactant solution. The vertices of equilateral triangle represent surfactant, oil, and water. Multiphase regions will be bounded by continuous binodal curve. Everywhere above bimodal curve single phase and that may change as composition changes. In the multiphase region, three-component systems involve two phases throughout: one is oil-external and the other is water-external. The following three types of microemulsion are formed:

- Winsor type-I surfactant forms oil-in water microemulsion. Aqueous phase containing brine and surfactant is external part. In this type of phase behaviour ultra low interfacial tension is achieved.
- Winsor type-II surfactant forms water in oil emulsion and surfactant is lost to oil phase. This is not a good case for EOR application.

- Winsor type-III surfactant separates microemulsion phase which is continuous in oil and aqueous phase. Ultra-low IFT 0.001 dynes/cm can be achieved. This is ideal surfactant type for EOR application for reducing the residual oil saturation. A schematic of these surfactant types are depicted in Figure 2.22. Phase diagram of surfactant characterization along with brine and oil is expressed using ternary diagram. Two phases and three phases regions are shown in Figure 2.23

Laboratory studies are carried out to understand the phase behavior and to study the impact of:

- Thermal stability of surfactant
- IFT reduction
- Microemulsion densities
- Surfactant and microemulsion viscosities
- Adsorption of surfactant
- Displacement studies for estimation of residual oil saturation
- Optimization of concentration of surfactant formulation
- Effect of salinity on surfactant efficiency

2.6 Gas-Based EOR

Gas-based EORs both in miscible and immiscible forms are widely used EOR techniques. The most popular techniques are as follows:

- Hydrocarbon gas flooding
- Miscible CO_2 gas flooding

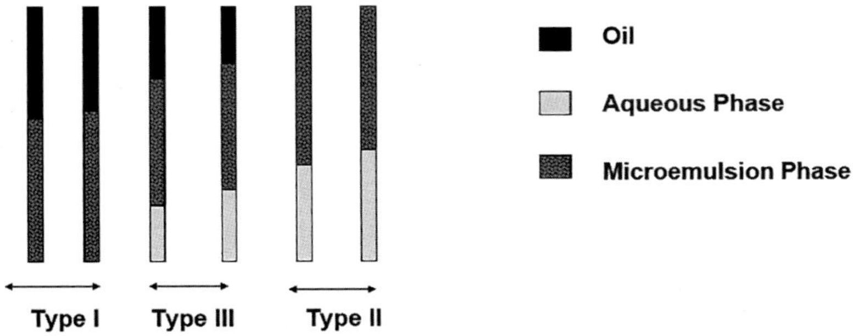

FIGURE 2.22
Microemulsion types with surfactant.

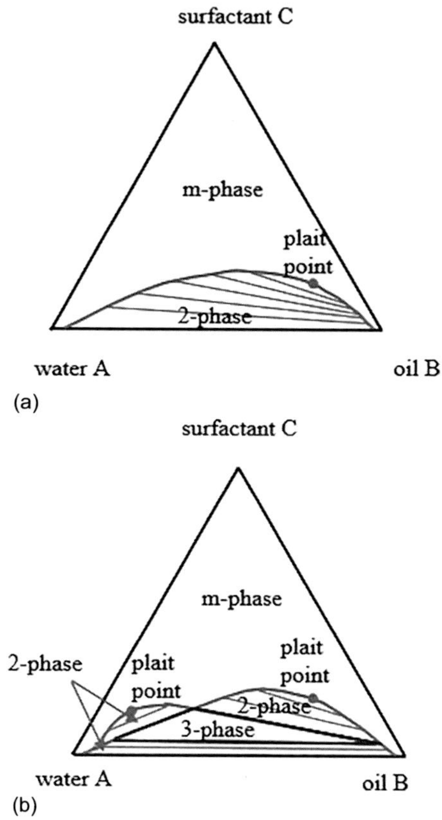

FIGURE 2.23
(a) Ternary diagram of oil, water, and surfactant. (b) Three phase region in ternary diagram.

- Immiscible water-alternating gas (WAG) injection
- Miscible WAG injection

Understanding of the phase behavior of injectant gases and their interaction with crude oil is vital for the design of scheme.

2.6.1 Immiscible and Miscible Water-Alternating Gas (IWAG) Injection

WAG injection process is a mixture of water and gas. They are injected alternatively in the oil column. It is a successful method for improving oil recovery. The combined mobility of the phases is less than that of the injected gas alone, and thus improves the mobility ratio as follows:

$$\text{Mobility ratio} = \frac{k_s/\mu_s + k_w/\mu_w}{k_o/\mu_o}$$

Water/gas injection ratios range from 0.5 to 4 volumes of water per volume of gas. Sizes of the alternate slugs range from 0.1% to 2% PV. Originally visualized that water and first-contact miscibility (FCM) gas injected simultaneously and flowed together without segregation. WAG practiced mostly in near-miscible and miscible floods in horizontal reservoirs gives better result. Realistic relative permeability curves are very important for evaluating WAG performance. In practice need a simulator to determine optimum WAG ratio. WAG reduces mobility and improve sweep when fluids are not segregated to appreciable degree. Even with appreciable segregation, there may be a benefit due WAG injection. The following factors impact the incremental oil recovery with WAG application:

- Reservoir heterogeneity (stratification and anisotropy)
- Rock wettability
- Fluid properties
- Miscibility conditions
- Gas trapped
- Injection technique
- Hysteresis effect
- WAG parameters—cycling frequency, slug size, WAG ratio, and injection rate

Gravity stable displacement is critical for better recovery where fluid properties plays an important role:

$$u_c = \frac{0.0439\left(\rho_o - \rho_g\right)\sin\alpha}{\left(\dfrac{\mu_o}{k_o} - \dfrac{\mu_g}{k_g}\right)}$$

where
u_c = darcy velocity, ft/d;
r_o = oil density, lb/ft^3;
k_o = oil permeability, darcies;
m_o = oil viscosity, cP;
m_g = gas viscosity, cP; and
a = angle of dip relative to horizontal.

2.6.2 Miscible Flooding

Miscible flooding with hydrocarbon gases or CO_2 gas injection works on the principle that these injectants becomes miscible with crude oil and improves the displacement efficiency of oil with no capillary resistance. When oil is mixed with a miscible fluid, there are no interfacial tensions or capillary forces and no interface exists. This process results in a reduction of residual oil saturation and oil swelling. Relative permeability of oil is also improved. Oil viscosity is also reduced and mobility of crude is improved. These two factors are the main ingredient for enhancing the recovery. The miscibility in reservoir can be obtained by using:

- FCM
- Multi-contact miscibility (MCM)

Minimum miscibility pressure (MMP) is the pressure at which injected gas dissolves in crude oil and forms single system. This is measured in lab using slim tube experiment. Miscibility with CO_2 gas is easily achieved compared to hydrocarbon gases. MMP can be lowered by enriching the hydrocarbon gas by adding propane and butane. This is called enriched gas injection. It is expensive and hence it is not used in field (Figure 2.24).

MMP is also calculated using empirical correlation if lab-measured data are not available.

Ternary diagrams are used to understand the phase diagram of gas-based EOR techniques. There are three vertices of equilateral triangle: light, intermediate, and heavy components of hydrocarbons. At the vertices they represent 100% saturations and 0% saturations to opposite lines. Variation in saturation can be understood by dropping a perpendicular to opposite line from vertices. Their percentage varies 100–0 from vertices to opposite line on the perpendicular line (Figure 2.25).

FCM process consists of injecting a small primary slug that is miscible with crude oil and conditions for miscibility are demonstrated by the use of

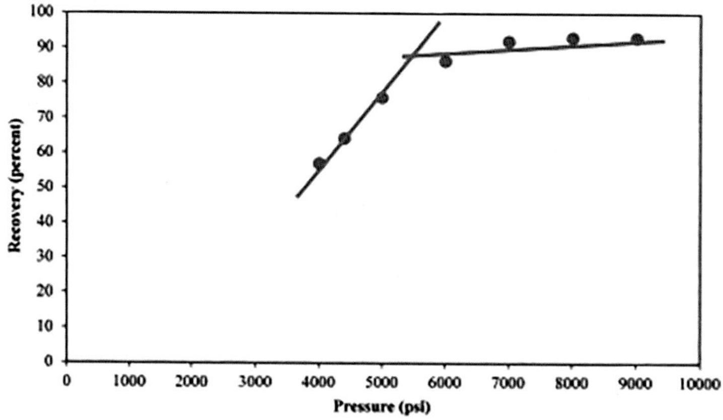

FIGURE 2.24
A schematic of minimum miscibility pressure (MMP).

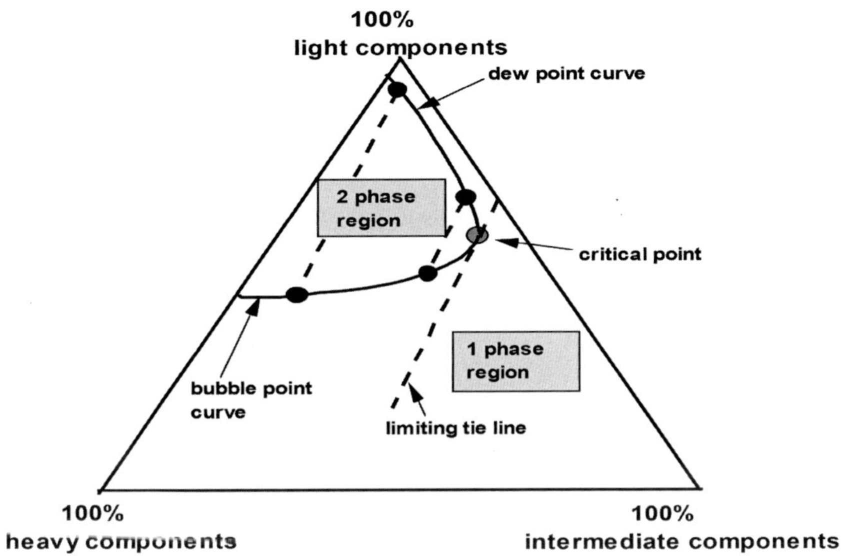

FIGURE 2.25
Typical ternary phase diagram.

the pseudo-ternary diagram. C1 content of oil does not affect MMP as long as composition of dead oil is outside all tie line extensions.

2.6.2.1 Multi-Contact Miscibility

Miscibility is generated in the reservoir through in situ composition changes resulting from multiple-contacts and mass transfer between reservoir oil and injected fluid. Multiple-contact processes are classified as follows:

- Vaporizing-gas (lean gas, N_2) displacements
- Condensing and condensing/vaporizing-gas (enriched gas) displacements

2.6.2.2 Vaporizing Gas Drive

In this miscible drive mechanism, lean gases like methane and nitrogen are injected into the oil reservoir to achieve the miscibility. When these lean gases are in contact with the oil, the light and intermediate MW hydrocarbons get vaporized and hence become the part of injected gas, thus vaporizing to gas. Formation of miscibility may require several contacts between gas containing vaporized components and fresh reservoir oil. If the injected gas becomes sufficiently enriched with vaporized components and miscibility results, then lean gas and oil will have MCM. A further multicontact laboratory test is carried out for the evaluation of vaporizing drive process.

In this test, the light and intermediate components are stripped from the oil by multiple contacts with the gas. The test also indicates that how many contacts are required before injected gas becomes miscible with crude oil. PVT simulation software is used to design the molar ratio with composition at each contact step. Required pressure for miscibility is high: >3,500 psia and gas may be injected continuously or as a very large slug (>50%). This process can occur with WAG injection also (Figure 2.26).

This ternary diagram indicates that the injection gas (Point G) includes mostly light components, and the reservoir oil (R) comprises a mixture of all components. After first contact, the phases equilibrate with a calculated mixture composition (M1). The first-contact liquid is removed isobarically and analyzed to determine its composition (L1). A small amount of gas is removed and analyzed to determine its composition (G1). The first-contact gas (G1) is equilibrated again with fresh reservoir oil (R), resulting in the second equilibrated mixture composition (M2); liquid composition (L2) and gas composition (G2) are evaluated. This process is repeated for several more contacts. In this case, miscibility is achieved after the sixth contact (G6). The point on the two-phase envelope, where the mixture forms one phase, is

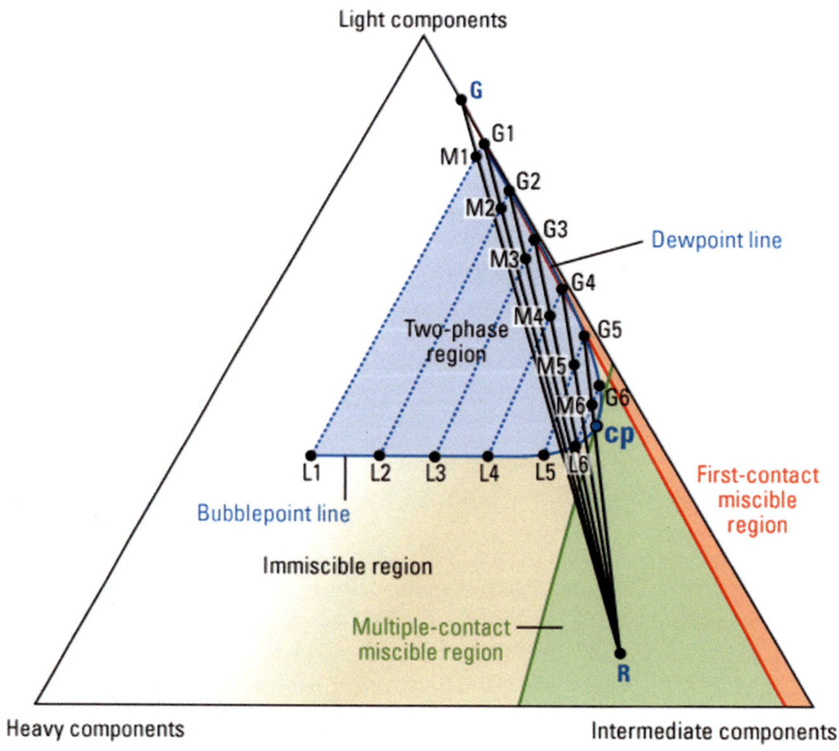

FIGURE 2.26
Vaporizing gas drive.

called the cp. The region in the ternary diagram to the right of the tangent line between the gas and the two-phase envelope indicates oil compositions that are miscible with the gas on first contact (pink). The region of oil compositions to the left of that line and to the right of the tangent passing through the cp contains multiple-contact miscible compositions (green). Oil compositions outside these two regions are immiscible with the gas (tan).

2.6.2.3 Condensing Gas Drive

When enriched gas containing intermediate gases like ethane, propane, and butane are injected in the oil reservoir, an MCM can be achieved. These intermediate gases condense into the oil phase on contact, making the oil lighter and less viscous. On repeated contacts, oil becomes light enough to become miscible with the solvent. Typically solvent slug is of the order of 10%–40% HCPV and is driven by a lean gas slug and/or water for economic reasons for continuous application. The important conditions for the success of the process include the lower temperature and higher pressure of reservoir and lower C_{7+} MWs and higher solvent enriched level. A backward multiple-contact test is carried out in laboratory to understand the process. This ternary diagram indicates that the injection gas contains only the light and intermediate components (Point G), and the reservoir oil (R) comprises a mixture of all components. After first contact, the phases equilibrate with a calculated mixture composition (M1). The first-contact gas is removed isobarically and analyzed to determine its composition (G1). A small amount of liquid is removed and analyzed to determine its composition (L1). The first-contact oil (L1) is equilibrated again with fresh injection gas (G) resulting in the second-equilibrated calculated mixture composition (M2), and gas composition (G2) and liquid composition (L2) are evaluated. This process is repeated for several more contacts. In this case, miscibility is achieved after the fifth contact (L5). The point on the two-phase envelope (blue) where the mixture forms one phase is the cp. The region in the ternary diagram below the tangent line between the oil and the two-phase envelope indicates injectant compositions that are miscible with the oil on first contact (pink). The region of gas compositions above that line and to the right of the tangent passing through the cp contains multiple-contact miscible compositions (green). Gas compositions outside these two regions are immiscible with the oil (tan) (Figure 2.27).

2.6.2.4 Condensing and Vaporizing Drive

Vaporizing and condensing gas-drive mechanisms occur in ideal scenario. In reality, combination of both vaporizing and condensing mechanism occurs in miscible gas flooding, which cannot be represented by pseudo-ternary diagrams. Composition path of the displacement may stay within the two-phase region without developing miscibility between gas and the oil but the

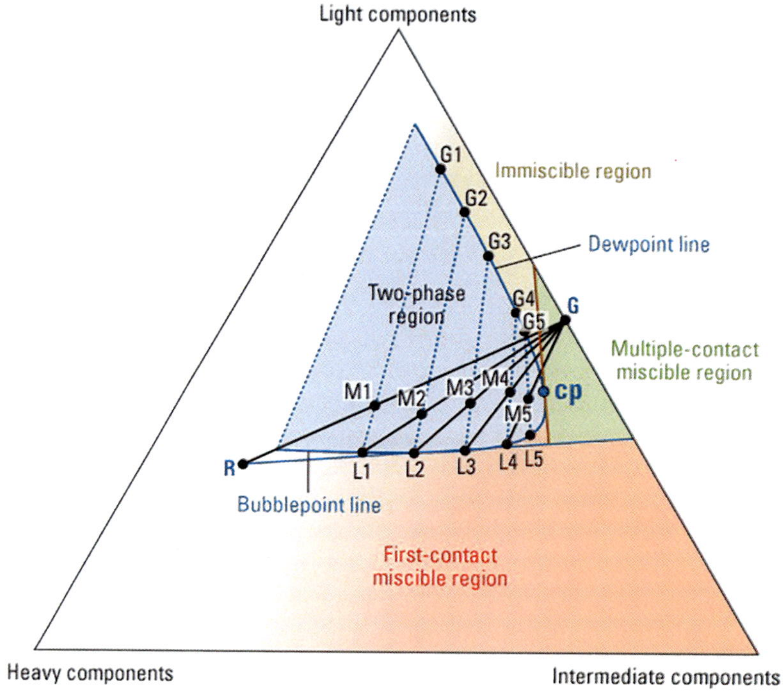

FIGURE 2.27
Condensing gas drive.

displacement process may be very efficient. Displacement behavior looks like condensing/vaporizing when pressure is few hundred psi lower and above of MMP and a lean gas bank precedes the displacement front, and a residual oil saturation is left. Oil swelling, viscosity reduction, and vaporization also improve oil recovery below the MMP. When reservoir pressure is well above the MMP, the phase behavior looks like vaporizing-gas drive except displacement may be liquid–liquid rather than gas–liquid at low $T<125°F$.

2.6.3 CO_2 Sequestration

Carbon dioxide capture and sequestration (CCS) in deep geological formations has emerged as an important option for reducing greenhouse emissions. CO_2 is a gas at ambient condition, but it becomes liquid at greater depth and supercritical fluid high temperature and pressure. The transition from one state to another depends on the geothermal gradient. In most of the sequestration scenarios, CO_2 is injected in liquid form (low T, modest to high P), but it transforms into a supercritical fluid as it is injected and warms to the temperature of the formation. In saline aquifers and oil reservoirs, CO_2 is less dense than the in situ fluids, so it rises to the base of the

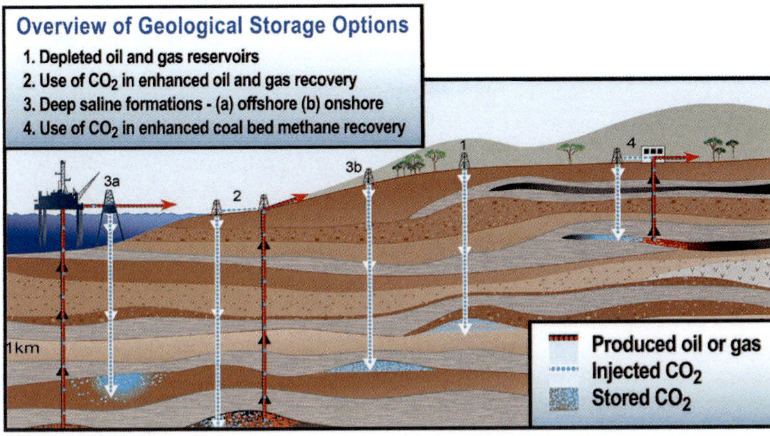

FIGURE 2.28
Schematic of CO_2 sequestration.[13]

seal. Clearly, maintaining an impermeable caprock is crucial to containing the buoyant CO_2. When CO_2 is injected into deep geological formations, it displaces the pore fluid. Depending on the fluid's properties, CO_2 is either miscible, that is, it can mix completely to form a single liquid phase, or immiscible, so the phases remain separate (Figure 2.28).

2.6.3.1 Trapping Mechanisms and Long-Term Fate of CO_2

- Structural trapping
- Capillary trapping
- Solubility trapping
- Mineral trapping

Phase behavior of CO_2 and its interaction with fluid present in formation where sequestration is planned must be understood.

2.7 Thermal EOR

More than 50% world oil resource is heavy to extra-heavy oil. The classifications of crudes are shown in Table 2.2. Primary recovery from such reservoirs is very small to nil due to very high viscosity. Viscosity of oil in these viscous fields sometimes is as high as a million cP (Figure 2.29). Heat is required to be injected into viscous reservoirs to reduce the viscosity and realize the

TABLE 2.2
Classifications of Crude

	Degree, API Gravity	Specific Gravity	Viscosity (cP)
Light oil	>31.1	0.87	<100
Medium oil	22.3–31.1	0.87–0.92	<100
Heavy oil	10–22.3	0.92–1.0	100–1,000
Extra-heavy oil	<10	>1.0	100–10,000
Bitumen	<10	>1.0	>10,000

Source: 12th World Petroleum Congress (WPC) 1987.

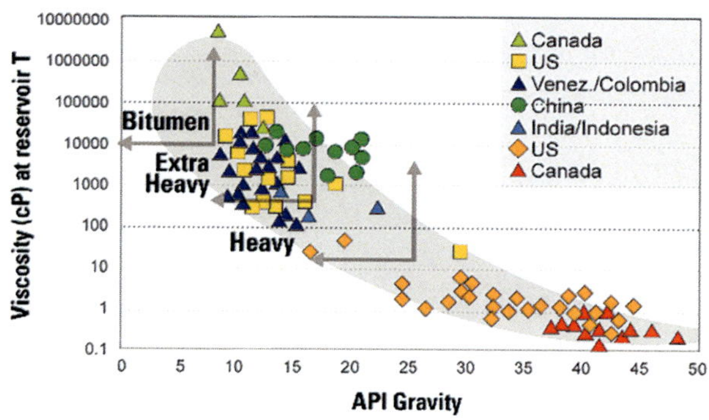

FIGURE 2.29
Degree API and viscosity.

flow in reservoir to well and to surface. The following are the three popular methods of thermal recovery:

- Cyclic steam stimulation
- Steam flooding
- In situ combustion

The fluid characterizations in terms of composition, viscosity, and thermal properties along with rock properties are to be understood before designing a suitable EOR. The change in relative permeability is found to be simulating the flow in the reservoir.

Thermal conductivities and steam properties

Aquathermolysis

2.7.1 Cyclic Steam Stimulation

This is a three-step thermal process. First step is to inject high-quality (80%) steam in the well and the second step is to stop the steam injection and allow the steam to soak in the reservoir. The third step is to produce the well. These steps vary depending up on the reservoir and fluid characteristics. This process is also called huffing and puffing.

CSS quickly stimulates the wells and increases the production rapidly and also provides the valuable data for steam injectivity, interaction of steam with rock and fluid, and generation of some revenues in the beginning as continuous steam flood projects have some lead times before first oil is realized through its application.

The analytical and numerical simulation techniques are used to estimate the incremental oil through CSS. The estimation of expected oil can be simplified as follows:

$$Q_{hot} \propto Q_{cold} = \frac{\text{Viscosity of the oil before steaming}}{\text{Viscosity of the oil after steaming}}$$

Thus, the main objective of thermal stimulation process is to allow maximum heat energy in the formation through well. During this process, benefit is not only reduction in oil viscosity but also due to removal of certain types of near wellbore damages, such as fine solids, asphaltic deposits, and paraffinic deposits, including sand-shale ratio, sand thickness, reservoir pressure, oil saturation, production mechanism, cold productivity index, formation depth, etc. In summary, the following points are favorable for the success of cyclic steam stimulation:

- Good reservoir characteristics with high porosity and permeability
- High oil saturation
- Higher oil viscosity
- High steam quality at wellbore
- High cold productivity index
- Permeability damage near the wellbore
- Thick sand with high net to gross ratio
- Low water–oil and GOR
- Weak aquifer support
- Shallow depth
- Higher reservoir pressure but below 1,500 psi for saturated steam injection

The main advantage of this process is that it can be applied easily in a number of wells and quick revenue can be generated due to oil production. One of the main advantages is reduction in reservoir pressure and making the reservoir more suitable for continuous steam flooding. Cycles can be repeated depending when oil rate becomes two low or sometimes it can be repeated on fixed frequency. This process is more effective in highly viscous oil filed with good permeability. Incremental oil can be estimated using thermal numerical simulation and empirical correlations. Variations in fluid properties with temperature are one of the most influencing parameters. Figure 2.30 shows the important parameters of a CSS project in North Africa.

It has been observed that steam injection with high rate is desirable as it provides the fast growth of heated radius and required injection is also completed in shorter period. In addition to this well is not on production for less time as injection and soaking period is reduced. With this background understanding, numerical simulation study is carried out for optimizing the steam injection rate, soaking period, and estimate of likely incremental oil gain. Simulation results guide to plan for the steam generation quantity, fuel requirement, availability of source water, treatment facility, and most importantly treatment size and cycle length optimization. The following parameters optimized in simulation study are shown in Figure 2.31.

- Treatment size
- Injection rate
- Soaking time
- Maximization of flowing bottom hole pressure

2.7.2 Steam Flooding

Steam is continuously injected in the reservoir in the pattern form. Injected steam forms a steam zone and increases gradually in the reservoir. Steam

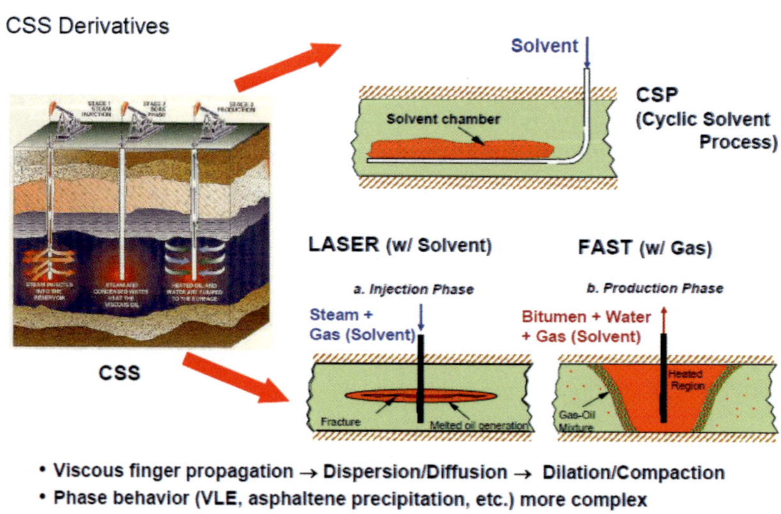

FIGURE 2.30
Schematic of CSS.

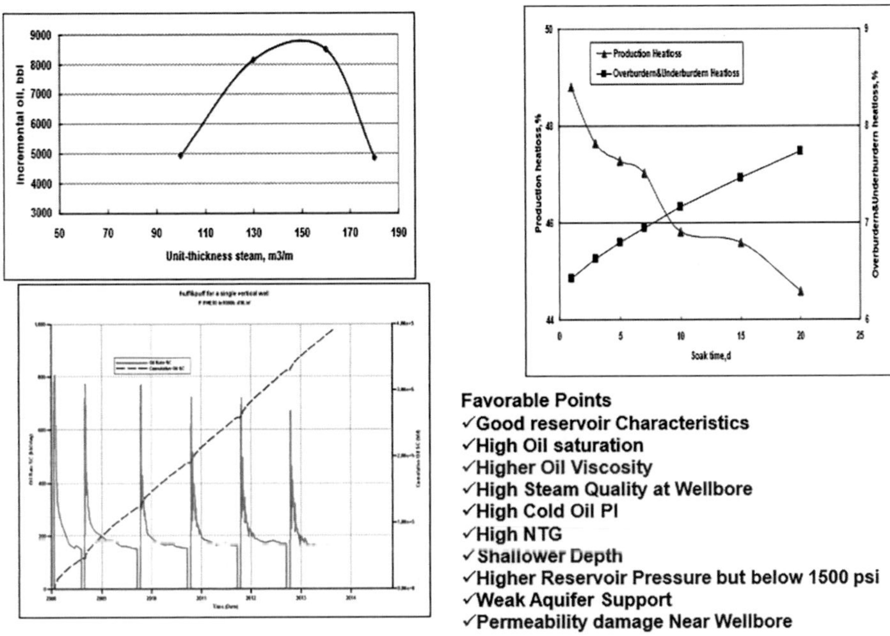

FIGURE 2.31
Parameters for CSS.

reduces the viscosity of crude oil and mobilizes the oil toward producers. The following phenomena take place in the steam flooding process (Figures 2.32 and 2.33):

Reduction of oil viscosity (Figure 2.34)

- Effect of temperature on ratio of viscosity of various oils to that of water
- Even after heating, water still fingers through oil because of the adverse mobility ratio

Changes in relative permeability (Figure 2.35)

- Relative permeability for oil flow tends to be increased during steam flooding, and the residual oil saturation is reduced
- A possible partial explanation is that water has a tendency to form water oil emulsions, within the reservoir, with bituminous oils under steaming conditions—(partial miscible)

The following phenomena take place with steam injection in the reservoir and are important to be measured in laboratory to be used in engineering calculations;

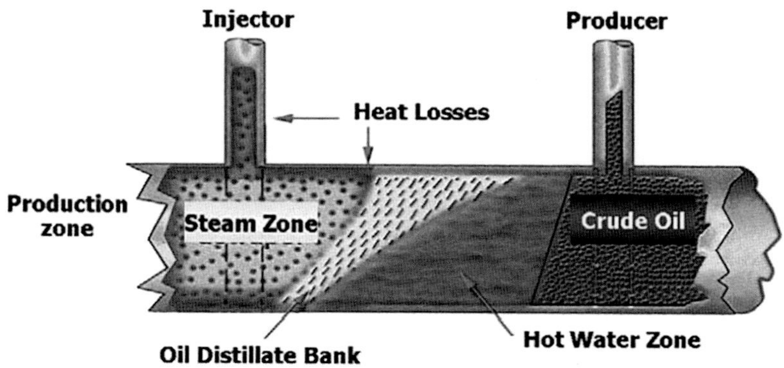

FIGURE 2.32
Schematic of steam flooding.

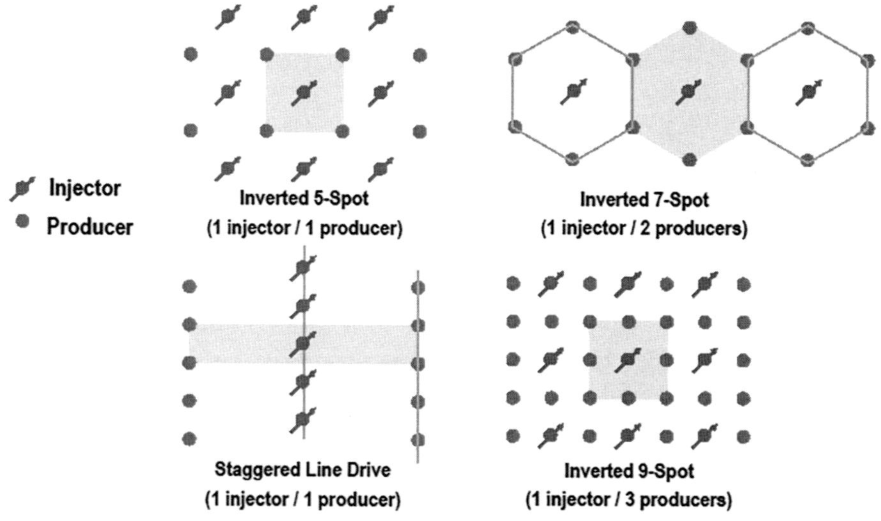

FIGURE 2.33
Pattern for steam flooding.

- The density of crude oils and gases decreases with increasing temperature
- The density of rocks changes slightly with temperature
- The viscosity of real gases increases with increasing temperature
- The viscosity of a liquid decreases very rapidly with increasing temperature
- Thermal conductivity of most natural rocks changes slightly with temperature
- Thermal conductivity of gases varies little with temperature

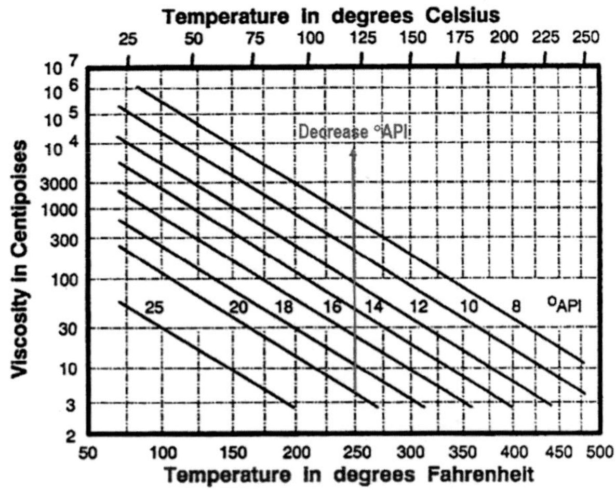

FIGURE 2.34
Viscosity vs. temperature relationship (after Farouq Ali 1983).

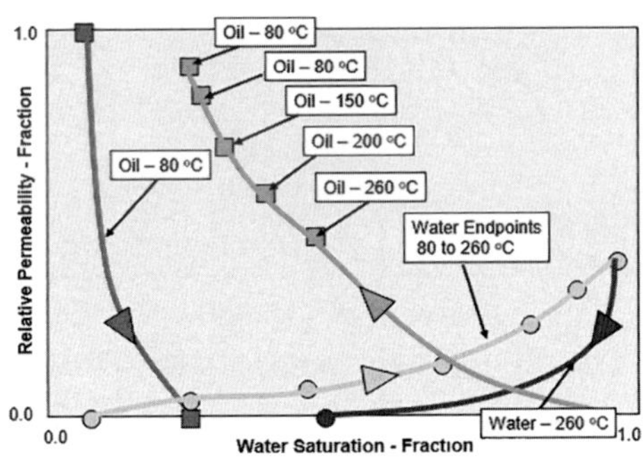

FIGURE 2.35
Modification of water–oil relative permeability with temperature.

- Thermal conductivity of liquids decreases with increasing temperature
- Thermal conductivity is not additive
- As a general rule, the head capacity per unit mass of solids, liquids, and gases increases with increasing temperature
- The heat capacity of a complex mixture is equal to the sum of the heat capacities of its constituent elements
- Relative permeability is modified due to increase in temperature. Residual oil saturation is reduced.

One of the major limitations is heat loss in injectors. Therefore, most of the successful steam flood applications are in shallower reservoirs so that maximum heat can be utilized. Thumb rule suggests that it will not give the desired result in a reservoir deeper than 1,000 m. In addition, aquifer or perched water within sand bodies bearing the heavy oil is also a major concern. Reservoir thickness should be good to avoid the heat loss to underburden and overburden rocks. Steam oil ratio also plays an important role. Successful projects have SOR 2.5–3.0. It means for every barrel of crude oil produced 2.5–3.0 barrel of cold water is used for steam generation. Higher ratios adversely impact the projects economics (Figures 2.36 and 2.37).

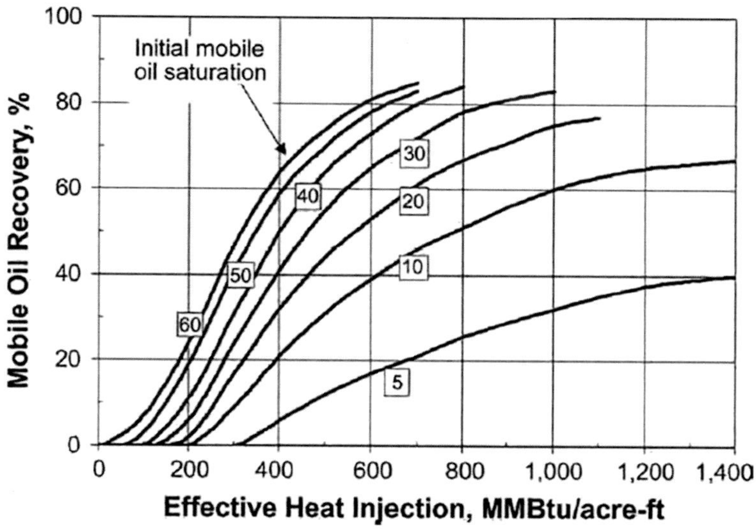

FIGURE 2.36
Recovery vs. heat injection.

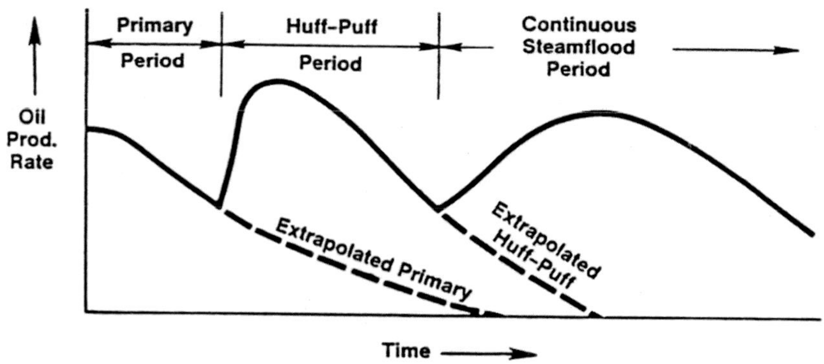

FIGURE 2.37
Steam flood recovery followed by CSS.

2.7.3 Steam Assisted Gravity Drainage (SAGD)

This is one of the derivatives of continuous steam flooding. In this process, a pair of two horizontal wells are used, as shown in Figure 2.38. Steam is injected in upper horizontal well. Injected steam forms a steam chamber which reduces the viscosity of crude oil and that oil moves to lower horizontal well with gravity. This is one of the successful thermal EOR process. There are certain limitations for this process. Some of these limitations are listed as follows:

- Reservoir thickness should be higher than 15 m
- Vertical separation between two horizontal wells should be 5 m
- Net-to-gross thickness should be good
- There should be enough clearance from aquifer to avoid the unnecessary heat loss

2.7.4 In Situ Combustion

The in situ combustion process involves creating a fireflood in the reservoir by injecting oxygen. Air is continuously injected to sustain the burning of some of the crude oil to for fire flood. Ignition is created spontaneously or artificially by injecting hydrocarbon gas in the beginning. Lighter components of crude oil start the ignition followed by heavier components of crude oil burning and forming the coke which sustain the fireflood. Nearly, 15%–20% crude is burnt to sustain the fireflood. This process does not have limitation of depth and thickness of reservoirs. Heat transfer in reservoir happens due conduction and convection process and reduces the crude oil viscosity. Combustion temperature is very high, nearly 600°C–700°C.

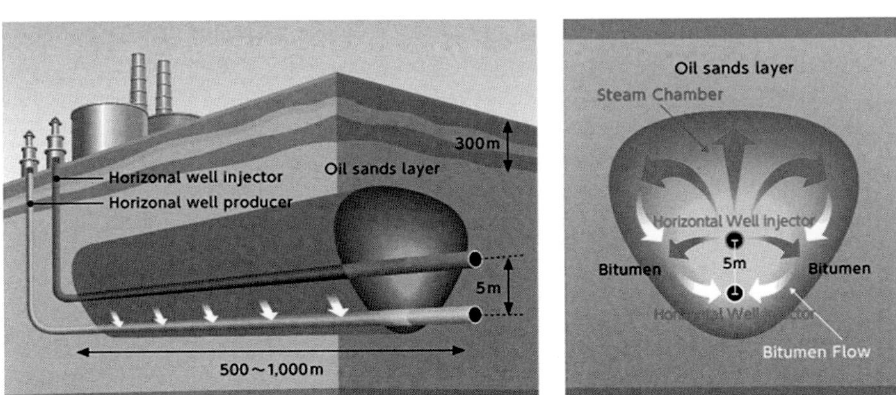

Oil Sands Development Using the SAGD Process

FIGURE 2.38
Schematic of SAGD.

Heavier components of crude also break down into small components due to thermal cracking and facilitate the viscosity reduction. Additionally formation water is converted into steam and flue gases are generated due to combustion which provides energy to displacement of crude oil in the reservoir. Continuous injection of air also maintains the reservoir pressure. The two major process of ISC are as follows:

Dry combustion

- Injection of compressed air and normally some thermal energy

Wet combustion

- Simultaneous injection of air and water
- Water acts as scavenger of heat and converts into steam

Additionally, ISC is applied either in forward or reverse combustion modes. In forward combustion, the reservoir is ignited in the vicinity of an air injection well, and the combustion front propagates away from the well. Continued injection of air drives the combustion zone through the reservoir to nearby producing well (Figure 2.39).

The reverse combustion process is started in the same manner as forward combustion, but after burning out a short distance from the ignition wells, air is switched to adjacent wells. This moves the oil toward the previously ignited well, while the combustion front moves in the opposite direction toward the adjacent wells. The process is developed as a method for improving recovery in reservoirs containing extremely heavy crudes (Figure 2.40).

In situ combustion has found less success than steam processes because of the difficulty in controlling the process. Fire front tends to advance much erratically than steam front, and it is much harder to obtain an even sweep of the reservoir. Nevertheless, there are successful fire flooding projects that are required high-quality tubing and cement (can stand high temperature). Gases like CO_2 in complete and CO due to incomplete scenario are produced. Both are gases and are not environment friendly (Figure 2.41).

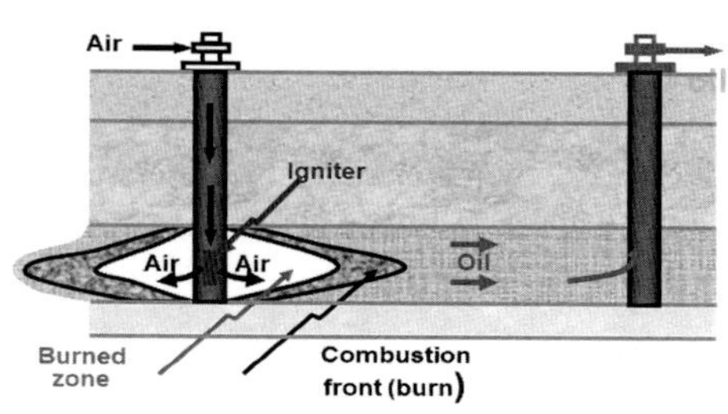

FIGURE 2.39
Forward combustion.

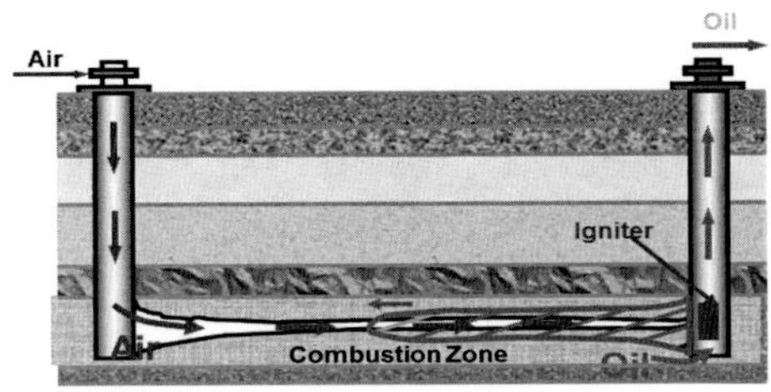

FIGURE 2.40
Revers combustion.

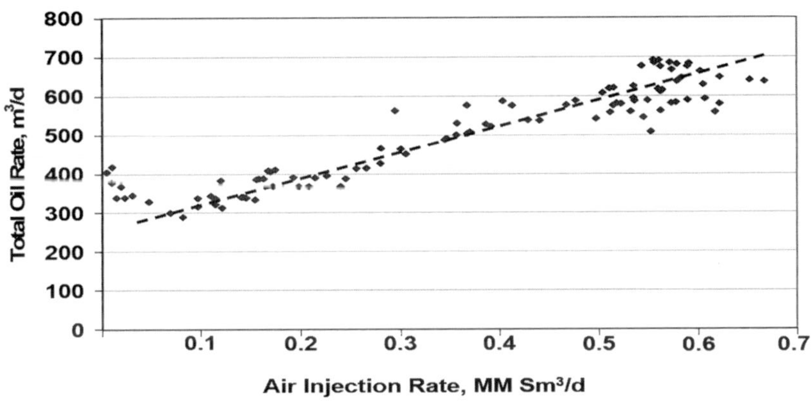

FIGURE 2.41
Air requirement per bbl incremental oil.

2.8 PVT Data for Thermal EOR Processes

- Low- and high-temperature oil property measurements (viscosity, density, formation volume factor, GOR, etc.)
- Vapor oil ratio crude of fluid for numerical simulation
- Aquathermolysis testing to evaluate levels of CO_2 and H_2S generation that may occur during thermal operations
- Low- and high-temperature relative permeability testing to ascertain residual
- Kinetics of crude oil and heat in in situ combustion
- Requirement of air for sustaining the fire front in ISC
- Oil saturations and injectivity and productivity for heavy oil applications

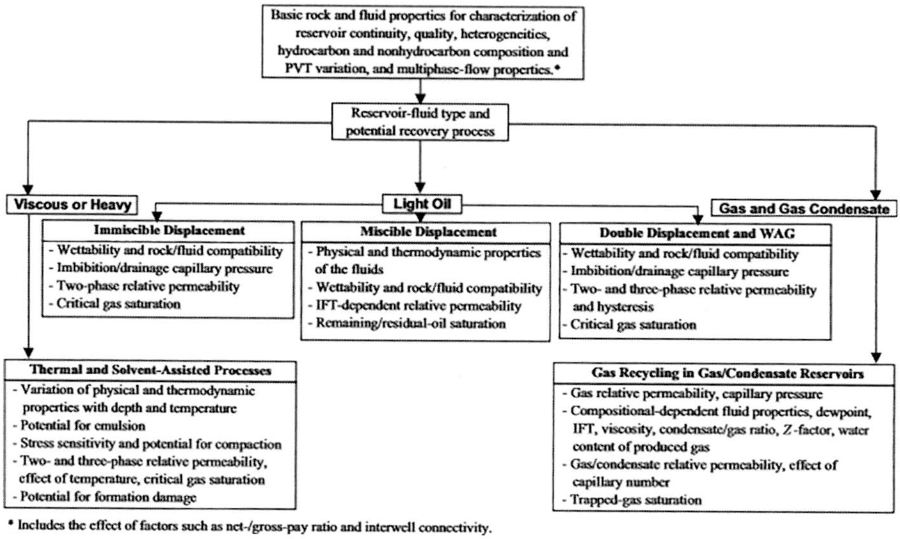

FIGURE 2.42
Data requirement for EOR (after Nagarajan et al.).

- Hysteresis testing to determine the variance in relative permeability character between injection and production cycles (CSS operations)
- Testing to evaluate possible permeability reductions due to thermal dissolution, migration, or mineral transformation issues
- Sand control and liner design for optimum performance of horizontal SAGD production wells (Figure 2.42)

References

1. Dake, L.P. 1983. *Fundamental of Reservoir Engineering*. Amsterdam: Elsevier Science.
2. Ahmed, T. 2010. *Reservoir Engineering Handbook*. Houston, TX: Gulf Professional Publishing.
3. Towler, B.F. 2002. *Fundamental Principal of Reservoir Engineering*. Richardson, TX: SPE.
4. Muskat, M. 1949. *Physical Principles of Oil Production*. New York: McGraw-Hill.
5. Cockcroft, P.J. and Moore, K.S. 1994. *Development Planning: A Systematic Approach*. Richardson, TX: SPE.
6. Sattar, A. and Thakur, G. 1994. *Integrated Petroleum Reservoir Management*. Tulsa, OK: Penwell Publications.
7. Mattax, C.C. and Dalton, R.L. 1990. *Reservoir Simulation*. Dallas: Richardson, TX: SPE.

8. Orr, F.M. 2007. *Theory of Gas Injection Processes*. Holte: The Line Publication.
9. Green, D.W. and Wllhite, G.P. 1998. *Enhanced Oil Recovery*. Richardson, TX: SPE.
10. Stalkup Jr., F.I., SPE, ARCO Oil & Gas Co. Summary. *Methods for Miscible Flooding have been Researched and Field Tested since the Early 1950's*. Richardson, TX: SPE.
11. Kelkar, M. 2008. *Natural Gas Production Engineering*. Tulsa, OK: Pennwell Publications.
12. Talbot, D. 2014. Carbon sequestration: Too little, Too late? MIT Technology Review.
13. Alvarado, V. and Manrique, E. 2010. Enhanced oil recovery. *Energies* 3: 1529–1575.
14. Lake, L.L. et al. 2015. *Enhanced Oil Recovery*. Richardson, TX: SPE.
15. Sheng, J. 2010. *Modern Chemical Enhanced Oil Recovery*. Burlington, MA: Elsevier Publication.
16. Ali Farouq, S.M. et al. 1997, *Practical Heavy Oil Recovery*. Edmonton: University of Alberta.
17. Butler, R.M. 1991. *Thermal Recovery of Oil and Bitumen*. Englewood Cliffs, NJ: Prentice Hall.
18. Intergovernment panel on climate change document 2006.
19. Honarpour, M.M. et al. 2006. Rock fluid characterization and their integration-implication on reservoir management. Richardson, TX: SPE.
20. Tewari, R.D. et al. 2011. *Successful Cyclic Steam Stimulation Pilot in Heavy Oilfield of Sudan*. Richardson, TX: SPE.

3

Advanced Fluid Sampling and Characterization of Complex Hydrocarbon Systems

3.1 Introduction

Representative fluid sampling and characterization are fundamental for exploration and development activities in the oil industry; accurate in situ fluid properties are used by multiple disciplines not only to design, optimize, and manage field development operations but also to understand fluid migration, identify flow assurance challenges, and provide realistic fluid in place estimations.

Criterion and caveats of sampling design vary depending on the hydrocarbon type, reservoir heterogeneity, development maturity, and cost.[1–3] Understanding and recognizing the effect of each of these elements on the sampling and characterization strategy is key for a successful description of the in situ reservoir fluid dynamics.

3.2 Hydrocarbon Sampling

Traditionally, fluid systems are interpreted to be in hydrodynamic equilibrium; that is, fluid properties follow the same distribution bounded only by changes in pressure and temperature across the reservoir. As such, reservoir hydrocarbons are interpreted to be either saturated (likely with two phases in the reservoir) or undersaturated (single phase). It is not uncommon to observe compositional gradients in reservoirs adding to the complexity of the fluids but still encompassing a single genetic reservoir fluid.

A second type of fluids are those where conditions or pressure and temperature are such that they operate beyond the mixture critical point. There are, other, special cases where fluid migrations augmented by geological heterogeneity may result on a deviation of the genetical material of the reservoir fluids, where a more complex interaction takes place and the assumption of simple pressure and temperature dependence is no longer valid.

Sampling and characterization need to be considered as being position, structure, and/or time dependent, requiring their own fit-for-purpose collection and analysis strategy.

Gravity and thermal gradients are responsible of most compositional gradients; thick reservoirs with considerable vertical permeability show an increase of lighter components toward the top of the reservoir (and therefore higher saturation pressures) in response to the gravity and thermal components in the reservoir. The reservoir fluid (albeit with different saturation pressures) remains at thermodynamical equilibrium.

3.2.1 Sampling Location Considerations

Conceptual understanding the hydrocarbon system/s of the reservoir/s is key to a successful sampling program. Sampling in and on itself should be designed to address key fluid uncertainties and further develop the reservoir/s thermodynamical interpretation. It is key, therefore, that compartmentalization, compositional gradients, fluid contacts, etc., be identified (albeit with higher uncertainty) prior to the sampling location design. That, however, does not imply that changes to the sampling program will not occur as a result of data acquisition during the sampling operation.

3.2.1.1 Pressure Data

Wireline formation testing is widely used in both exploration and development stages, offering a higher resolution (vertical) of the hydrocarbon pressures and formation mobility across the reservoir. Several authors have discussed the advantages and challenges of the technology application for reservoir characterization,[4,5] particularly for low permeability and highly laminated reservoirs. Pressure data, when adequately used, provide an insightful appreciation of the compartmentalization and compositional changes in the reservoir, making it fundamental for hydrocarbon sampling design.

Pressure gradient analysis by far remains one of the most useful tools not only to identify and quantify phase changes (GOC, FWL) but also enables identification of potential compositional changes in the oil column as well as selective pressure depletion. Use of pressure vs. depth plots can easily identify fluid gradient changes (basis of contact recognition) but may (as function of pressure resolution) obscure subtle changes on the oil column gradient. It is not uncommon to attribute subtle deviations of a straight line on the pressure vs. depth plot to measurement issues, and while those need to be considered and their effect on the pressure analysis identified, small gradient changes may be an indication of higher reservoir heterogeneity. Several techniques have been developed thru the years to address the validity of pressure gradients. Fluid head method[6,7] is one of the more straightforward techniques, which simply involves the difference between the measured pressure and the pressure expected from the weight of a fluid between the datum and the pressure depth. This technique allows for the identification of small pressure changes and ultimately potential compositional changes along the vertical column. Other techniques, albeit more sophisticated, involve the use of statistical gradient analysis (student *t*-test[8]) to determine the statistical probability that the observed gradient changes is a function of the reservoir and not responding to the measurement operations.

One example of the application of pressure data on the identification was presented by Fujisawa et al.,[6] on a north sea exploration well where a fairly homogeneous 100mt formation was sampled with 25 formation pressure tests (see Figure 3.1). A first pass analysis identified three phases, gas, oil, and water; a single gradient may be inferred from the oil column (albeit

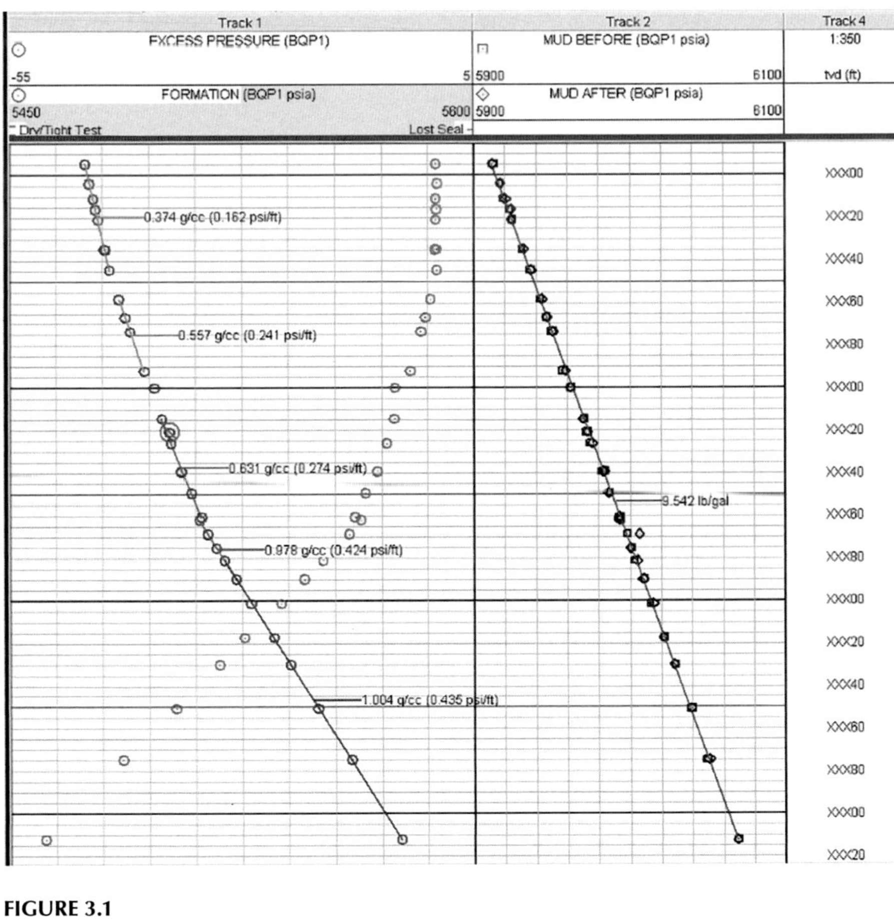

FIGURE 3.1
Pressure gradients.[6]

there are subtle deviations). A second, more detailed investigation (using both the fluid head and *t*-test methods[8]) revealed five different gradients within the oil column. All of these gradients are statistically different (*t*-test method results), most likely indicating the existence of a compositional gradient, with a curved pressure profile and not a straight line. The details of the interpretation and analysis can be found in Fujisawa et al.[6]

3.2.1.2 Compositional Gradients and API

Changes of composition due to gravity and thermal effects have been widely documented in the industry[7,9,10] and are often associated with lighter API hydrocarbons (>35 API) where vertical migration of the light and heavy components is less restrictive than in the heavier oils. These changes are naturally augmented if the reservoir conditions are near that of the critical point of the hydrocarbon mixture. Heavier oils (20–30 API) also exhibit changes in composition with depth, less related to the fugacity and diffusion but more dependent on gravity segregation, particularly when asphaltenes are present. These oils will typically exhibit a large variation of viscosity (increasing with depth) and are frequently associated with the existence of a "tar-mat" near the water interface. Biodegradation and loss of light ends may also cause a gradient on heavy oils occurring in shallow formations.

Gravity segregation models allow for an easy calculation of the compositional variation by accounting the changes of chemical potential of each of the hydrocarbon components. A typical calculation (not accounting for capillary forces) involves using a simple regression to (based on fugacities) estimate the changes with depth under a stable temperature gradient (assuming single phase and therefore $y_m = z_m$):

$$R_m^{l+1} = R_m^l \frac{F_m^{v1}\left(P^1, T^0, y_m^{l+1}\right)}{F_m^{v0}\left(P^0, T^0, y_m^0\right)} \exp\left[\frac{Mw_m\left(D^1 - D^0\right)}{144RT}\right]$$

Compositions are iterated upon to that the R_m operator is minimized:

R_m is the gauss-newton operator

F_m is the fugacity of the liquid (l) and vapor phase (v)

y_m is the mole fraction of each component, for single phase equivalent of Z_m

P is the reservoir pressure at corresponding depth D

T is reservoir temperature at corresponding depth D

R is the universal constant for ideal gas law

There are, of course, more considerations to be accounted for in the modeling of such reservoirs. Hydrocarbon migration history, pressure potential changes, diffusion, convection, etc., do affect the dynamic distribution of the hydrocarbons and pose a challenge for fluid characterization and most importantly predictive power of any EOS (equation of state).

Al-Shaheen reservoir in Qatar is an example[11,12] of lateral compositional changes, far greater than those attributed to the effect of thermal, gravity, and chemical potential changes due to structure, where reservoir API gravity (and more importantly viscosity) changes dramatically within a short areal distance, making fluid characterization a key challenge on the successful selection of the conceptual field development as well the potential enhanced oil recovery (EOR) application. While the origin of changes on the hydrocarbon quality of Al-Shaheen is under investigation, it is clear that sampling and nonequilibrium characterization was fundamental on the evaluation of the ongoing waterflooding as well as building the foundations for the upcoming EOR application.[13] This characterization program included a comprehensive acquisition of drilling samples (particularly useful given the length of the horizontal wells), subsurface, surface, and, most importantly, a constant monitoring of API and viscosity at the well head. Thus, hydrocarbon heterogeneity was effectively incorporated onto the development planning. The details on the characterization of Al-Shaheen reservoir may be found in Lindeloff et al.[11]

The existence of a tar mat, along with compositional changes on heavier oils, has been widely reported in Middle Eastern reservoirs, and historically have posed a challenge on any EOS characterization particularly when dealing with asphaltene deposition prediction. Recent advancements in understanding asphaltene nanoscience particularly the Yen–Mullins model[14,15] have contributed to the better characterization of asphaltenes, their molecules, and nanocolloidal structures. Overall the model states that the molecular form of asphaltenes is dominant at low concentrations (lighter oils), and as the concentration increases nano aggregates form and ultimately, for the higher concentrations nanoaggregates form clusters (see Figure 3.2). Once the size of the asphaltene molecules is understood, the effect of gravity is

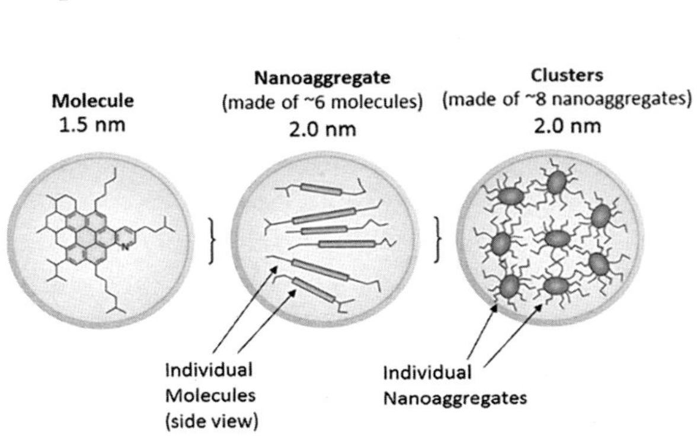

FIGURE 3.2
Asphaltene molecules.[15]

accounted for on a more precise manner (unavailable on the traditional EOS), thus improving the predictive power of the model.

Laboratory characterization and sampling for asphaltenes requires, as expected, different fit-for-purpose approaches; as such DFA (downhole fluid analyzer) techniques have been widely used to help characterize asphaltenes and aid on the decision of sampling for subsequent laboratory analysis. One such example was described in Mullins et al.,[15] where DFA (downhole fluid analysis) was used to determine fluid gradients and more importantly fluid discontinuities that, in turn, resulted in a fit-for-purpose, more representative sampling of the reservoir. The presence of heavier fluids on the upper part of the oil column was an indication of potential vertical barriers—and therefore compartmentalization; an observation which otherwise would not have been plausible should fluid densities, not have been measured along the oil column—see the caveats of single-point sampling.

There is often the misconception of disequilibrium associated with reservoir connectivity, while it is true that reservoir discontinuity will result on pressure isolation, pressure communication does not warranty flow communication. Thus, fluid equilibrium state on a reservoir needs to be taken in context before flow continuity may be assumed (or rather the premise that fluids that are in disequilibrium are isolated). The presence of asphaltenes in the system only exacerbates this situation and may lead to erroneous conclusions if they are not properly characterized. It is, therefore, highly recommended that an appropriate EOS must be selected for the characterization of the fluid (prior to making any observations in terms of communication).

3.2.2 Sampling Challenges

There are several considerations when collecting, transferring, and transporting hydrocarbon samples. Changes on pressure and/or temperature may cause the sample to be compromised from the moment it is produced at the reservoir to the time it arrives at the laboratory[1], particularly when sample sizes are small, whereas changes on pressure and/or temperature may trigger a compositional change compromising the integrity of the sample. Presence of contaminants such as H_2S and CO_2 must be considered during the design of the sampling protocol, as they not only require additional HSE considerations but also special bottles (corrosion resistant) to be safely transported to the laboratory.

The main objective of sampling is, in simple terms, to collect the designed volume of in situ hydrocarbon fluid, that is, with a composition (and phase) representative of that of the reservoir without any contaminants. It follows that formations with limited conductivity (where fluids do not flow easily), pressure, geomechanically unstable, with commingled production, and with two or three phases flow, pose the greatest challenges for a proper sampling. Near-critical reservoirs and those of high temperature pose a different problem with phase stability and changes when the fluids are brought to the lower surface pressure and temperature. It is, therefore, imperative that a detailed sampling addressing the foreseeable reservoir and operational issues should be in place prior to any sample collection activity.

While several advances have been made in the field of sampling, oil industry relies mainly on two different methods: downhole and surface sampling. The former allows for a fresh, uncontaminated (if sampled properly), localized sample (higher vertical resolution), typical volumes ranging from 0.25 to 4 L and the latter allows for a cheaper albeit less representative sample (operational challenges). The following sections will discuss in detail the advantages and challenges of either techniques, highlighting the latest advancements on sampling technology.

3.2.2.1 Downhole Sampling

Downhole is often preferred over surface sampling, given the lesser risk of contamination (either by other fluids and/or unstable two/three-phase samples at surface). They reduce environmental and safety risks, are easier to plan, and overall require less lead time. Downhole sampling is flexible, allowing for multiple reservoirs/zones to be sampled during the same operation, highly coveted in laminated formations and those with compositional gradients.

Majority of the challenges for downhole sampling are related to sample contamination from mud cake and filtrate (in the case of open hole), two phase flowing and sanding for formations requiring a large drawdown, not to mention any potential operational issues related to packer sealing and loss of reactive components to the tubulars among others.

Much like pressures, downhole samples are best taken close to the reservoir, and majority of wireline sampling tools have been designed to address the issue (Saturn, MDT of Schlumberger, Armada of Halliburton and RCX Sentinel of Baker to mention a few). A typical sampling operation may be summarized in five stages, namely, location selection, tool positioning, sampling, transport, and analysis.

Technical and operational considerations need to be accounted for in the sample location decision; geological sequence, structure, containment, mobility, fluid contacts, known compositional gradients, reservoir pressure, permeability, existence of contaminants (CO_2 and/or H_2S), and sample volumes are among the main considerations. Single-phase sampling (both at reservoir and downhole) is preferred. Operational considerations include tool selection, well deviation, hole conditions, formation stability and maximum allowable drawdown, and wellbore fluids including drilling filtrate, mud cake, sample depth, pressure, temperature, and sample transport.

3.2.2.1.1 Formation Considerations

Formation heterogeneity plays an important role on the sample location selection process, understanding its potential impact on the reservoir fluid distribution. Existence of a single fluid in hydrodynamic equilibrium, vertical compositional gradient, and lateral changes on composition (either by different migration and/or reservoir compartmentalization) are key to determine the number and location of the samples necessary to achieve the objective of the sampling operation.

As the sampling operation will likely require to adapt to the conditions found downhole, particularly for the open hole operations, it is important that a rigorous pre-sampling investigation is performed[16] to clearly understand the implications and more importantly design mitigation plans of deviations on the expected in situ conditions. A comprehensive wireline-log characterization of the sample formation may significantly increase the probability of success of the operation. Traditionally, resistivity and nuclear logs have been used to infer fluid types, while caliper logs provide a measure of hole stability and suitability for the downhole sampling. Use of more sophisticated logs such as nuclear magnetic resonance (NMR) has been documented in the literature to aid not only permeability estimations (to determine the type of sampling and infer potential challenges during the operation) but also to provide means of identifying and quantifying the radial saturation of fluids in the near wellbore area, i.e., to determine the type and size of filtrate invasion (including OBM), hydrocarbon type (oil, gas) and well as to infer potential changes on reservoir fluid composition (compartmentalization, compositional gradient) and overall to provide a continuous sampling of the complete reservoir interval to complement the discrete sampling.[17] Furthermore, it is not uncommon to combine NMR measurements with in situ fluid analysis[18] in order to better understand oil density and viscosity relationships. The main advantage, however, is possibly to identify and measure the oil-based-mud filtrate, with applications not only relevant for sampling depth selection but also to help on the characterization of intervals which were too difficult or too expensive to sample. The application of these techniques, of course, is case dependent.

It follows that success of a downhole operation relies on the fact that a "clean" uncontaminated sample of reservoir fluid gets in contact with the sample probe; that is, mobility (or permeability) at the formation level is high enough that a single phase flow may be achieved and maintained long enough to remove any contaminants from either completion and/or drilling fluids from the tested formation. Volumes and timing are a function of the formation permeability, saturation, pressure, viscosity, and of course drawdown. Low mobility, highly heterogeneous formations require a more comprehensive real-time monitoring, along with a specialized reservoir characterization tool in order to facilitate "on the fly" decisions for sampling location based on the encountered hole and reservoir conditions. A pressure survey will be part of the sampling operation to not only provide insight on the reservoir pressure, fluid gradients, and mobility but also to identify supercharging and any other challenges likely to affect the sampling tool position. When sampling highly heterogeneous formations with a probe-based tool, it is imperative that both reservoir pressure (mobility) and fluid sampling occur at the same station (should mobility be high enough) as it would be very difficult (highly improbable) that the same spot be tested with the probe after the tool has moved.

The effects of supercharging (build-up of pressure due to hydrostatic mud filtrate head) may be observed on both fluid gradients and contacts. Pressures as expected are not representative of the formation but rather mud gradients and are therefore not suitable for any analysis. Consequently, samples obtained under such conditions are likely to be heavily contaminated with drilling fluids. Supercharging is far more common on low mobility formations (rule of thumb < 1 md/cP). It is possible, however, for supercharged intervals to yield a pressure point consistent with the reservoir gradient, and as such they require further validation, a quality check of the shape and time of the build-up, extracted volumes, and repeated tests are among the industry practices.

3.2.2.1.2 Sampling Challenges

Containment and isolation of the formation and of course sampling tool from the wellbore are crucial on any operation. Furthermore, the sampling tool must have continuous uninterrupted contact with the formation to reduce both contamination risk and sampling time, and this applies to both probe and dual-packer sampling devices. The decision of tool specifics and sampling protocol is often designed with the aid of numerical models where optimum withdrawal rates (no to cross saturation pressure) and volumes are evaluated against different formation, mud filtrate, and wellbore-operating conditions.

Traditionally, unconsolidated formations and highly viscous fluids are challenging when using downhole samplers. Pressure changes associated with sampling along with the expected surge of fluids may cause unconsolidated formations to exceed critical velocities and start transporting fines (not to mention emulsions) toward the sampler while also causing formation stability problems. Low mobility on the other hand does require a larger pressure drop to mobilize the oil. The high viscous oil may have relatively low invasion profiles and have the risk of disturbing both formation and fluid (crossing the saturation pressure) as well as increasing cleanup time. Flow rates are therefore key for the sampling operation and are also a function of the type of tool selected for the job. Several studies are available to investigate the effect of either parameter on the tool selection, considering different effective flow areas (EFAs),[19] but it is clear that each solution needs to be tailored to the specific fluid/formation conditions in order to reach a meaningful conclusion.

Use of numerical models for sampling planning purposes has been extensively documented in the literature[16,20] with a clear emphasis on both high-resolution reservoir characterization thru wireline logs and core (whenever possible), tool geometry, and fluid chamber-filling techniques. These models differ from the traditional sector and single well not only due to finite difference grid resolution (detailed enough to capture the probe, invasion zone, and expected flow regimes during sampling) but also due to the necessary corrections to deal with capillary end effects and other numerical artifacts which may overshadow the formation responses. These models, when properly built and calibrated, enable the evaluation of sensitivities to quantify the effect of the remaining uncertainties on the overall operational risk, enabling a fit-for-purpose pre-sampling characterization. Given the complexity and high dependence on operations, this design is often carried out with the sampling service provider ensuring that specific tool strengths and limitations are considered.

Several advances on wireline sampling have been made since its introduction in the late 1990s, such as low-shock sampling which minimized the pressure disturbance by using a pump to mobilize the fluid onto the sample chamber against the hydrostatic pressure, reverse low-shock sampling, and packer filters to reduce mobilization of fines into the sample chamber to downhole segregation (SEG*), among others. As discussed before, one of the key design parameters for sampling is the flowing pressure, which affects not only the quality of the sample (single phase if higher than saturation pressure) but also impacts the sampling time (higher production rate of necessity will require shorter time to flush the near formation contaminants and hence speed up the sampling). Consolidated, high mobility/permeability formations allow for a larger drawdown and therefore are suited for either probe or conventional dual-packer tools; unconsolidated formations, nearly saturated fluids and high viscosity hydrocarbons, on the other hand are limited to smaller drawdowns to preserve formation stability (and seals) on the former and as result of low mobility on the latter.

Drawdown of course is also a function of the flow area available at the ports of the sampling tool, limiting the velocity at which fluids may enter the sampling chamber. As the effective area increases, so the flow rates increase for a given pressure drop. The largest available EFA ranges near 80 $in.^2$ enabling the sampling of formations/fluids which otherwise would have been beyond the scope of a traditional sampling.

3.2.2.1.3 *Downhole Fluid Analysis*

Once the formation, tool, and operational issues have been considered, the next step is the sampling operation itself. During this stage, every effort is made to minimize the sample contamination and retrieve a fluid as close as possible to the one saturating the reservoir. Sample contamination may not be avoided (particularly when dealing with post-drilling samples) but it certainly needs to be minimized to ensure representativeness.

When fluids saturating the near wellbore area are immiscible (such as hydrocarbon oil and water-based mud filtrate), fluid segregation is easier and thus pure hydrocarbon samples are easier to obtain. There is the risk, however, of emulsions and or hydrocarbon precipitation depending on the composition of the filtrate, but it may be minimized thru proper planning. Separation is more complicated when the fluids are likely to be miscible (case of oil-based mud), particularly in cases—such as gas condensates—where small changes on the composition and/or mix of any condensed hydrocarbon with the filtrate result in a much different fluid than the one saturating the reservoir. Extensive research has been conducted on the topic of sample contamination, aiming to determine permissible levels of mixed filtrate on the sample that still yields consistent PVT behavior at the laboratory. One rule of thumb is for given set of reservoir oil, the deviation of predicted gas–oil ratio (GOR) and saturation pressures was less than 5% when the mud contamination of the sample was less than 20% in weight. This level of tolerance, however, does change depending on the nature of laboratory work and, as expected, on the specific composition of the reservoir oil, with permissible contamination ranges being much lower for highly volatile oils and lower still for gas condensates.

Given the importance of recognizing the quality of the sample, it is not uncommon to introduce real-time composition monitoring during the sampling operation, aiming not only to obtain a clean sample but also aid with reservoir characterization and sample quality control at the surface.

Optical fluid analysis provides an alternative, using visible and near-infrared absorption spectrometry to recognize the nature of the fluid passing thru the tool measuring the combined effect on the light transmission and optical density (OD) of scattering and absorption. Crude oil can absorb light of a given wavelength, showing a distinct OD (vs. wavelength) depending on its composition. For shorter wavelengths,[6] the OD distribution is smooth and is a function of the aromatic content of the hydrocarbon oil. The lower the content of aromatics, the lower the OD and will be perceived as clear, as opposed to the high content aromatics which will be perceived as dark. For higher wavelengths, the shape of the OD distribution changes yielding several peaks, consistent with the strength of the H–C bonds on the alkane chains, shorter wavelength associated with CH_4 (methane) and the longer ones with more complex alkane chains.

The strength of the process relies mainly on the clear differentiation of the fluids saturating the near wellbore area. Water as expected does have a significantly different OD response as the mud filtrate does, making it plausible to recognize in situ the type of fluid flowing thru the sample at any given time. Scattering on the other hand refers to a non-absorbing process where the photons (of the beam of light) interact with the particles and molecules of the reservoir fluid and get deflected, thus reducing optical transmission. The magnitude of scattering is proportional to the particle surface, with clay particles (often present in mud solids) generating a high level of scatter, thus aiding on the fluid characterization.

This technique, however, has the underlying assumption that the mixture between the different components is limited and given sufficient cleaning up time representative reservoir samples may be collected. That is to say uncertainty (albeit minimized) still exists on the quality and composition of the sample; several authors have proposed the use of probabilistic methods where the OD tool, environmental and operational issues are accounted for on a systematic manner in order to provide a more comprehensive confidence level on the sample.[21] These are of particular interest when characterizing complex reservoirs as they allow for the statistically probability that any two samples are genetically different and compositional changes observed during the monitoring and sampling are not likely to be caused by sample contamination. These techniques are highly relevant for exploration wells where compartmentalization estimation plays a big role on the evaluation of the prospect. Application of such analysis is often done while the sampling operation is ongoing so that it may be tailored to suit the new findings.

Maintaining single phase inside the sampling chamber is extremely important. This includes limiting if not avoiding any foreign solids from entering the chamber and of course ensuring that the sample conditions of pressure and temperature are always in the single-phase region. It should be noted that while pressure within the chamber may be adjusted, temperature drops are often harder to control as the chamber surfaces. Empirical data suggest that should phase segregation occur in the chamber, recombining the components is often a very difficult process, particularly if asphaltenes and/or waxes drop out. Thus, sample preservation is crucial for a successful characterization.

There are a few rules of thumb in order to validate the quality of the sample once it reaches the laboratory,[22] such as documenting the pressure/temperature at the wellsite and laboratory before opening the PVT chamber, taking samples from top and bottom of the chamber (~5 mL) to identify potential compositional changes, measuring sample density (a higher value than the reservoir density may indicate water contamination for example), and carrying out a gas chromatography analysis which as compared with the drilling fluid fingerprinting as well as DFA may reveal either contamination or compositional changes due to poor handling of the sample at surface.

3.2.2.2 Surface Sampling

Surface recombination involves the sampling oil and gas samples from the outlets of a conventional production separator; these samples are then recombined following the relative amounts observed at the reservoir, which is the gas–oil-ratio the well/s exhibit when produced at steady-state conditions after the cleanup period (for undersaturated reservoirs). There are several assumptions involved in any surface sampling operation, namely: the stable GOR observed in the production test is consistent with that of the saturated fluid (i.e., a single phase is present at the sand face), the measurement and translation (accounting for separator and well conditions) of the GOR to reservoir is accurate, and lastly sampled fluids are properly recombined at the laboratory (avoiding any precipitation of either gas and/or liquid phases).

Sample integrity is, as with downhole sampling, a priority for any surface operations. Generally, surface sampling is less affected by near-wellbore fluids (compared to the downhole sampling) given the volume of fluids produced before the sample operation (longer cleanup), but it requires the well to be properly conditioned to avoid contamination and to produce at a steady rate for a long period of time such that the ratios of fluids (oil and gas) are stable both at the formation and at the separator. It follows that drawdown limitations are similar to those explained on the downhole sampling to avoid two-phase occurrence on the formation. The production rates must be closely monitored to ensure that steady-state flow is achieved, and any foaming, emulsification, etc., product in the flow line and the separator must strictly be avoided.

Given the dependence of separation conditions and GOR, separator fluid levels need to be maintained as stable as possible (and constantly monitored). The pressure and specially temperature changes need to be minimized, given the larger impact on the GOR behavior and subsequent sample validity.

Rules of thumb are also available for validation of separator samples.[22] Concerning both the gas and oil chambers, gas cylinder is often heated about the separator conditions and contents examined. Any liquid present in the chamber may suggest contamination, liquid carry over, and/or issues with the temperature sampling. It follows that the sampling pressure for the liquid equals its saturation pressure at the separator temperature. As such a partial CCE (constant composition expansion) should yield a similar saturation pressure if the sample is valid. Density measurements similar to those carried out on the downhole sample may also reveal (although less likely) any water contamination.

3.3 Field Sampling

Hydrocarbon sampling strategy, as explained in the previous sections, is as much a function of the hydrocarbon type as it is of the reservoir. Several case studies are available in the literature. Some of these case studies address fluid type, while the others address the formation issues. Downhole sampling on low permeability and unconsolidated formations has been particularly challenging given the restrictions on drawdown, cleanup, mud cake, packer deployment, and sealing, among others. The application of new technology in Thailand reported a successful wireline sampling in an otherwise prohibitive low mobility formation,[23] where delayed mudcake caused sealing issues on conventional formations. It was reported that six stations were successfully deployed on mobilities ranging from 0.3 to 17 mD/cP. Cleanup times were reported to be in the range of 45–114 min. Another example is the sampling of an unconsolidated laminated reservoir in the Gulf of Mexico,[24] where single-phase uncontaminated samples were collected with minimum cleaning up time (~2.5 h). This operation took advantage of the larger surface flow area (79.44 in.^{2}) and targeted sealing of the sampling interval, which not only aided the cleaning up but also provided support to the formation enabling a single-phase flow with a 35 psi drawdown.

Nagarajan[2] discussed a few sampling operations, particularly on sampling considerations of gas condensate and near-critical fluids. Near-critical fluid of the Oso field in Nigeria where temperatures and pressures were of 232 F and 6,300 psia, respectively, used downhole sampling for the characterization of the compositional gradient as well as near-critical behavior. The critical factor to the success of the sampling was the isolation of each station to ensure a proper compositional gradient imaging.

References

1. Lawrence, J.J., Chorneyko, D.M., Smith, C.K., and Nagarajan, N.R. Representative reservoir fluid sampling: Challenges, issues, and solutions. IPTC 12401 presented at the *International Petroleum Technology Conference in Kuala Lumpur, Malaysia*, December 2008.
2. Nagarajan, N.R., Honarpour, M.M., and Sampath, K. Reservoir fluid sampling and characterization—Key to efficient reservoir management. SPE 101517 presented at the *International Petroleum Exhibition and Conference in Abu Dhabi, U.A.E.*, November 2006.
3. Bon, J. and Sarma, H.K. Reservoir fluid sampling revisited—A Practical Perspective. SPE 101037 presented at the *Asia Pacific Oil & Gas Conference and Exhibition in Adelaide*, September 2006.
4. Jackson, R.R., Carnegie, A., and Dubost, F.X. Pressure measurement and pressure gradient analysis: How reliable for determining fluid density and compositional gradients. SPE 111911 presented at the *Annual Technical Conference and Exhibition in Abuja, Nigeria*, August 2007.
5. Dubost, F.X., Carnegie, A., Mullins, O.C., O'Keefe, M., Betancourt, S., Zuo, J.Y., and Eriksen, K.O. Integration of in-situ fluid measurements for pressure gradient calculations. SPE 10849 presented at the *International Oil Conference and Exhibition in Veracruz, Mexico*, June 2007.
6. Fujisawa, G., Betancourt, S.S., Mullins, O.C., Torgersen, T., O'Keefe, M., Terabayshi, T., Dong, C., and Eriksen, K.O. Large compositional gradient revealed by in-situ optical spectroscopy. SPE 89704 presented at the *Annual Technical Conference and Exhibition in Houston, USA*, September 2004.
7. Montel, F., Bickert, J., Hy-Billot, J., and Royer, M. Pressure and compositional gradients in reservoirs. SPE 85668 presented at the *27th Annual SPE International Technical Conference and Exhibition in Abuja, Nigeria*, August 2003.

8. Kabir, C.S. and Pop, J.J. How reliable is fluid gradient in gas-condensate reservoirs? SPE 99386 presented at the *SPE Gas Technology Symposium in Calgary, Canada*, May 2006.
9. Whitson, C.H. and Belery, P. Compositional gradients in petroleum reservoirs. SPE 28000 presented at the *University of Tulsa Centennial Petroleum Engineering Symposium in Tulsa, USA*, August 1994.
10. Canas, J., Pop, J., Dubost, F., and Elshahawi, H. Advanced compositional grading analysis. SPE 115429, presented at the *SPE Annual Technical Conference and Exhibition, Denver*, 21–24 September 2008.
11. Lindeloff, N., Mogensen, K., Van Lingen, P., Do, S.H., Frank, S., and Noman, R. Fluid-phase behaviour for a miscible-gas-injection EOR project in a giant offshore oil field with a large compositional variation. SPE 115970 presented at the *Annual Technical Conference and Exhibition in Denver, USA*, September 2008.
12. Hoch, E., Ohrt, H.B., Brink, D.I., Flikkema, J., and Noman, R. Pushing the limits for field development. SPE 138301 presented at the *International Petroleum Exhibition & Conference in Abu Dhabi, U.A.E.*, November 2010.
13. Frank, S., Pedersen, M.H., Lingen, P., and Noman, R. water flooding efficiency in a low-permeability mixed-wet to oil-wet carbonate reservoir; A case study from the Kharaib B reservoir in the Al-Shaheen Field, Offshore Qatar. IPTC 13330 presented at the *International Petroleum Technology Conference in Doha, Qatar*, December 2009.
14. Mullins, O.C. 2010. The modified yen model. *Energy & Fuels* 24: 2179–2207.
15. Mullins, O.C., Andrews, A.B., Pomerantz, A.E., Dong, C., Zuo, J.Y., Pfeiffer, T., Latifzai, A.S., Elshahawi, H.,Barré, L., and Larter, S. Impact of asphaltene nanoscience on understanding oilfield reservoirs. SPE 146649 presented at the *Annual Technical Conference and Exhibition in Denver, USA*, October 2011.
16. Canas, J.A., Low, S., Adur, N., and Teixeira, V. Viscous oil dynamics evaluation for better fluid sampling. SPE/PS-CIM/CHOA 97767 PS2005–347 presented at the *International Thermal Operations and Heavy Oil Symposium in Calgary, Canada*, 2005.
17. Awaoud, A., Zaggas, J., Haddad, S., and Xia, C.G. Formation fluid sampling from heterogeneous carbonate reservoirs. Presented at the *SWPLA 49th Annual Logging Symposium in Edinburgh*, Scotland, May 2008.
18. Mihn, C.C. and Weinheber, P. et al. Using the continuous NMR fluid properties scan to optimize sampling with wireline formation testers. SPE 115822 presented at the *Annual Technical Conference and Exhibition in Denver, USA*, September 2008.
19. Asly, M., AS, V.N., Sibbald, G., and Orrell, W. Formation testing and sampling in low-mobility formations: an example of new technology solutions. Presented at the *SWPLA 57th Annual Logging Symposium in Reykjavik, Iceland*, June 2016.
20. Kristensen, M., Ayan, C., Chang, Y., Lee, R., Gisolf, A., Leonard, A., Corre, P., and Dumont, H. Flow modeling and comparative analysis for a new generation of wireline formation tester modules. IPTC 17385 presented at the *International Petroleum Technology Conference in Doha, Qatar*, January 2014.
21. Venkataramanan, L., Elshahawi, H., McKinney, D., Flannery, M., Hashem, M., and Mullins, O.C. Downhole fluid analysis and fluid comparison algorithm as an aid to reservoir characterization. SPE 100937 presented at the 2006 *Asia Pacific Oil & Gas Conference and Exhibition, Adelaide, Australia*, 11–13 September.

22. Shaikh, J.A., Sah and Sah, P. 2011. Experimental PVT Data Needed to Develop EOS Model for EOR Projects. SPE 144023 presented at the *SPE Enhanced Oil Recovery Conference in Kuala Lumpur, Malaysia*, July 2011.
23. Schlumberger 2013 case study "Saturn 3D Radial Probe Samples Fluids from Tight Formations, Gulf of Thailand" retrieved from www.slb.com/services/characterization/reservoir/wireline/saturn.aspx?t=4&libtab=2.
24. Schlumberger 2015 case study "Saturn 9-in Probe Samples Near-Saturation Hydrocarbon in an Unconsolidated Laminated GOM Reservoir" retrieved from www.slb.com/services/characterization/reservoir/wireline/saturn.aspx?t=4&libtab=2.

4

Planning of Laboratory Studies

4.1 Introduction

As important as fluid sampling is to preserve the integrity of the reservoir fluids, so is the planning of sequence and type of laboratory experiments to aid on the fluid and displacement process characterization. While standards for laboratory experiments are clearly defined with the expectation that all fluid laboratories adhere to such, there are challenges that arise when trying to characterize fluids and processes that somewhat differ from the day-to-day operations. Any laboratory planning needs to start with a clear definition of objectives and resolution; that is, use of the laboratory results on the reservoir characterization and most importantly on the development decisions needs to be properly defined and recognized by the development team to avoid misrepresentations and ultimately sub-utilization of the laboratory results.

Impact of planning and clear strategy for the use of laboratory experiments is particularly important for miscible and near-miscible flooding (along with critical and retrograde systems) where laboratory results help to further calibrate equations of state that subsequently are used to predict reservoir performance and recovery. Traditionally miscible or near-miscible processes are explained through either vaporizing or condensing mechanisms—the former where the light components of the oil help enriching the gas phase and the latter where a reverse mechanism occurs where the light components from the gas phase enrich the oil. The laboratory experiments have been designed to fit this interpretation. While the results of the experiments do fit the mutually exclusive vaporizing/condensing processes, a combination of both is more likely to occur at reservoir conditions; implications of which go from the different approach of tuning EOS to the quality checking of the laboratory experiments where laboratories may omit data points that do not conform to the existing understanding. While efforts in the industry are being made to improve these shortcomings, it is important that engineers involved on planning and evaluation of laboratory experiments for miscible flooding must be aware of such limitations when using the results for prediction purposes.

4.2 Conventional Oil and Gas Experiments

PV cell experiments have been extensively used in the industry for the characterization of hydrocarbon fluids under the assumption of thermodynamic equilibrium. The main sets of experiments include constant composition expansion (CCE) often called constant mass expansion (CME), where the change of volume of the undersaturated fluid is recorded as a function of pressure; no hydrocarbon is removed from the cell—hence called constant composition/constant mass—during the process. The experiment relies on the changes of compressibility of the overall system to determine the saturation pressure (P_{sat}). At pressures higher than P_{sat}, overall compressibility of the system follows that of the oil phase, however, upon crossing the saturation pressure, evolved gas will significantly increase the compressibility of the system, changing the slope of the pressure vs. relative volume chart; saturation pressure is then defined as the pressure where the change of slope occurs.

CCE experiments are followed by either differential liberation (DL) and/or constant volume depletion experiments (CVD)—the former often performed in black oil reservoir and the latter for volatile, condensate reservoirs. DL follows a similar process to that of the CCE, oil at P_{sat} is equilibrated in the PV cell at reservoir temperature, and pressure is decreased at fixed steps; however, unlike the CCE, any evolved gas is removed from the cell at the end of each pressure step (upon reaching equilibrium), and hence composition of the hydrocarbon is constantly changing with the fluid getting progressively heavier as the lighter components migrate toward the gas phase. At each step, volumetric factors such as Bo, Z, oil, and gas viscosities are measured and reported, as well as the evolved gas volumes. Due to the nature of the experiments (series of two phase flash), a larger number of intermediate components are found in the gas phase—when compared with a one/two flash experiments and separator, and therefore volumetric factors and solution gas (R_s) are pessimistic and need to be corrected with separator information prior to be used in any subsequent analysis.

Constant volume depletion (CVD) is somewhat in-between the CCE and DL experiments, where the volume of the cell (and not composition) is maintained constant through each of the pressure drops performed during the experiment; compositions of the liberated gas are systematically measured as well as the properties of the hydrocarbon fluids left in the cell. CVD experiments arguably do represent the physics of the reservoir better than the DL and CCE, where pressure and saturation are changing in response to the fluid production where the actual reservoir volume remains the same. Results, much like the DL, need to be further corrected to account for separator conditions.

In addition to the pressure depletion experiments, single, multi-flash experiments are conducted to represent the separator conditions of the field, to optimize them as well as to provide means of correcting DL and/or CVD results to field conditions. It is expected that through this optimization a larger proportion of the low-mid weight hydrocarbon components remain in the liquid phase, thus improving the overall value of the asset.

Several authors have discussed the importance of complete laboratory experiments for the proper characterization of hydrocarbons under primary and secondary production.[1,2] A set of laboratory experiments, however, does not ensure a successful fluid characterization, requiring a complete set (see Table 4.1) for the analysis. Furthermore, laboratory data need to be validated with field observations (see Chapter 3) and should be consistent within on itself.

Single-phase composition is one of the main objectives of any laboratory studies, to either complement or validate the in situ measurements taken during the sampling activity and/or to characterize the reservoir fluid for the first time. While laboratory equipment measurements of composition have been substantially improved in the past decade (with ability to characterize heavy fractions well beyond the traditional $C7^{+}$ carbon group), these experiments are not without uncertainty depending both on the accuracy of the laboratory equipment and the concentration and nature of the components (smaller fractions of components are much difficult to measure than larger ones, for example). Traditionally,[3] the accuracy of the compositional measurement has been estimated as follows:

$$R_{x_i} = 0.07 \cdot x_i^{0.43}$$

where R is the repeatability and x is the fraction of component i.

Overall, a 0.5 mol% is the higher limit for repeatability for higher concentrations and near 0.01% for lower ones, N_2 has an absolute detectability of nearly 0.3 mol%.

Many techniques have been proposed through the years to test the validity of the laboratory data: Y function,[4] density, R_s tests, and inequality tests for Black oil reservoirs, and Y function, Bashbush,[5] Whitson[6] and Hoffman crump for volatile and condensate reservoirs, among others.

TABLE 4.1

PVT Experiment Requirements Per Fluid Type

Experiment	Black Oil	Volatile Oil	Dry Gas	Condensate
Composition	√	√	√	√
CCE	√	√		
DL	√	√		
CVD		√	√	√
Viscosity	√	√	√	√
Separators	√	√	√	√

Use of "Y" function is very common in the industry. It mainly uses the pressure and volume measurements during CCE to further validate the saturation pressure estimation. It should be noted that saturation pressures are not measured directly, rather they are estimated based on other observations, in the case of CCE, saturation pressure estimation relies on the considerable change of compressibility as two phases are present in the PVT cell. Recognition of the "two slopes" is easier on black oil samples where the oil compressibility is significantly lower than the two phase, but it does become more difficult on volatile oil samples where the compressibility contrast is lower. Experimental errors due to equipment and laboratory experience are not uncommon and are mostly associated with measurements being taken prior equilibrium has been reached (particularly when two phases are present). The concept of "Y" was introduced by one empirical equation[4]:

$$Y = \frac{(P_{sat} - P)}{[(P_{abs}) \cdot (V_r - 1)]}$$

where

P_{sat} is the saturation pressure,

P is the experiment pressure,

P_{abs} is the same pressure "P" but in absolute units, and

V_r is the relative volume at pressure "P" or volume at pressure "P" divided by the volume at "P_{sat}".

A smooth curve of "Y" vs. pressure is expected, and any deviations should be investigated to determine source and take appropriate action. The main objective, however, is to validate the selection of the saturation pressure (to which "Y" function is most sensitive). It follows that if the curve not be smooth or change direction (in the exclusion of laboratory measurement error) there are errors on the saturation pressure estimation. The pressure should then be changed and "Y" function calculated again until the curve is smooth. It must be noted that any changes on P_{sat} prompt changes on V_r as they are mutually linked.

Material balance has been used to both validate the accuracy of compositional PVT experiments and to extract valuable data from the experimental results prior to any equation of estate calibration. Whitson and Torp[6] described the process in detail as follows:

$$n_{tk} = n_{Lk} + n_{Vk}$$

and

$$n_{tk} \cdot z_{jk} = n_{Lk} \cdot x_{jk} + n_{Vk} \cdot y_{jk}$$

where

n indicates the number of moles, the subscripts L and V indicate the liquid and vapor phases, respectively, z, x and y are the molar composition of component "j" for total, liquid and vapor, and k refers to each pressure stage during the CVD experiment. Only y composition is measured and the others need to be inferred. Considering 1 mol of initial fluid:

$$n_{tk} = 1 - \sum_{i=2}^{k} \Delta n_{pi}$$

and

$$n_{tk} \cdot z_{jk} = z_{j1} - \sum_{i=2}^{k} \Delta n_{pi} \cdot y_{ji}$$

where

Δn_{pi} is the incremental moles of vapor produced from the cell. Using real gas law, the remaining moles in the cell can be calculated as follows:

$$V_{cell} = \frac{Z_d \cdot R \cdot T}{P_d}; \quad \text{for gas condensate and}$$

$$V_{cell} = \frac{M_b}{\rho_b}; \quad \text{for volatile oils}$$

where

M_b and ρ_b are bubble-point molecular weight and mass density and Z_d/P_d are dew-point compressibility and pressure, respectively. The liquid saturation is reported at every pressure stage of the CVD experiment and can be expressed as follows:

$$V_{Lk} = S_{Lk} \cdot V_{cell}$$

and

$$V_{Vk} = 1 - V_{Lk}$$

and

$$n_{Vk} = \frac{P_k \cdot V_{Vk}}{z_k \cdot R \cdot T}$$

It follows that

$$x_{jk} = \frac{n_{tk} \cdot z_{jk} - n_{Vk} \cdot y_{jk}}{n_{tk} - n_{Vk}}$$

where K values can be calculated as follows:

$$K_{jk} = \frac{y_{jk}}{x_{jk}}$$

A plot of K values with pressure should give a smooth monotonicity curve.

The product of $K{\cdot}P$ (K values times experiment pressure) may also be plotted against the component characterization factor "F" defined by Hoffman and Crump[7] where the lines for each pressure depletion stage should be smooth and should not cross each other. In addition to these, there are other simple QC[8] that have been widely used in the industry such as plotting composition (x and y) vs. pressure on a logarithmic scale, where curves should be smooth and without any "jumps" in the data. The factor F[7] is defined as follows:

$$F = \frac{\log\left(\frac{P_c}{14.7}\right)}{\left(\frac{1}{T_B} - \frac{1}{T_c}\right)} \cdot \left(\frac{1}{T_B} - \frac{1}{T}\right)$$

where

P_c and T_c are the component's critical pressure and temperature (in psia and degR), T_B is the boiling pressure in degR, and T is the experiment's temperature (degR).

A similar procedure may be applied to validate the separator experiments, where composition (using the material balance described above) may be calculated at each stage of the separation and used as feed for the next one; smoothness and monotonicity are expected on the Hoffman and Crump plots as well as the K values vs. pressure plot.

Quality of laboratory data and key considerations during the experimental exercise have been documented extensively,[9] where some of the practical challenges during the laboratory procedures are highlighted, ranging from compositional changes when the PVT cell is bled or hydrocarbon fluid is added to match the saturation pressure on DL/CVD experiments (where the initial composition after the addition of fluid may not be measured and reported), to data gaps on the measurements around the saturation pressure where not enough fluid (gas for oil samples and condensate for gas) is collected to perform a meaningful analysis, hence having to draw to a lower pressure and rely on extrapolation.

Application of these techniques in ranking and evaluating laboratory results has been discussed in the literature. Nnabou et al.[10] used sample validity checks, recombination accuracy, and overall experimental consistency, which were investigated for a sample in the Niger delta field. Use of material balance and Hoffman factors are also explained.

Quality of PVT data was also investigated by Stephen et al. (2008) where the authors used a set of identical fluid samples (dry gas-3, live oil-1 and stock tank-3) to several PVT laboratories around the world and compared the quality of the data—ranked against a reference value. Some of the deviations range from up to 3% in the measurement of C1 in the gas, to 17 wt% in mud base oil content of the stock tank liquid and –11–20 wt% in the C36+ characterization. Errors were attributed to flawed laboratory procedures, inaccuracy of the quantification of the low volatility materials in C36+ and failure to follow laboratory procedures, highlighting the human error exposure on any laboratory analysis.

While these experiments are standard for green and brown developments, further characterization work is required when enhanced/improved oil recovery activities are considered, particularly if the EOR agent is likely to interact with both liquid and gas phases of the reservoir hydrocarbon.

4.3 Special Laboratory Experiments

4.3.1 Swelling Test

This test is performed on a PV cell (often after having performed a CCE experiment) where a given volume of reservoir fluid is equilibrated at reservoir pressure and temperature conditions, a mixture of gas is then injected to the PV cell and pressure increased until a single phase is observed. Sample volume is measured and the incremental (swelled) volume is determined. The mixture then undergoes a second CCE to determine the new saturation pressure along with liquid shrinkage. The process is repeated (adding gas) to match the expected field conditions. It follows that with the sequential addition of hydrocarbon gas, the hydrocarbon composition within the cell changes, thus gradually increasing the concentration of mid-light weight components and transforming the fluid from a black oil to volatile, critical fluid, and gas condensate. Swelling tests are useful to determine the potential interaction between the injection agent (gas, CO_2, solvent, etc.) and the reservoir fluid providing valuable information in terms of PVT property changes (swelling, stripping, changes on viscosity, etc.) that can be used for EOS calibration and more importantly to further substantiate the injection potential of the tested agent. It is common to perform multiple swelling tests with different injection agent concentrations (close to the ones expected to be available on the field) to shortlist the most attractive ones and decide upon the viability of further laboratory tests.

4.3.2 Minimum Miscibility Pressure

The minimum miscibility pressure (MMP) is defined as the minimum pressure at which a single phase exists between the injected solvent (hydrocarbon gas, CO_2, etc.) and the hydrocarbon fluid at reservoir conditions. There are a couple of measurements available for determining MMP. The most popular is the slim tube test where a long stainless-steel tube (typically 60 ft long and ¼ in diameter) is filled by either sand or glass beads and used to perform a series of displacement tests at different pressures (while keeping temperature constant). Porosity, permeability, and most important volumes (including dead volumes) are measured prior the displacement tests to ensure consistency, the tube is heated to reservoir temperature (typically with toluene inside) and then displaced with reservoir oil until reaching stable conditions. The basic PVT properties of the reservoir fluid are then measured to confirm representativeness and to ensure all the toluene has been displaced prior start of the test. Experiment pressure is controlled by applying back pressure at the outlet. Gas or solvent is injected at a fixed rate (to resemble the expected frontal velocity at the reservoir) while continuously measuring the fluids at the outlet, where volumes, density, and rates are measured nearly every hour. It is common that the injection rate is increased during the experiment. Gas/solvent breakthrough is constantly monitored and tests are typically carried out to inject between 1.2 and 1.5 PV upon which injection stops and backpressure is removed. Produced fluids are then used for material balance purposes to ensure consistency on the experiment. The experiment is then repeated for different pressures. A plot of oil recovery at 1.2 PV injected vs. backpressure is used to determine the MMP. It is expected that the plots have two different slopes (one at lower pressures where miscibility is not yet achieved and the other at higher pressures where displacement is miscible). The intersection of these two slopes (which is again expected to correspond to a higher recovery ~90%) defines the MMP (Figure 4.1).

It is common for the laboratory to measure and report:

- Slim tube downstream pressure
- Slim tube injection pressure
- Slim tube differential pressure
- Oven temperature
- Injection solvent temperature
- Produced oil volume
- Produced gas volumes at ambient conditions
- Produced gas composition
- Visual observations of produced fluids

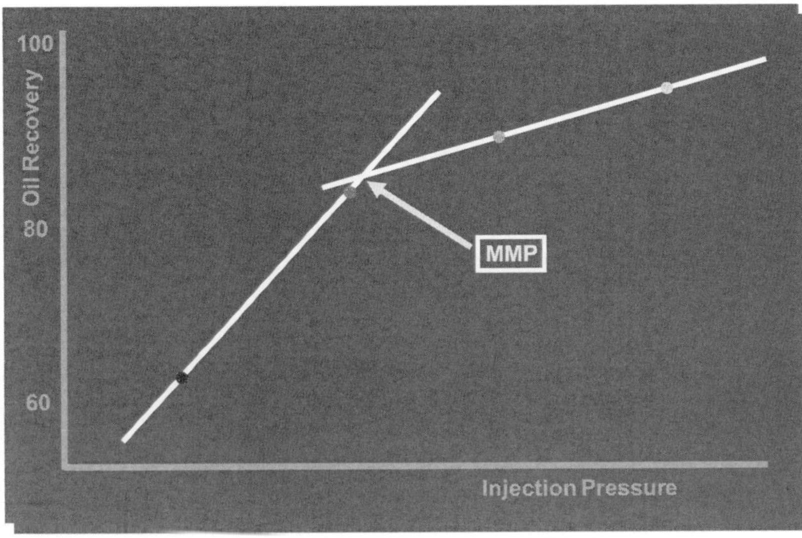

FIGURE 4.1
MMP calculation example.

Pore volume and inlet/outlet dead volumes of the slim tube need to be measured and duly reported prior the start of the experiment, along with the permeability, porosity, length, diameter type of tube packing material and particle size distribution. A portion of the reservoir fluid should be flashed out of the slim tube at the gas injection pressure and temperature conditions to assess certain parameters such Bo and GOR.

Recovery trends should be consistent, requiring an oil material balance to be performed to validate each slim tube run. There are four methods of measuring the recovery at 1.2 system volumes available at each pressure: (1) the produced oil, (2) the pore volume minus the residual oil volume from expansion of helium gas before and after each run (Boyle's Law expansion), (3) the pore volume minus the weight of residual oil divided by its density, and (4) the pore volume minus the oil volume recovered from the solvent wash. Any three of the four methods can be used to estimate the standard deviation around the average measure of the recovery at 1.2 system volumes at each pressure. In turn, the standard deviation can be used to assess the uncertainty in the MMP measurement.

Another method for defining miscibility pressure is the rising bubble apparatus (RBA) test where miscibility is defining visually injecting (on a clear PV chamber) gas/solvent at the bottom and watching it raise at different pressures. The injection system includes a small hand pump and a needle. The hand pump displaces injection gas into the bottom of the glass tube through the needle. The needle is placed at the bottom of the glass tube. The key component is the visual cell with a glass tube in it. The cell body has two windows on either side, which allows for observation. The glass tube is about 30-cm long and immersed in the confining fluid: water. The top of the tube is connected to the oil cylinder and the bottom is open to the confining fluid. The needle is just inside the glass tube about 1-cm deep. The tracking system includes a video camera mounted on a motor-driven support. The motor speed is controlled to keep track of the bubble movement. A VCR records the entire movement of the bubble from the bottom to top. The visual cell with the glass tube is first filled with water and equilibrated at the specified testing pressure and temperature. Then fresh oil is pushed by the positive displacement pump into the top of the glass tube, when the oil water interface reaches a point just above the needle, the pump stops.

The criterion to determine the MMP is as follows: at pressures below the MMP, the solvent bubble may hold its shape when it enters the oil column and may reach the top of the tube intact. Above MMP, the gas bubble may disintegrate, dissolve, or disappear into the oil immediately after it contacts the oil or it may disintegrate more slowly, becoming fuzzy and developing miscible tails and eventually disappearing prior to reaching the top of the tube.

The test results on their own do provide substantial evidence of the injection potential (miscible vs. non-miscible) of the tested EOR agent. Fluid compatibility issues (such as asphaltene precipitation) should be de-risked prior embarking on the slim test experiment—given it complexity, duration, and of course cost, such that test run smoothly. Use of the results of the slim tube experiments to calibrate EOS will be discussed on a subsequent chapter of this book.

4.3.3 Multiple Contact Miscibility

Multiple contact experiments simulate, what is perceived, a more realistic case where the injected agent (gas, CO_2, etc.) contacts the reservoir oil multiple times allowing dynamical compositional changes to take place. There are two main mechanisms for multiple contact miscibility: namely vaporizing and condensing drives.

4.3.3.1 Vaporizing Gas Drive

On a vaporizing drive, the injected agent attracts the light components of the oil enriching as it moves through the reservoir up to the point where miscibility is achieved between the enriched gas and the reservoir oil. The process is often recognized when the oil composition lies to the right of the limiting tie line on a ternary diagram and the injectant to the left. Miscibility on a vaporizing gas drive occurs at the front of the advancing gas. Figure 4.2 shows a ternary diagram with a vaporizing gas drive:

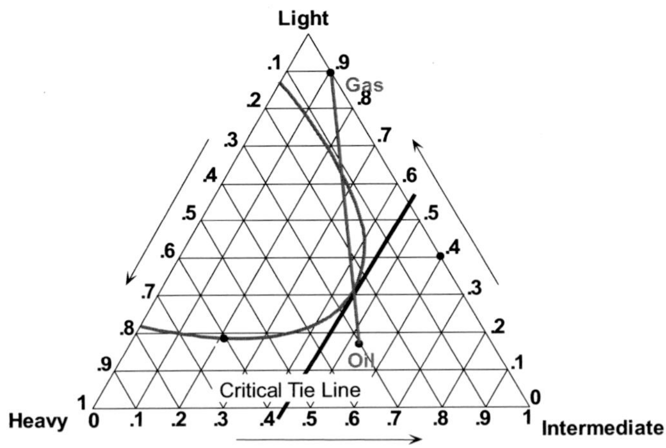

FIGURE 4.2
Vaporizing gas drive ternary diagram.

The experiment designed to represent the vaporizing gas drive is often referred to as forward contact, where a portion of the reservoir fluid is charged onto the PVT cell and equilibrated at reservoir conditions (pressure, temperature), injection gas is introduced to the cell and its contents mixed and allowed to equilibrate (often laboratories may perform the mixing at pressures close to first contact miscibility and then lower it to the reservoir or multicontact pressure). This enriched gas is then transferred to another cell where it encounters fresh oil and the process repeated until a single phase is observed in the cell (typically designed for four contacts). Through this process, compositions and concentrations of the enriched gas and oil are monitored, as well as densities and volumes. Partial CCEs are performed on the remaining oil of every cell to determine their saturation pressure, thus validating the migration of the lighter components to the gas phase.

4.3.3.2 Condensing Gas Drive

Much like on the vaporizing drive, condensing processes rely on the interchange of light-med weight components between the injecting agent and the reservoir oil, in condensing drives it is the oil (unlike vaporizing) that gets enriched by the interchange, thus getting lighter as the injecting agent passes through the pore structure. In theory to achieve a condensing gas drive, oil composition should lie to the left of the critical tie line and the injectant either at, or to the right. Miscibility is achieved at the injection point (Figure 4.3).

The experiment designed to represent the condensing gas drive is often referred to as backward contact, which is very similar to the forward contact with the only difference that it is the oil that is brought into contact with fresh injectant gas upon equilibration of the PVT cell. Figure 4.4 shows a schematic of both processes.

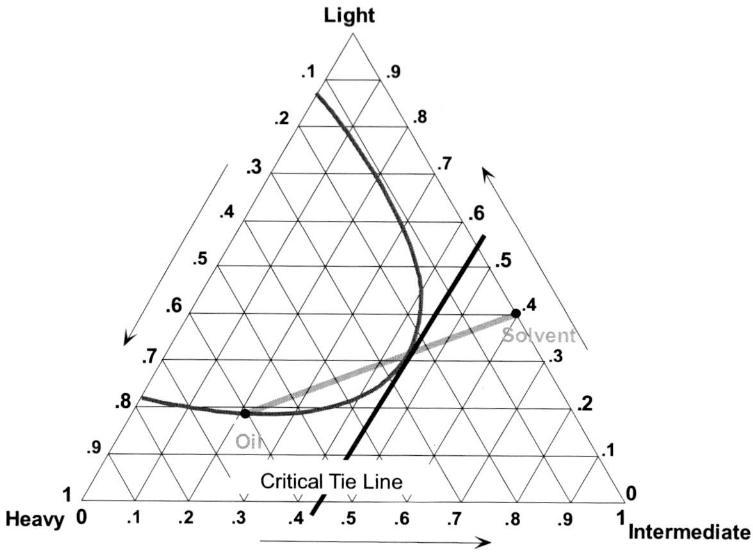

FIGURE 4.3
Condensing gas drive ternary diagram.

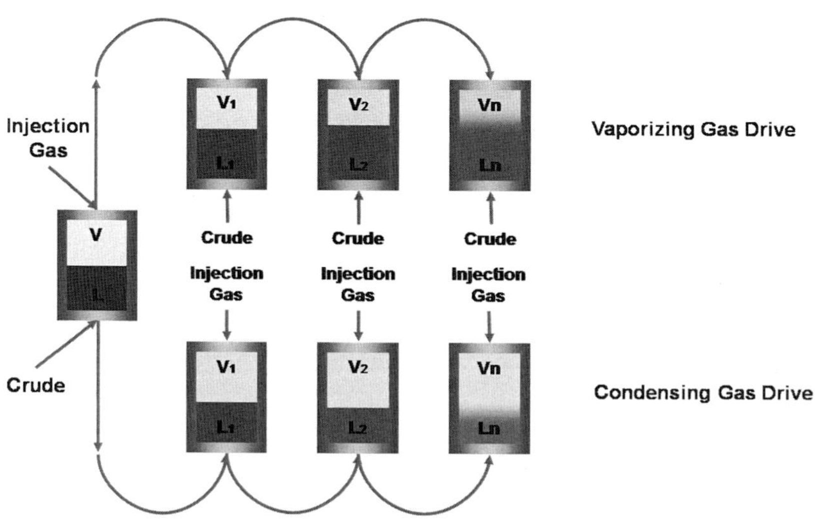

FIGURE 4.4
Vaporizing and condensing gas drive experiment setup.

4.3.4 Fluid Compatibility Studies

Fluid compatibility, particularly when introducing a foreign fluid onto the reservoir, is key for the success of any injection project. Wax and asphaltene precipitation are among the most common issues experienced during EOR injection. CO_2 has the ability to reduce the asphaltene precipitation pressure causing an undesired effect in the reservoir as it not only significantly reduces formation injectivity but also triggers compositional changes.

4.3.5 Asphaltene Precipitation Envelope

Asphaltene phase envelope is used both for surface and subsurface processes design to minimize/mitigate the deposition of asphaltene by ensuing that the fluid follows a proper thermodynamic path. A typical asphaltene phase envelope is shown in Figure 4.5 where the upper locus is the boundary of asphaltene onset. As the pressure decreases below this point, the severity of asphaltene precipitation increases and maximizes at the saturation pressure. Below the saturation pressure, asphaltenes tend to re-dissolve in to the solution as the pressure decreases. Depending on the oil, asphaltene may dissolve completely, as indicated by the lower locus, or may suspend in the solution even if the pressure is reduced to ambient. In this case, the lower asphaltene locus may not be determined.

Asphaltene precipitation is often determined in the laboratory using "injection gas compatibility test" where the onset is mapped at different pressures and compositions (to resemble likely frontal conditions). There are, however, a few steps to follow to minimize measurement uncertainty, firstly preliminary screening using de Boer plots,[11] validation of the sample quality, modeling of the observations using an appropriate EOS—looking for reproducibility of the experimental results, and finally use the field observations to further validate the predicted precipitation conditions—whenever available.

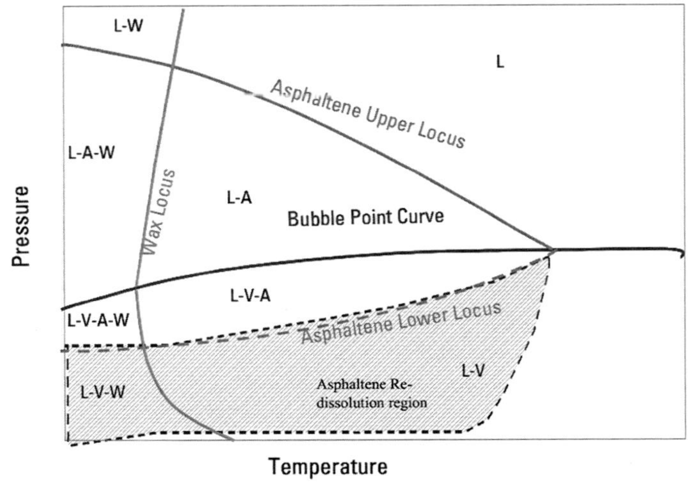

FIGURE 4.5
Typical asphaltene phase envelope.

Special experiments for stock tank oil, do include, asphaltene content, saturated–aromatic–resin–asphaltene (SARA) and compositional analysis. Live oil experiments are run concurrently with CCE to validate saturation pressures.

References

1. McCain, W.D. 1990. *The Properties of Petroleum Fluids.* Tulsa, OK: PennWell Publishing Company.
2. Paredes, J.E., Perez, R., and Perera, L.M. Screening criteria and methodology for quality check and PVT selection for reservoir studies. SPE 170695-MS presented at the 2014 *SPE Annual Technical Conference and Exhibition in Amsterdam,* October 27–29.
3. Eyton, D.G.P. Practical limitations in obtaining PVT data for gas condensate systems. SPE 15765, presented at the 1987 *SPE Middle East Oil Show, Bahrain,* March 7–10.
4. Williams, J.M. Why Y? SPE 146394. Presented at the 2011 *SPE Annual Technical Conference and Exhibition in Denver, Colorado,* November 2011.
5. Bashbush, J.L. 1981. *A Method to Determine K-values from Laboratory Data and its Applications.* Dallas, TX: SPE.
6. Whitson, C.H. and Torp, S.B. 1981. *Evaluating Constant Volume Depletion Data.* Dallas, TX: SPE.
7. Hoffman, A.E., Crump, J.S., and Hocott, C.R. 1953. Equilibrium constants for gas-condensate system. *Transactions, AIME* 198: 1–10.
8. Samaniego, V.F., Bashbush, B.J., Leon, G.A., Mazariegos, U.C., Corona, B.A., and Casticllo P.P. On the validation of PVT compositional laboratory experiments. SPE 91505 presented at the 2004 *SPE Annual Technical Conference and Exhibition, Houston, TX,* September 26–29.
9. Drohm, J.K., Trengove, R.D., Murdoch, U., and Goldthrope, W.H. On the quality of data from standard gas-condensate PVT experiments. SPE 17768, presented at the 1988 *SPE Gas Technology Symposium, Dallas, TX,* June 13–15.
10. Nnabou, N.N., Okafor, I.S., and Ubani, C.E. Interpretation of laboratory PVT analysis results (A Case study of a Niger delta field). SPE-172412-MS, presented at the 2014 *SPE Nigeria Annual International Conference and Exhibition, Lagos,* August 5–7.
11. De Boer, R.B., Leerlooyer, K., Eigner, M.R.P., and van Bergen, A.R.D. 1992. *Screening of Crude Oils for Asphaltene Precipitation: Theory, Practice, and the Selection of Inhibitors.* Dallas, TX: SPE.

5

Phase Behavior of Petroleum Reservoir Fluids in the Dense Phase or Supercritical Region

5.1 Definition of Dense or Supercritical Phase

In the context of normal or broad reference made to petroleum reservoir fluids, terms such as vapor, gas, or liquid are commonly used because of a particular state of existence in the rock pore spaces, production tubing, surface processing facilities, and pipelines. Obviously, this state of existence being a vapor or gas *phase* or liquid *phase* is primarily dictated by the prevailing temperature, pressure, overall composition, and chemical makeup of the system. For example, it is possible for the same pure or single component or mixture to exist as either a vapor or gas phase or a liquid phase depending on the prevailing conditions. The phase descriptors simply refer to how close or how far the molecules are packed, and thus alternate terms such as less dense and more dense referring to the vapor or gas phase and liquid phase, respectively, are also employed.

We now take a closer look at these phases for a pure component and mixture system from the standpoint of phase behavior in order to highlight the importance of the so-called dense phase as far as its application in miscible displacement-based enhance oil recovery (EOR) and transportation applications is concerned. Given the importance of carbon dioxide (CO_2) as a recognized gas that contributes to climate change, its importance from a geological sequestration standpoint and its popularity as an injection gas for EOR processes, we will use the phase behavior of CO_2 as an example to describe the dense and other phases. Figure 5.1 shows the phase diagram of CO_2 predicted by the Peng–Robinson (PR) equations of state (EOS).[1]

As can be seen in Figure 5.1, the vapor pressure curve or the temperature and pressure points on the vapor pressure curve marks the equilibrium between the CO_2 vapor and liquid phases, respectively. These vapor and liquid phases can also be referred to as less and more dense phases, respectively. CO_2 exists as a vapor phase and liquid phase, respectively, below

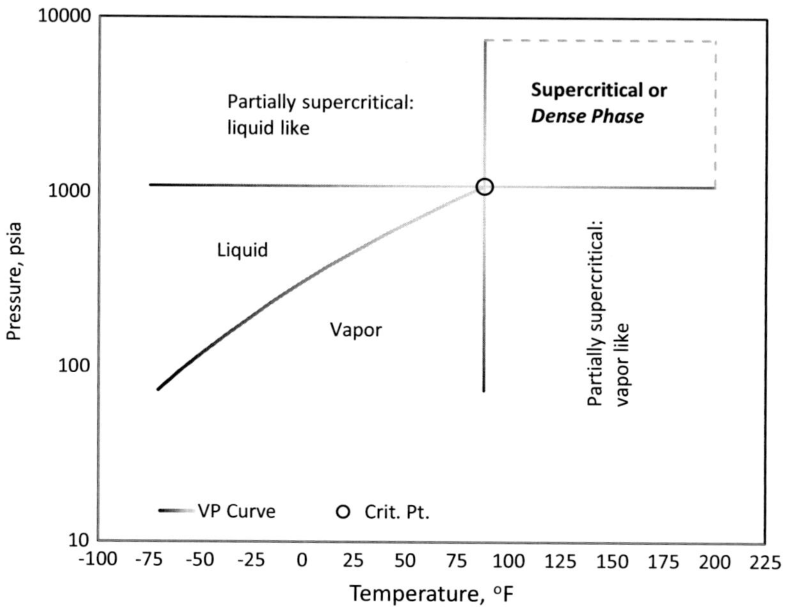

FIGURE 5.1
Phase diagram of carbon dioxide (CO_2) showing the distribution of phases based on the temperature and pressure conditions.

and above the vapor pressure curve. However, at temperatures exceeding the critical temperature, properties of CO_2 are partially supercritical vapor like and the converse is true when pressures exceed the critical pressure. All these regions are identified in Figure 5.1. The end of the vapor pressure curve is the critical temperature and pressure of CO_2, collectively known as the critical point (88°F and 1,085 psia), which also is the limit of the temperature and pressure at which CO_2 vapor and liquid phases can coexist. The lighter shading intensity of the various lines around the critical point signifies similarity in the phase properties, such as density. As the prevailing conditions move away from the critical point, the darker shading intensity of the lines indicates a relatively higher density. Clearly, if both the temperature and pressure are above the critical temperature and pressure, CO_2 cannot be distinguished as a vapor or a liquid and is simply referred to as a *supercritical or dense phase*. The hypothetical envelope of dense phase conditions is shown by a box in Figure 5.1; the broken horizontal and vertical lines are in no way limiting but are shown to somewhat simply represent typical operating condition limits (220°F and 7,500 psia) for various applications such as sequestration, supercritical extraction, and EOR. Given the near-ambient critical temperature and somewhat low critical pressure, practically speaking CO_2 will almost always be in a supercritical or dense phase for most applications. This particular dense phase of a pure component possesses some rather unique properties that are beneficial and advantageous in various applications (see Sections 5.2 and 5.5).

Let us now consider the phase behavior of a somewhat rich (in terms of intermediate components) natural gas type mixture, which is shown in Figure 5.2, as predicted by the PR EOS model. Obviously, unlike a pure component, instead of a single vapor pressure curve there is a broad region known as the phase envelope that encompasses variety of temperature and pressure conditions at which vapor and liquid phases in equilibrium are present. The bubble- and dew-point curves meet at the critical point at which the intensive properties of the liquid and vapor phases are identical (signified by the light shading of the curves as they approach the critical point). The temperature and pressure corresponding to the critical point are the critical temperature and pressure, respectively. The definition of critical point as it applies to a pure component is not the same as far as the mixture is concerned, i.e., this point for a mixture will vary (as well as the location of the saturation curves on the temperature–pressure plot) depending on the components making up the mixture and their individual compositions. Temperature and pressure conditions on the far left and far right indicate the presence of the mixture in a single-phase liquid and vapor, respectively (Figure 5.2).

The concept of dense phase for hydrocarbon mixtures (especially natural gas type) was first suggested by Hankinson and Schmidt,[2] based on which the dense phase boundary is portrayed in Figure 5.2. The lower and upper temperature limits of the dense phase region are bounded by the critical temperature and the cricondentherm (highest temperature on the phase

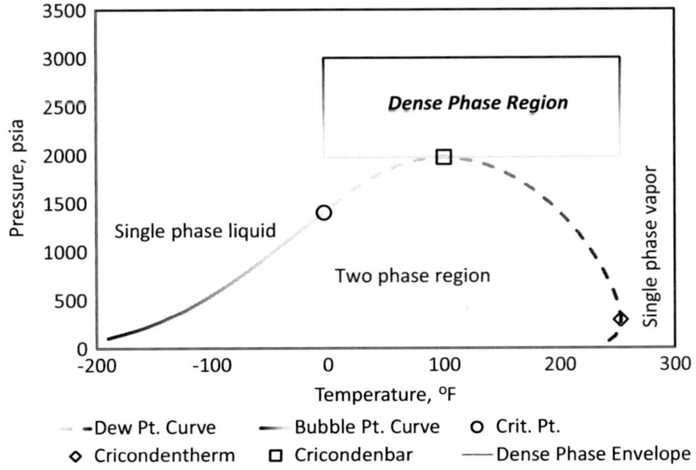

FIGURE 5.2
Phase envelope of a rich natural gas mixture showing the distribution of phases based on the temperature and pressure conditions.

envelope), respectively. As far as the lower limit of pressure for the dense phase region is concerned, it is bounded by the cricondenbar (highest pressure on the phase envelope). However, the upper limit is not fixed or well defined, but a reasonably hypothetical value (some type of average operational condition pressure) of 3,000 psia is shown just as an example to complete the dense phase loop or the box in Figure 5.2. For example, if this value were to be 4,000 psia, then the dense phase loop would just stretch upwards slightly. As mentioned before, this particular dense phase of a mixture possesses some rather unique properties that are beneficial and advantageous in various applications (see Sections 5.2 and 5.5).

The behavior described above for CO_2 and the mixed system is "qualitatively universal" in that the same concepts and definition apply to other pure components or (compositionally similar) mixtures as well—the only difference being the magnitudes which are different. Note though that in the context of phase behavior of the five reservoir fluids (black oils, volatile oils, retrograde gases, wet and dry gases) the concept and practical significance of the dense phase region is mostly valid or applicable to lighter mixtures, i.e., retrograde, wet and dry gases. This is because of the fact that these fluids: (1) are relatively much rich in methane; (2) have relatively small or narrower phase envelopes with critical temperature not much greater than methane; (3) have the critical point on the left of the cricondenbar; and (4) also have a much lower cricondentherm. On the other hand, in the case of black and volatile oils, although methane is present (in smaller quantities), the phase envelope is much broader on the temperature scale due to the presence of large heavy hydrocarbon molecules, which results in very high cricondentherms (in the range of 700°F–900°F), and critical points much on the right of the cricondenbar. McCain,[3] however, states that the appearance of critical point on the right of the cricondenbar is normally the case with fluids that are deficient in intermediates or those that have considerable dissolved nitrogen. Nevertheless, for these fluids the dense phase region of any practical significance would occur in a relatively much higher temperature range that is either not operationally encountered or achievable. The supercritical or dense phase region is, however, of particular significance especially for gas condensates or retrograde gases and wet gases. In their original state of existence in the reservoir, these fluids are under supercritical or dense phase conditions, or in other words, the typical original reservoir temperature and pressure conditions are located within the dense phase region. Therefore, given the excellent solvating power of dense phase fluids, the heavier molecules (although in small proportions) are present in a dissolved state or are "held" in the supercritical phase, which eventually drops out as retrograde condensate as the pressure is reduced, i.e., when the reservoir fluid is no longer in the supercritical state and can no longer keep the heavier molecules dissolved. *This is the primary reason why a retrograde (opposite of normal behavior) phenomenon is observed in the case of these fluids in which the initial reservoir conditions start out in the supercritical region.*

A normal behavior is of course expected in the case of oil systems (black oil and volatile oil), i.e., gas coming out of solution with pressure reduction (bubble point) because the initial reservoir conditions for these fluids is on the left side of the dense phase region.

5.2 Variation of Dense Phase Fluid Properties with Temperature and Pressure

We now consider the effect of temperature and pressure on two specific physical properties, i.e., density and viscosity of pure CO_2 and the rich natural gas mixture in the dense phase region and compare them with other regions to highlight the differences. The reason we consider only density and viscosity is because of their unique characteristics that are of particular practical importance when transporting fluids in the dense phase region. Additionally, we also examine the variations in the density of another pure component (propane) as a function of temperature at two fairly contrasting isobars. The density and viscosity data shown in several plots in the following paragraphs are predicted by the PR EOS[1] and the Lohrenz–Bray–Clark (LBC)[4] methods, respectively, as both are reasonably reliable predictors for lighter or less dense fluids. However, in order to eliminate any prediction biases, in comparing data at various conditions, reduced density ($\rho_r = \rho/\rho_c$; ρ_c is the critical density) and viscosity ($\mu_r = \mu/\mu_c$; μ_c is the critical viscosity) are plotted rather than absolute values.

Figure 5.3 shows the reduced density of propane plotted as a function of temperature at two different isobars of 500 and 2,000 psia, respectively. In both cases, the temperature span crosses the propane critical temperature of 206°F. Starting at the lowest temperature, the abrupt break observed at 500 psia (below the critical pressure) and 185°F in the reduced density is basically indicative of the transition from a liquid phase to vapor phase to partially supercritical vapor like (after crossing 206°F). In other words, this marks the horizontal intersection of the propane vapor pressure curve, i.e., at the vapor pressure of 500 psia at 185°F, and the high and low reduced densities (upper and lower ends of the vertical line) corresponding to the equilibrium liquid and vapor phases coexisting at these conditions. In contrast when the data trend at 2,000 psia is now considered, the reduced density change with increasing temperature is rather smooth because of the very high pressure that is over three times the critical pressure, i.e., the vapor pressure curve is never crossed and it merely represents the transition from partially supercritical liquid like to fully supercritical. Specifically, this intersection point is the 206°F critical temperature of propane and 2,000 psia. What is noteworthy though is the comparison of all reduced densities for both pressures but at all temperatures greater than

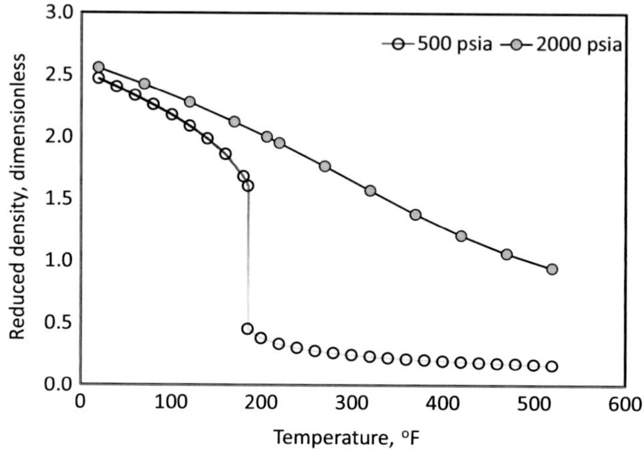

FIGURE 5.3
Reduced density of propane vs. temperature at two isobars of 500 and 2,000 psia. Critical temperature, pressure, and volume of propane are 206°F, 619 psia, and 4.54 cm^3/g, respectively.

206°F, which indicates that the densities in the dense phase region (2,000 psia) are on an average six times higher than those in the partially supercritical vapor like region. As discussed later, such high densities achieved in the dense phase region is one of key characteristics desired when transporting fluids.

Figure 5.4 shows the reduced density–temperature relationship for CO_2 in four different phase regions, namely the equilibrium liquid and vapor

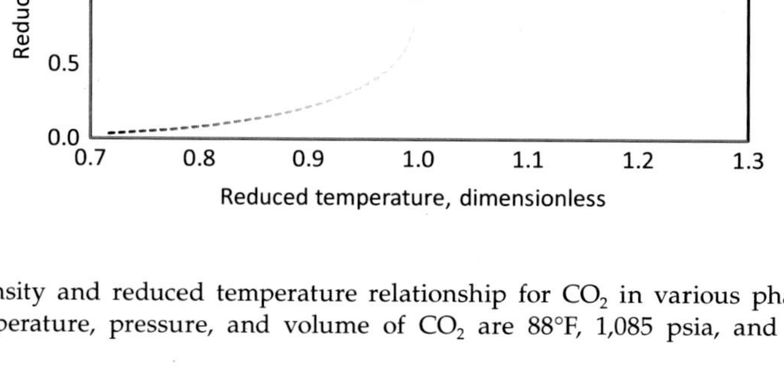

FIGURE 5.4
Reduced density and reduced temperature relationship for CO_2 in various phase regions. Critical temperature, pressure, and volume of CO_2 are 88°F, 1,085 psia, and 2.14 cm^3/g, respectively.

along the saturation or the vapor pressure curve, the critical and finally the dense phase region. Although not explicitly shown but the pressures (or the reduced pressure, P_r) corresponding to the reduced temperature T_r on the x-axis up to the critical point are the vapor pressures, i.e., $T_r = P_r = 1$ at the critical point. In other words, if reduced pressure and reduced densities are plotted instead then the data would show similar trend as in Figure 5.4. Obviously, both the densities of the equilibrium phases converge toward a common value at the critical point, i.e., reduced density of 1. Beyond the critical point (fully supercritical region) in order to show the variation in the reduced density as a function of reduced temperature ($T_r > 1$ as well as $P_r > 1$), a pseudo extension of the vapor pressure curve is obtained thus diagonally entering the dense phase box such as the one shown in Figure 5.1. CO_2 density at the corresponding temperature and pressure values is then predicted and expressed as reduced density of the dense phase (diamonds in Figure 5.4). Again, somewhat similar to the propane data shown in Figure 5.3., the dense phase density values, even at relatively high temperatures, appear to be consistently quite high and liquid-like, as compared to the vapor phase.

For the same conditions of reduced temperature (and pressure), Figure 5.5 shows the reduced viscosities of the four different phases for CO_2. Qualitatively, the data trends are almost identical to the one seen in Figure 5.4; however, what is noteworthy is the relatively low reduced viscosity of the dense phase CO_2. Alternatively, if the differential between the reduced density and the reduced viscosity of the liquid and dense phases is compared, then this indicates that the dense phase densities are somewhat

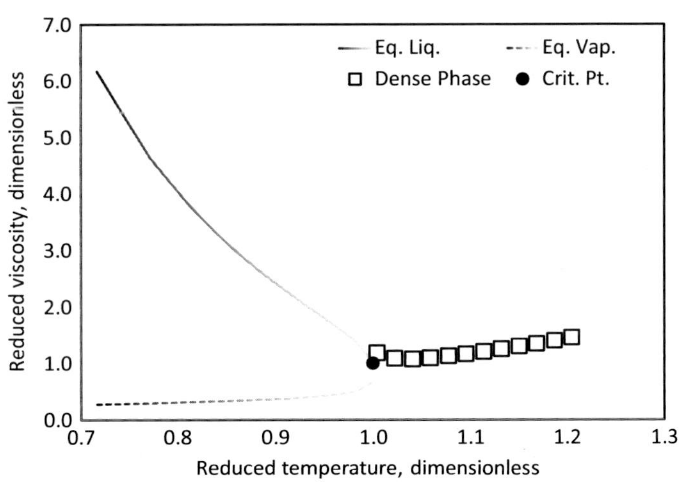

FIGURE 5.5
Reduced viscosity and reduced temperature relationship for CO_2 in various phase regions. Critical viscosity of 0.04059 cP is interpolated from data presented by Fenghour et al.[5]

closer to the liquid phase but the viscosities, on the other hand, are relatively quite different (lower). Basically, what this means for a dense phase is relatively higher density but lower viscosity, which is of significance in applications such as transportation, as shown later.

Following a somewhat similar approach as above, Figures 5.6 and 5.7 show the plots of reduced temperature (*x*-axis of the phase envelope) vs. reduced density and reduced viscosity, respectively, for a natural gas mixture of fixed overall composition in four different phase regions: single-phase liquid, vapor, critical, and the dense phase. The single-phase liquid and vapor reduced densities and viscosities are at the reduced temperature and reduced pressure (implicit, not shown on plots) conditions just at the exterior of the phase envelope, i.e., the liquid phase thinly separated outwardly from the bubble-point curve and the vapor phase from the dew-point curve, respectively. Tracing these paths or the curves from the left to the right and vice versa, both the reduced density and viscosity curves meet at the critical point such that $\rho_r = \mu_r = 1$ at $T_r = 1$. The dense phase region is covered next, which, in principle, is the box shown in Figure 5.2 with the temperature and pressure conditions selected along the diagonal for density and viscosity comparison with the single-phase liquid and vapor regions. This is identified by the diamonds and the squares in Figures 5.6 and 5.7 for reduced density and viscosity, respectively. Again, considering the differential between the reduced density and the reduced viscosity of the liquid and dense phases, conclusions similar to that of CO_2 can be drawn meaning liquid-like dense phase densities but relatively low viscosities, which are beneficial in transport applications.

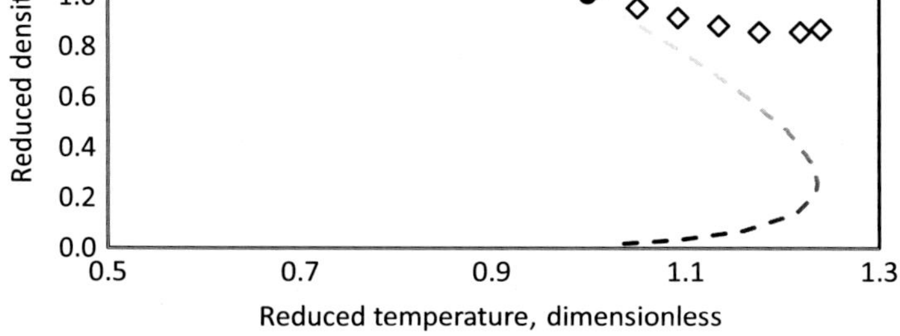

FIGURE 5.6
Reduced density and reduced temperature relationship for a natural gas mixture in various phase regions.

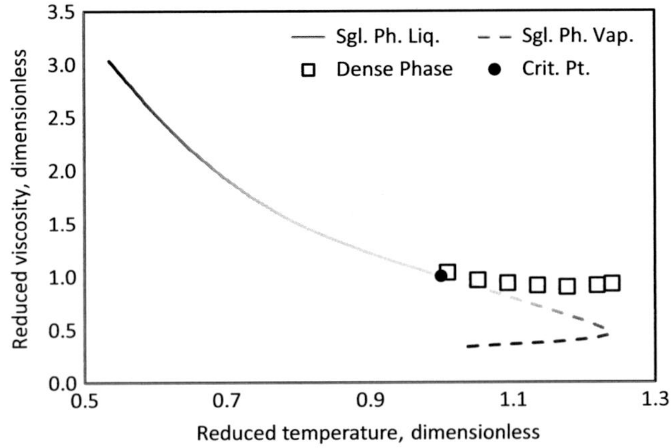

FIGURE 5.7
Reduced viscosity and reduced temperature relationship for a natural gas mixture in various phase regions.

5.3 High Temperature High Pressure (HTHP) and/or Hyperbaric Reservoir Fluids

In a conventional sense, the average upper limit of the temperatures and pressures that are typically encountered in petroleum reservoirs are 300°F and 7,500 psi.[6] However, the quest for hydrocarbons have led to accumulations or deposits existing at unusually high temperature and/or pressures for which a collective term high temperature–high pressure (HTHP) is commonly used. According to the Schlumberger Oilfield Review,[7] the HTHP system classification begins at 300°F, i.e., the upper limit of conventional fluid systems, and 10,000 psi, followed by ultra HTHP with temperatures and pressures in excess of 400°F and 20,000 psi, and finally the extreme HTHP with 500°F+ and 35,000 psi+ pressures. According to this classification, it is not necessary that both the temperature and pressure has to be high; rather it could be either one or both. On a plot presented by Zaostrovski[8] on several HTHP fields around the world, the lower and upper limits of the temperature and pressure starts at 212°F (100°C) and 4,351 psi (300 bar) and 464°F (240°C) and 20,300 psi (1,400 bar), respectively. This range basically somewhat covers the HTHP and ultra HTHP as per Schlumberger's classification. A review of the plot presented by Zaostrovski[8] indicates that many fields lie on a 45° line, i.e., HT as well as HP. However, there are fields such as Arun from Indonesia where the temperature is rather high and the pressure is relatively low and Thunder Horse in the US Gulf of Mexico in which pressures are in excess of

17,000 psi but the temperature is rather low. At least based on the collection of data by Zaostrovski,[8] the highest temperature as well as pressure appears to be in the Mobile Bay gas field in the US with formation intervals below 20,000 ft having bottomhole temperatures and pressures in excess of 400°F and between 10,000 and 20,000 psi.[9]

Other authors[10–12] have used an alternate term called "hyperbaric" to define the HTHP fluids. In this particular definition, they also include the compositional as well as phase behavior peculiarities that are associated with these fluids. Specifically, these fluids have a very asymmetric fluid composition, which means the presence of a substantial amount of the lightest hydrocarbon (methane) and long chain *n*-alkanes reaching a carbon number of 40[6]; 60[10] or even 80.[13] The methane mole fraction present in these fluids has been stated as anywhere between higher than 0.4[11] and 0.6[6], respectively. Because of the asymmetric composition, these fluids are prone to the precipitation of solids (waxes in particular) during production[6,10,11] and exhibit retrograde behavior at high temperatures and become bubble-point system at moderate ones.[11] Ungerer et al.[6] have stated that in some cases, the critical temperature of the fluid lies between reservoir temperatures and flowline temperatures, viz. transition from dew- to bubble-point system along the curved reservoir to surface production path. A review or comparison of the characteristics (either composition and/or phase behavior) of other HTHP fluids, however, indicates that they do not fit the profile of hyperbaric fluids as has been described[6,10–12]. For example, the compositions of the Mobile Bay gas field report[14] nearly 90 mol% methane and negligible amount of C_{6+} fraction, although the gas does contain a fairly high proportion of both CO_2 and H_2S. The oil from Thunder Horse field is reported to be sweet having an ~ API gravity of 30° and a GOR of 1,000 Scf/STB,[15] which can be considered as typical black oil characteristics.[16] The phase envelope of the fluid from the Kristin gas condensate field in Norway also does not display any unusual features and shows what is typical of a retrograde fluid. Given the foregoing, it is therefore perhaps best to consider the hyperbaric fluids having these special compositional and phase behavior characteristics as a special class of generally defined HTHP fluids.

In the context of dense phase, what is interesting about the HTHP or hyperbaric fluids is generally the large degree of undersaturation, i.e., a very vast difference between the reservoir pressure and the saturation pressure (dew or bubble). In other words, the HTHP fluids can be considered as existing in the reservoir under supercritical or *super dense phase* conditions as compared to conventional reservoir fluids. Arnold et al.[15] report a bubble pressure of the Thunder Horse oil as less than 5,500 psi, which is 7,500 psi below the 13,000 psi reservoir pressure. Incidentally, such large degree of undersaturation is a somewhat common feature among many deepwater (and thus HTHP) Gulf of Mexico fields.[16] In the case of Kristin gas condensate field the spacing between the reservoir pressure and the dew-point pressure is ~7,300 psi,[8] meaning a large degree of undersaturation. It is also interesting to note that at the initial reservoir conditions of 338°F and 13,200 psi the fluid (gas) has a relatively high density[8] of 0.45 g/cm^3—typical dense phase characteristics. The phase envelope and in particular the reservoir conditions reported for the Kristin gas condensate truly depict what a super dense phase fluid in a reservoir would be like. Finally, for a hyperbaric fluid reported by Daridon et al.,[12] the dew-point pressure at the reservoir temperature of 366°F is 4,772 psi, whereas the reservoir pressure is 16,100 psi, again indicating a very large degree of undersaturation. Therefore, the preceding discussion thus reveals one common feature as far as HTHP or hyperbaric fluids are concerned, and that is the wide spacing between the reservoir pressure and the saturation pressure or unusually large undersaturation not seen in typical petroleum reservoirs. The existence of these fluids at such extreme temperature and pressure conditions is obviously conducive to their behavior as super dense phases for the most part of the recovery by a simple depletion type process.

5.4 Measurement and Modeling of Dense Phase Fluid Properties

The most comprehensive database and report on the measurement and modeling of dense (mostly HTHP or super dense) phase fluid properties, mainly density and viscosity, is published in the final technical report of the European Union sponsored EVIDENT (Extended VIscosity and DENsity Technology) project.[17] This effort was a multi-university partnership with each having specific sets of expertise, i.e., measurement and modeling. Detailed results from this project have also appeared in other publications[18,19]; however, a summary of the experimental results and the recommended methods for modeling density and viscosity are included here.

Extensive experimental data on viscosity and density were measured as well as their accuracy evaluated. The tested fluids included six pure alkanes, eight binary mixtures, two ternary mixtures, three multicomponent model systems, and seven real gas condensate systems from the North Sea area. The HTHP limits for density and viscosity measurements were up to just under 400°F and 22,000 psi, respectively—conditions corresponding to the HTHP Mobile Bay gas field mentioned earlier. The measurement of density was accomplished by a pycnometer (uses simple mass/volume concept) and viscosity by capillary tube (uses Poiseuille's law relating flowrate, pressure drop, and viscosity). It is interesting to note that cross-comparisons or round-robin tests between various partner measurements on viscosity and density of a common reference binary system were used to establish the accuracy and reliability. The reference system composed of 60 mol % methane and 40 mol% *n*-decane was tested at 200°F and 4,350–20,000 psi. Based on the PR EOS predictions, this system exists as a super dense phase at these conditions because the bubble-point pressure at this temperature is ~3,500 psi. Following this verification, dense phase fluid property measurements were carried out on several custom-made binary systems representing numerous data collected in the range of dense to super dense phase regions. Density and viscosity experimental uncertainties in such a wide temperature and pressure range are reported as ±0.5% and ±1%, respectively, which is quite remarkable. The measurement techniques used and the reported uncertainties thus serve as a guideline for obtaining dense phase density and viscosity data for other fluids.

Figure 5.8 shows the partial phase envelope for a 19 component North Sea gas condensate on which two bubble- and two dew-point pressures[17] and HTHP viscosities[20] were measured. Since reliable and accurate EOS-based phase behavior predictions are generally not guaranteed for systems such as these, which was the case here as well, the PR EOS was tuned to the experimental bubble- and dew-points at 122°F and 369°F, respectively. The tuned PR EOS was then used to predict the cricondentherm, cricondenbar, and the critical point for this fluid to delineate the dense phase region. In this particular case, the cricondenbar and the critical point appear to overlap each other and by earlier definition the true dense phase region is only on the right of the cricondenbar/critical point, while the region on the left up to the lowest temperature can be considered as a pseudo (*liquid like*) dense phase. The viscosity measurements reported[20] in this region are sketched by a pentagonal box in Figure 5.8. Next, in Figure 5.9 a comparison between the measured and predicted viscosities using the LBC and the principle of corresponding states one reference method[21] (referred to as PCS-1 ref.) at conditions along the pentagonal box are shown. The functional forms of both LBC and PCS-1 ref. models can be found in Pedersen et al.[21] Since the LBC method requires densities, and experimental values were not reported, these were predicted by the Soave–Redlich Kwong (SRK)[22] EOS in the true dense phase region and the Alani and Kennedy method[23] in the pseudo (liquid) dense phase region, respectively, based on the conclusions drawn by Dandekar.[24] Clearly, viscosity predictions in the pseudo dense phase region appear to deviate much more, particularly for the LBC method. In the case of the LBC model, its functional form as a sixteenth-degree polynomial in the reduced density or in other words strong density dependence is a major drawback that may result in large underpredictions in this particular region if the densities are not accurate—reliable predictions of which from an EOS are a major issue.[24] The PCS-1 ref. model on the other hand has one significant advantage in that it is independent of the fluid density. However, strictly speaking from the standpoint of true dense phase region as defined earlier, i.e., viscosity predictions at 302°F in this case, even the LBC model predicts the viscosity reasonably well despite the use of SRK EOS calculated density. Note that, overall, the PCS-1 ref. model does a very good job, particularly resulting in the highest accuracy in the dense phase regions. Similar conclusions were also was drawn by Dandekar[24] in 1994.

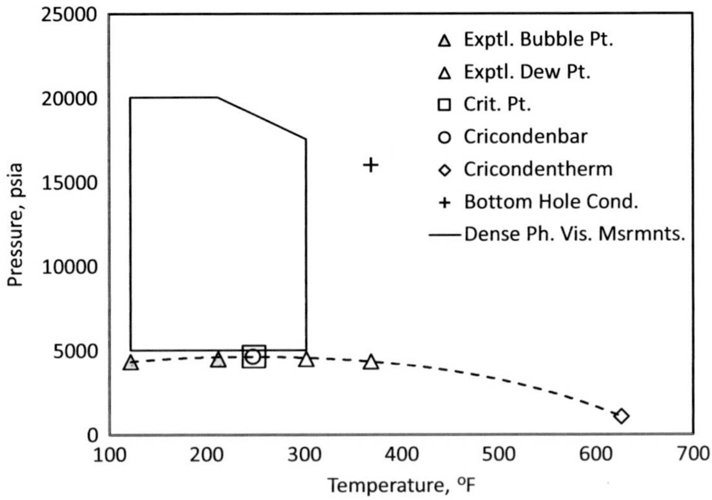

FIGURE 5.8
Partial phase envelope of a North Sea gas condensate showing four experimental[17] saturation pressures, tuned PR EOS predicted cricondenbar, cricondentherm, critical point, and bottom hole reservoir conditions. The dashed curve represents a best-fit of all the data points. Note that the PR EOS is regressed to the bubble point at 122°F and the dew point at 369°F, respectively. The boxed envelope indicates the temperature and pressure ranges of dense phase viscosity measurements.[20]

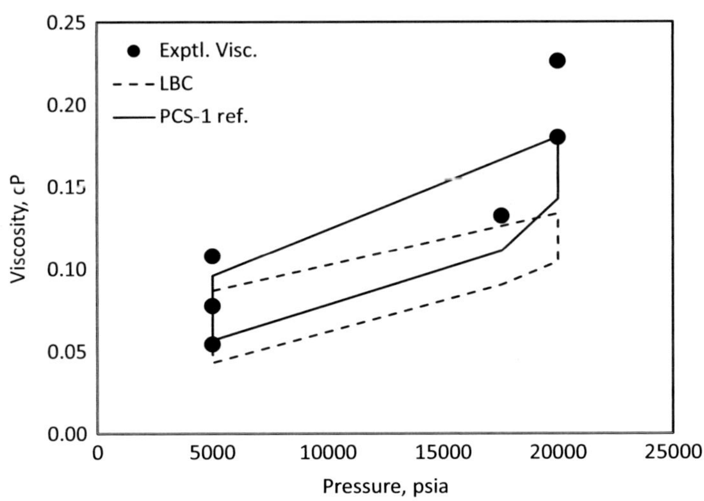

FIGURE 5.9
Comparison of experimental[21] and predicted viscosities of the North Sea gas condensate at conditions identified in Figure 5.8. Note that since this is not a surface plot, the temperatures are implicit and correspond to those shown in Figure 5.8.

As part of the EVIDENT project, Gozalpour et al.[25] reported comprehensive density and viscosity data on two lab-constituted real-gas condensate fluids representative of the North Sea for temperatures and pressures up to 302°F and a little above 20,000 psi, respectively. Both gas condensates (Fluids A and B) have similar compositions and exhibit dew-point behavior at the tested temperatures of 122°F, 212°F, and 302°F. The reported dew point for Fluid A is ~ 5,445 psi, presumably at the reservoir temperature of 212°F. Given the phase behavior characteristics and compositions of typical gas condensate fluids, which these two apparently are, all the single-phase data reported fall into the (true) dense phase to super dense phase regions (also see Figure 5.2 as a guideline). Choosing Fluid A at 212°F, as an example, Figures 5.10 and 5.11 show a comparison of SRK EOS[22] predicted density, and LBC and the PCS-1 ref. model predicted viscosities, respectively. In handling the fluid composition, the single carbon number (SCN) fractions C_6–C_{19} and the plus fraction C_{20+} were lumped according to the Whitson[26] lumping scheme, thus forming four multiple carbon number groups. The required properties of these groups were estimated based on empirical correlations that were used by Dandekar.[24] All calculations were carried out without any tuning, i.e., all are strictly pure predictions. Figure 5.10 also shows a comparison of experimental and SRK EOS predicted densities of a conditioned Statfjord gas at 35°F. In Figure 5.11, LBC-predicted viscosities using the measured densities[25] are also shown to highlight the sensitivity of the model to the density used.

As far as SRK EOS density predictions are concerned, incidentally, both comparisons produce an average absolute deviation (AAD) of ~1.8% for temperatures as diverse as 35°F and 212°F and pressures within such as a wide range

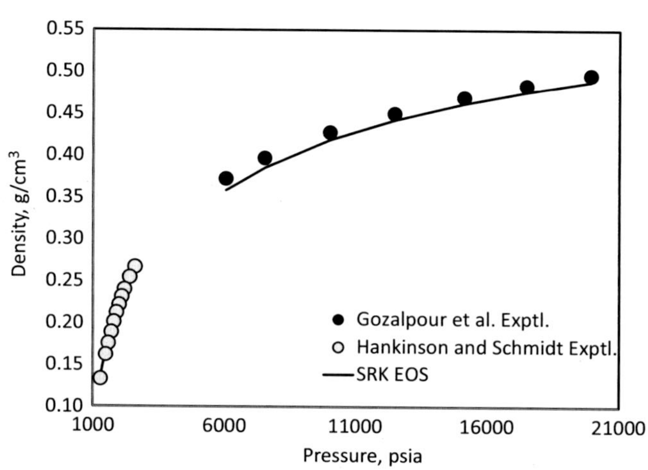

FIGURE 5.10
Comparison of experimental[2,25] and SRK EOS predicted densities of a Statfjord conditioned gas[2] and North Sea gas condensate (Fluid A)[25] at 35°F and 212°F, respectively, and various pressures in the dense phase region.

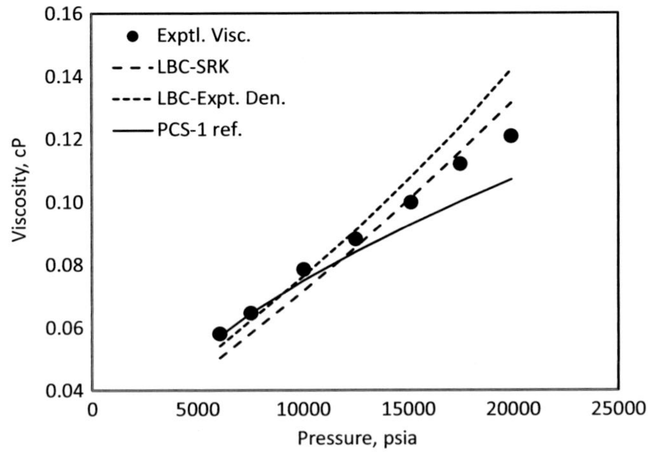

FIGURE 5.11
Comparison of experimental[25] and predicted viscosities of a North Sea gas condensate (Fluid A) at 212°F and various pressures in the dense phase region.

from as low as 1,300 to 20,000 psi. It is remarkable to note that despite the presence of SCN's and the plus fraction, where a significant amount of uncertainty in terms of critical properties and acentric factor exists as compared to well-defined components, the AAD of 1.8% is quite good. For viscosity predictions the AAD's are less than 6% for the PCS-1 ref. and 7%–7.5% for the LBC method using the SRK EOS predicted and experimental densities, respectively, which is quite good in light of the presence of SCN's and the plus fraction. The sensitivity of the density dependence is clear but interestingly the fluctuation (between using measured and SRK EOS predicted densities) is not very significant in the tested range. Finally, given the foregoing plots (especially Figures 5.10 and 5.11) and the discussion, as a guideline or recommendation for density and viscosity prediction in the dense phase region is concerned, the SRK EOS for density and PCS-1 ref. model for viscosity is the most appropriate and adequate combination. However, given a somewhat simpler functional form, popularity and availability in most software, the LBC model may also be used in conjunction with the SRK EOS, albeit with some caution. Clearly, these types of conclusions that were drawn back by Dandekar[24] appear to be still valid.

5.5 Practical Significance of Dense Fluid Phase in Transportation and EOR

The practical significance of dense phase fluid has been alluded to earlier; however, the specifics and some examples are provided in this section. Fluids in this particular state are single phase and have unique properties in that the densities are close to that of a liquid phase, which means more mass per unit volume, but the viscosities are similar to that of gases, which means lower frictional losses. Figures presented in Section 5.2 demonstrated these density and viscosity characteristics. Given these reasons, long-distance pipelines have been designed and built to transport CO_2 and natural gas mixtures in the dense phase region.[27,28] Since the state of the fluid in the dense region means single phase, liquid formation in the pipelines and slug flow, and thus the use of pigs can be avoided. For this reason, hydraulic calculations also are much more simplified, instead of computing two-phase pressure drops from empirical correlations, which may introduce uncertainties.

Operating in dense fluid phase conditions is however not necessarily always trouble-free. The problem is mostly caused by the presence of water in free-phase or otherwise, which may potentially condense on cooling. Potentially, a common problem for all types of fluids is the possibility of hydrate formation and something that is specific to CO_2 (e.g., pipelines), in addition to hydrates, is the formation of carbonic acid (H_2CO_3) that may be conducive to corrosion. However, Eldevik[29] states that there is a degree of uncertainty if free water in dense phase CO_2 will form hydrates before carbonic acid; the main variables playing a major role are temperature and pressure conditions and the water content (see Figure 5.12 for CO_2 hydrate and the accompanying discussion). The higher pressure favors hydrate formation risk, whereas the lower pressure means higher likelihood of corrosion.[29] Therefore, drying or dehydrating the fluid is rather an important step in pipeline transport. Chapoy et al.[30] state that the risk of hydrates is low if the water content of the CO_2 stream is below 250 ppm. The topic of corrosion is beyond the scope of this book; therefore, it is not covered here but the reader is referred to authoritative works on this area.[31] However, the formation conditions of hydrates are covered in more details here since it falls within the realm of phase behavior.

As characterized by Sloan,[32] gas hydrates are crystalline compounds that form when typically smaller hydrocarbon molecules or for that matter even components such as CO_2 and N_2 on their own or present in natural gas mixtures, and water come in contact at relatively lower temperatures and higher pressures. Note that the reference to water here is pure water and not containing salt because that influences hydrate phase behavior by bringing in an inhibitor effect. Such scenario is rather commonly encountered in long-distance pipelines operating in colder climates and deepwater flowlines, surrounded by colder seawater, carrying produced fluids. Due to their non-flowing nature hydrates tend to form plugs blocking the pipelines, thus having disastrous consequences on the flow infrastructure. Formation of gas hydrates in pipelines is thus considered as a major flow assurance issue. Similar to the vapor pressure curve, gas hydrate phase behavior is also portrayed on a pressure–temperature (PT) diagram which represents equilibrium between gas, water, and the hydrate phase. The

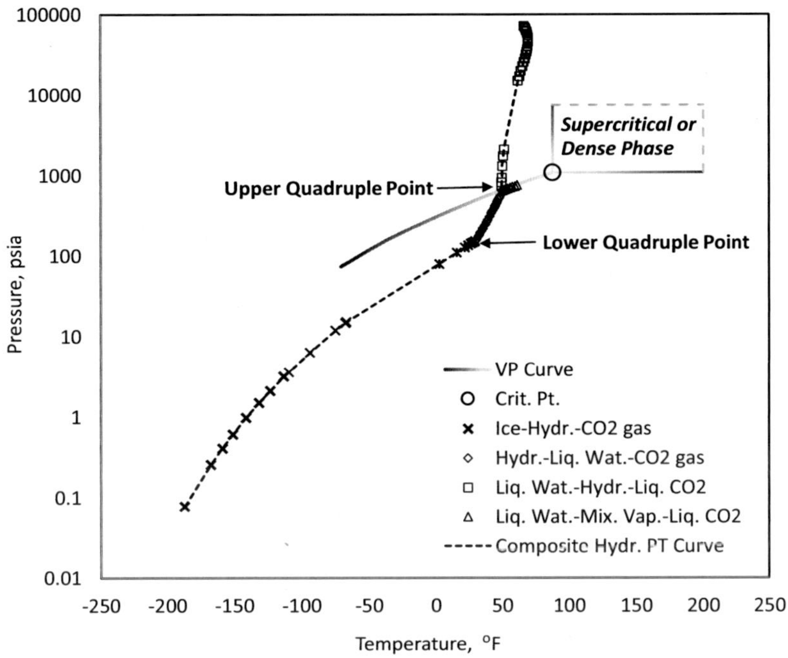

FIGURE 5.12
CO_2 pressure–temperature diagrams spanning the vapor pressure curve, dense phase region, vapor–liquid–liquid equilibria (VLLE), and various branches of hydrate phase equilibria. All hydrate and VLE data are experimental for which the sources corresponding to the legend are listed as follows: "Ice–Hydr.–CO_2 gas"—Falabella,[34] Larson[35] and Miller and Smythe[36]; "Hydr.–Liq. Wat.–CO_2 gas"—Larson[35]; "Liq. Wat.–Hydr.–Liq. CO_2"—Ng and Robinson,[37] Nakano et al.[38]; and "Liq. Wat.–Mix. Vap.–Liq. CO_2" or VLLE—Ohgaki et al.[39]

curve that is of primary importance in typical petroleum reservoir fluid handling is the hydrate equilibrium curve that separates the hydrate forming and non-forming conditions. Simply speaking, points on the right of the PT diagram represent "hydrate-free" conditions or simply equilibrium between the gas and the water phase, whereas points on the left represent equilibrium between hydrate phase and gas. From a detailed phase behavior perspective, there are other elements and branches that include ice–hydrate–gas; ice–liquid water–gas; ice–liquid water–hydrate and lower and upper quadruple (not applicable for methane, nitrogen) points marking four phase equilibria between ice–liquid water–hydrate–gas and liquid water–hydrate–gas–gas in liquid state, respectively. The lower and upper quadruple points on the hydrate PT diagram are analogous to the triple and critical points on the vapor pressure curve. Recalling Gibb's phase rule, $F = C - P + 2$, where F is degrees of freedom, C is number of components, and P is the number of phases. Every single component system will result in 0 degrees of freedom at the triple point ($F = 1 - 3 + 2 = 0$), i.e., this is a unique point. Similarly, with four phases in equilibrium at the quadruple points, the degrees of freedom will also be 0, i.e., $F = 2 - 4 + 2 = 0$ meaning a unique point and an invariant system; note that the components are two because one is the pure hydrate former and the other is water. Sloan and Koh[33] have covered these facets in more details.

Figure 5.12 shows an interaction between the CO_2 phase behavior, dense phase region, and the CO_2 hydrate equilibrium. In assembling Figure 5.12 and others that follow, the authors acknowledge the vast collection of experimental data on hydrate phase equilibria of various systems from diverse literature sources that Dr. Sloan and Dr. Koh[33] has put together in *Clathrate Hydrates of Natural Gases*. As mentioned earlier, the hydrate curve of practical importance is the one bounded by the lower and upper quadruple points. As shown in Figure 5.12, this curve represents the equilibria between the hydrate phase, liquid water, and CO_2 in the gas phase; the presence of CO_2 in gas phase being quite logical in that the PT conditions are below the vapor pressure curve. A pseudo extension of this curve going into the very low temperature and pressure region marks the equilibria between ice, hydrate, and CO_2 in gas phase; the gap (which also exists in the hydrate–liquid water–liquid CO_2 equilibria) is indicative of lack of experimental data but the transition or connection between the different data sources appears to be smooth. Note that the very low end of this curve may be of limited practical significance and is also the farthest from the dense phase region, but is shown for completeness. As one approaches the critical point, the upper quadruple point marks the bifurcation into two different equilibria: one is the hydrate, liquid water, and liquid CO_2 and the other is the VLLE between liquid water, liquid CO_2, and a mixed vapor phase (mostly composed of CO_2). Again, logically speaking the hydrate, liquid water, and liquid CO_2 equilibria initiate and continue on above the vapor pressure curve, i.e., CO_2 will be in a liquid state. The digression or reversal of the trend shown by Nakano et al.[38] for the hydrate–liquid water–liquid CO_2 equilibria at 69.53°F is an important finding in that regardless of how high the pressure is increased, CO_2 hydrates cannot exist in the temperature above 69.53°F, which they refer to as "maximum temperature point." Also note that this occurs at ultra-high pressures (of the order of 47,000 psi) that are not of any practical significance. Nevertheless, this means that there will be a minimum spacing of 18.36°F from the critical temperature of CO_2, thereby maintaining its distance from the dense phase boundary. As one moves to the right of the upper quadruple point along the VLLE region, there are no hydrates given the increasing or higher temperatures. Finally, joining all the three hydrate equilibria, i.e., hydrate with ice, gaseous CO_2, liquid water, and liquid CO_2 a composite hydrate PT equilibrium curve for CO_2 in its entirety for a wide range of conditions can be established—as shown by the dashed curve in Figure 5.12. Therefore, simply speaking, if CO_2 and water come in contact at the conditions on the composite PT curve or on the left, then hydrates will form, whereas on the right of the curve hydrate formation is thermodynamically not feasible. Clearly, as seen in Figure 5.12, the supercritical or dense phase boundary of CO_2 remains completely in the non-hydrate zone, i.e., hydrate formation strictly in the dense phase region is not a risk; however, drastic process or pipeline operational upsets causing a sudden drop in the temperature and pressure may result in crossing the hydrate boundaries increasing the risk of blockage. The overall (hydrate) behavior as seen in Figure 5.12 for CO_2 has been termed by Katz and Lee[40] as "hydrate forming conditions for a subcritical forming gas." Additionally, they[40] state that the upper quadruple point and the divergent VLLE behavior is exhibited only by subcritical gases such as CO_2.

Let us now look at the phase behavior of methane in similar details. As stated by Whitson and Brule,[41] two of the lighter components of natural gas, i.e., methane (CH_4) and nitrogen (N_2) do not have the upper quadruple point, which means that there is no upper temperature limit for hydrate formation and an absent VLLE region. Therefore, for CH_4 (and N_2) the two only hydrate equilibrium curves that exist are the ice–hydrate – CH_4 gas and hydrate–liquid water–CH_4 gas, respectively, with the lower quadruple point placed at the intersection between the two, which is similar to CO_2 as shown in Figure 5.12. Given Katz and Lee's terminology and the lack of the upper quadruple point, gases such as methane and nitrogen can thus be called as "supercritical forming gas" for most commonly encountered operational conditions. This is also somewhat obvious simply comparing the widely different critical temperatures of CO_2 and CH_4, which are 88°F and –116.6°F, respectively. Alternatively, it can be stated that CO_2 hydrates can form only at temperatures less than 88°F, but in the case of CH_4, hydrates can form at temperatures lower as well as higher than –116.6°F as shown later. This is also the reason why methane hydrates are found in nature (both arctic and marine sub-surface settings) at depths where the entire geothermal gradient results in a temperature range that far exceeds –116.6°F. It is the PT conditions on the hydrate–liquid water (typically saline to represent the pore water) – CH_4 gas equilibrium curve that are expressed in terms of temperature vs. depth (assuming a certain pore pressure gradient) and mapped against the geothermal (sub-surface temperatures) gradient to determine what is known as the hydrate stability zone (HSZ) or its thickness. Therefore, it is readily apparent that, unlike CO_2, CH_4 hydrates can form even in the dense phase region. A plot similar to Figure 5.12 that maps the hydrate forming conditions for CH_4 is shown in Figure 5.13, which basically summarizes the preceding discussion. In Figure 5.13, given the very low critical temperature of CH_4, the elongated dense phase region on the temperature side is shown to include temperatures (100+°F) that are of practical operational significance. Unlike CO_2, the hydrate PT equilibrium curve in the case of CH_4 cuts across the corner of the drawn dense phase region. This means that if CH_4 and water unite at any conditions on the equilibrium curve or on the left of it even in the dense phase region, hydrates will form. However, if the dense phase conditions are on the right of the equilibrium curve then there is no

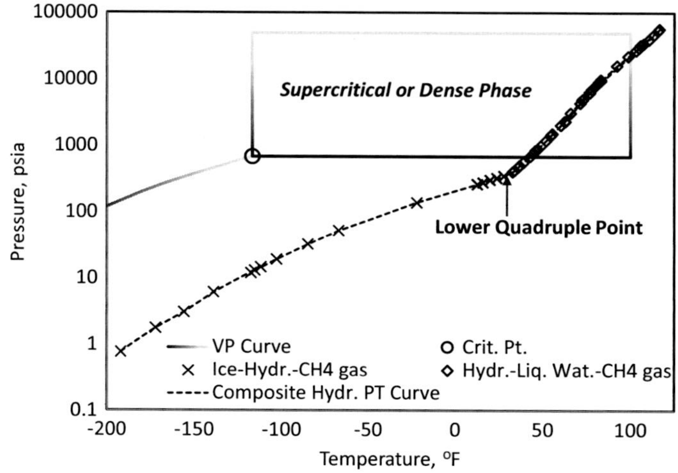

FIGURE 5.13
CH_4 pressure–temperature diagrams spanning the vapor pressure curve, dense phase region, ice–hydrate–CH_4 gas and hydrate–liquid water–CH_4 gas equilibria curves. All hydrate data are experimental for which the sources corresponding to the legend are listed as follows: "Ice–Hydr.–CH_4 gas"—Deaton and Frost,[42] Falabella,[34] and Makogon and Sloan[43]; "Hydr.–Liq. Wat.–CH_4 gas"—Deaton and Frost,[42] Kobayashi and Katz,[44] McLeod and Campbell,[45] and Marshall et al.[46]

risk of hydrate formation, but it should be noted that any unwanted or unexpected reduction in pressure and temperature may result in entry into the hydrate forming region.

Methane is the most dominant component present in a natural gas mixture from a compositional standpoint; thus, we now investigate the placement of the hydrate equilibrium curve with respect to the phase envelope and the dense phase region for four diverse multicomponent systems. The molar compositions and other details of these systems are provided in Table 5.1. Note that most natural gas type mixtures will typically contain CH_4 of the order of 90 mol%; however, others are purposefully selected in light of the widely different behavior observed with pure CO_2 and CH_4.

In Figures 5.14–5.17, unless otherwise stated the two-phase envelopes are modeled using the PR EOS but the hydrate equilibrium curves corresponding to the pertinent mixtures represent the experimental data from the listed sources in the individual figure captions or Table 5.1. Since all these mixtures are mainly composed of well-defined components, i.e., no SCN or plus fractions where most uncertainty exists, it is believed that the predicted phase envelopes are adequately accurate representations of the respective systems. Note that given the very small concentration of pentanes+ in Gas B, it is treated as *n*-pentane.

Figures 5.14 and 5.15 show the results for two mixtures (natural gas type) that obviously have high proportion of methane, which is reflected in the phase envelopes for both mixtures, i.e., positioned on the relatively low

TABLE 5.1

Molar Composition (all values in mol%) of Mixtures Corresponding to Figures 5.14–5.17. Column 2–4 Headings are Data Source References, and Parentheses Indicates the Mixtures so Named by the Investigators

Component	Wilcox et al.[47] (Gas B)	Bishnoi and Dholabhai[48] (Ternary Mixture)	Ng and Robinson[49] (Mixture III)	Ng and Robinson[49] (Mixture VI)
Nitrogen	0.64	0.00	0.20	0.00
Carbon dioxide	0.00	20.00	0.00	25.5
Methane	86.41	78.00	21.9	0.00
Ethane	6.47	0.00	24.7	17.0
Propane	3.570	2.00	40.8	38.6
i-Butane	0.99	0.00	12.4	18.9
n-Butane	1.14	0.00	0.00	0.00
n-Pentane	0.00	0.00	0.00	0.00
Pentanes+	0.78	0.00	0.00	0.00

temperatures, and lower cricondentherms. Note that it is possible for the ternary system to exist in single-phase conditions due to the higher mol% of CO_2. The entire hydrate–liquid water–natural gas (Gas B and Ternary mixture) equilibrium curve, which is of practical importance, is placed to the right of the phase envelope and the dense phase envelope. This means that if these mixtures were brought to be in contact with water in the dense phase region, hydrate formation is a possibility, which is influenced by the high proportion of CH_4 (see Figure 5.13 in conjunction with this).

The phase behavior similar to the natural gas type mixtures is also shown in Figures 5.16 and 5.17 for the two mixtures studied by Ng and Robinson[49]; however, the hydrate phase behavior is quite different. First, note the compositional differences—both are relatively rich in intermediates with CH_4 content ranging from a small ~22 to 0 mol%. It is due to this molar distribution the phase envelopes are slanted toward the right with a spread across the temperature scale with the critical point appearing on the corner or the top, which means the cricondenbar and the cricondentherm in the vicinity of the critical point. Note the location of the cricondenbar on the left of the critical point as opposed to the right, which is generally the case. Given these characteristics, the *pseudo* (called as such due to the cricondenbar placement) dense phase envelope is rather narrow. Second, the lack of influence of CH_4, owing to the small to none mol%, on the hydrate phase equilibria is also clear. In the case of these two mixtures (and others that are compositionally similar), mainly three different hydrate equilibrium curves are of importance, namely: (1) hydrate–liquid water–hydrocarbon vapor or gas phase, (2) hydrate–liquid water–hydrocarbon vapor or gas phase–hydrocarbon liquid phase, and (3) hydrate–liquid water–hydrocarbon liquid phase. The

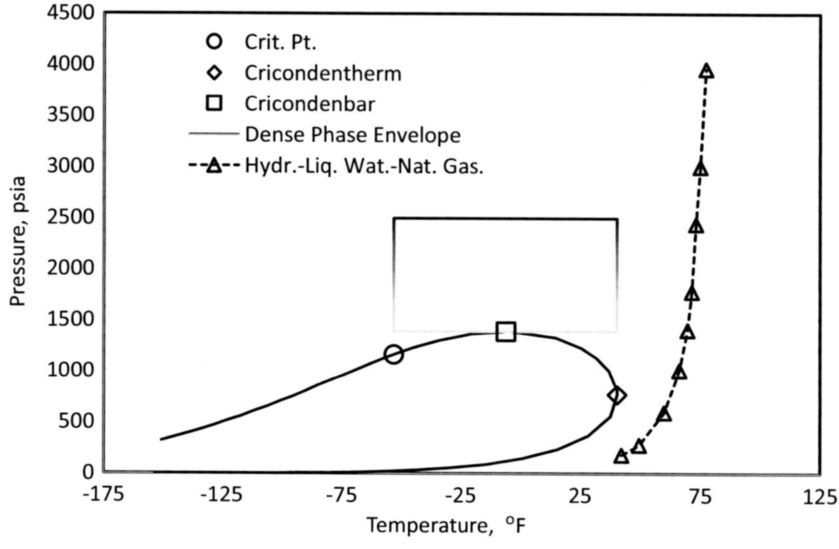

FIGURE 5.14
Gas B[47] phase envelope, dense phase region, and the hydrate–liquid water–natural gas (Gas B) equilibrium curve.

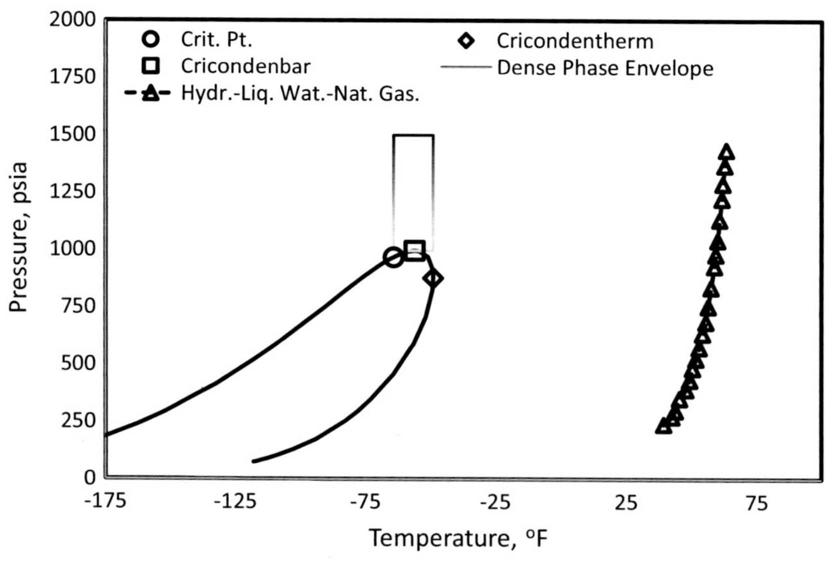

FIGURE 5.15
Ternary mixture[48] phase envelope, dense phase region, and the hydrate–liquid water–natural gas (Ternary mixture) equilibrium curve. The hydrate equilibrium curve is digitized from the simulated values presented by Pedersen et al.,[21] which basically mirrors the experimental data points reported by Bishnoi and Dholabhai.[48]

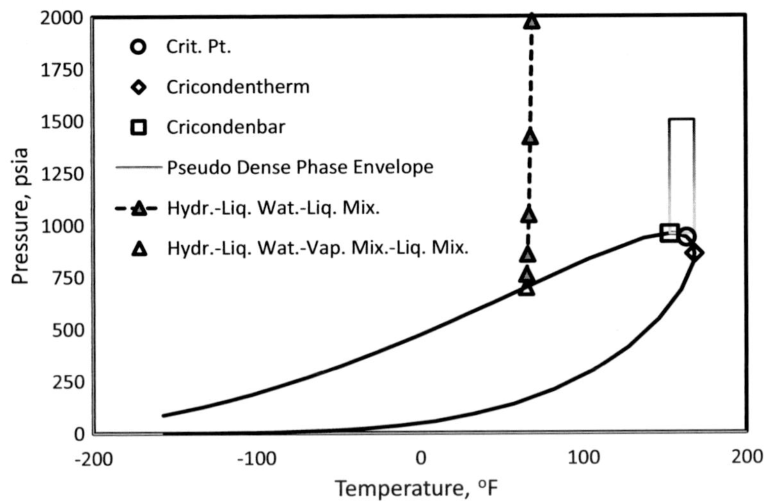

FIGURE 5.16
Mixture III[49] phase envelope, *pseudo* dense phase region, and the hydrate–liquid water–liquid phase mixture (Mixture III) equilibrium curve (solid triangles). The open triangle represents the *second upper* quadruple point on the bubble-point curve or the equilibria between hydrate–liquid water–vapor phase (resulting from Mixture III)–liquid phase (resulting from Mixture III).

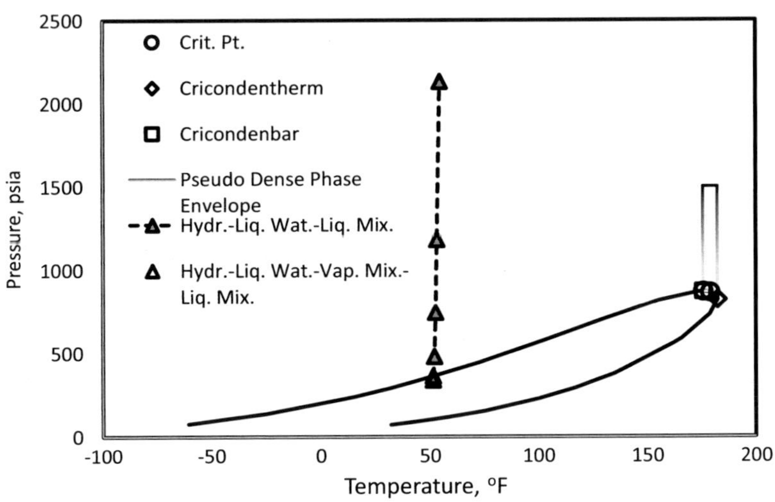

FIGURE 5.17
Mixture VI[49] phase envelope, *pseudo* dense phase region, and the hydrate–liquid water–liquid phase mixture (Mixture VI) equilibrium curve (solid triangles). The open triangle represents the *second upper* quadruple point on the bubble-point curve or the equilibria between hydrate–liquid water–vapor phase (resulting from Mixture VI)–liquid phase (resulting from Mixture VI).

placement of these curves is as follows: Curve (1) below the mixture dew-point curve; Curve (2) within the two-phase region, i.e., inside the mixture phase envelope; and Curve (3) above the mixture bubble-point curve. Obviously, these curves are placed as such and the equilibria they represent are because below the dew and above bubble-point curves the entire mixture is single-phase vapor and liquid, respectively, while within the phase envelope the vapor and liquid phases (splitting from the overall mixture and both having different compositions) coexist in equilibrium. The "Curve (1)—dew-point curve" and "Curve (3)—bubble-point curve" intersections are called the *second lower* and *second upper* quadruple points, respectively.[33] In other words, both mark the starting and ending points of Curve (2). Note that using the references of Figures 5.12 and 5.13, the second here means the upper quadruple point of which two exist in the case of these mixtures. Basically, we can sketch a somewhat composite hydrate equilibrium curve, i.e., a combination of all three, with inflections at the lower and upper second quadruple points, which will transect the phase envelope. Another transection of the phase envelope will also occur at relatively low temperatures with only one lower quadruple point below the dew-point curve, which is discussed by Sloan and Koh[33] in detail. Figures 5.16 and 5.17 show the experimental[49] *second upper* quadruple point and its continuation as Curve (3) is shown. Therefore, unlike Gas B and the Ternary mixture, if Mixtures III and VI are brought to be in contact with water in the (pseudo) dense phase conditions, hydrates cannot form given the location of the dense phase envelope on the right of Curve (3). However, qualitatively speaking upsets in the operating conditions resulting in somewhat minor temperature and pressure reductions may split the mixture into two phases (complicated vapor–liquid flow), and in cases of severe reductions Curve (2) may be approached and crossed causing hydrates to form.

CO_2 is a chemically stable and nontoxic[41] component that is typically present in a gaseous state at normal temperature and pressure conditions. The importance of CO_2 as far as EOR operations are concerned has been long recognized, dating back to the early 1950s.[50] In current times, there is renewed interest in sequestering CO_2 in oil reservoirs, thereby enhancing oil recovery, given its potential for causing climate change or global warming. Much has been written on these two somewhat controversial terms on which numerous literatures exist that the reader can refer to elsewhere. Given the somewhat near-ambient critical temperature of CO_2 and low critical pressure, the state of CO_2 in EOR applications will be supercritical or dense since most reservoir temperature and pressure conditions will exceed the critical point of CO_2. The existence of CO_2 in the dense phase in EOR applications is of a particular significance when it comes in contact with the oil in the porous media. It has been stated that substances in dense phase will effuse[51] or flow through easily in porous rocks like a gas and will dissolve in materials, i.e., oil in case of EOR like a liquid solvent. In other words, Holm and Josendal[52] state the CO_2 is "transported chromatographically through porous rock."

The behavior of CO_2 when brought in contact with the oil can be characterized as immiscible and miscible. In immiscible behavior, CO_2 causes oil swelling and viscosity reduction; oils that are typically amenable for this are viscous, heavy oils, and low pressure reservoirs producing less than 30° API stock tank oils.[41] The efficacy of swelling is expressed in terms of swelling factor. Generally, the miscibility of CO_2 in the oil is numerically characterized by a certain pressure that is denoted as the minimum miscibility pressure (MMP) that can be achieved either via a first contact or multiple contacts with the oil in question. The specific EOR benefits of CO_2 in the dense phase region, from a component exchange and phase behavior standpoint, have been highlighted and supported by experimental data, by Holm and Josendal.[53] In the case of live oil (containing solution gas), CO_2 in the gaseous or non-dense or subcritical phase region vaporizes only the lighter components such as C_6 and below, whereas in the dense phase region it selectively extracts and solubilizes components from the reservoir oil as heavy as C_{30} and even higher. This results in a dramatic shrinkage in the original oil volume because the significantly extracted C_5–C_{30} components become part of the CO_2 rich phase, which clearly means CO_2 in the dense phase is a much better solvent for oil.[54]

Whitson and Brule[41] have stated that the main parameter influencing the MMP of CO_2 is its density at the reservoir conditions, which for the most part will be supercritical or dense. Holm and Josendal[53] presented a graphical (empirical) correlation relating CO_2 density at MMP with the amount and distribution of extractable C_5–C_{30} components in the oil—note that the reservoir temperature (being a given) in this correlation is somewhat implicit. Their correlation suggests that CO_2 densities between 0.65 and 0.55 g/cm^3 are adequate for developing miscibility with typical oils containing 70–80 wt% C_5–C_{30} fraction.[41,53] These CO_2 densities are typical of the dense phase regions as shown in Figure 5.4. The CO_2 MMP data for various oils at different temperatures reported by Holm and Josendal[53] and Yellig and Metcalfe[54] are shown in Figure 5.18 in reference to the dense phase boundary for CO_2, i.e., all MMPs and the corresponding temperatures are at conditions at which CO_2 is in the dense phase. The EOR significance of CO_2 in the dense phase region is also portrayed in Figure 5.19. While the data and the caption of Figure 5.19 are self-explanatory, it supports the earlier statement on CO_2 densities that are adequate for developing miscibility for most typical oils, as well as the fact that these densities span the dense phase region.

Next we consider the advantages of transporting fluids in the dense phase region because of the three[27] favorable characteristics, namely: (1) liquid-like high densities (see Figures 5.4 and 5.6), which means more mass or mole per unit volume, (2) correspondingly low gas like viscosities (see Figures 5.5 and 5.7), resulting in less frictional losses and thus smaller pressure drops, and finally (3) single-phase fluid, which means simplified hydraulic calculations, ability to "carry" or transport intermediate components such as ethanes, propanes, and butanes over long distances to petrochemical complexes, and

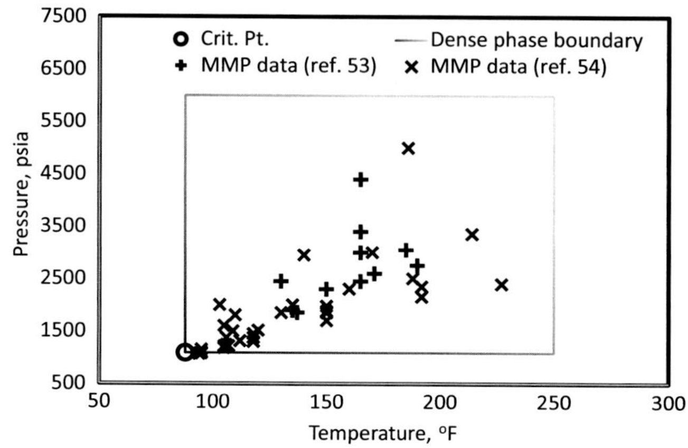

FIGURE 5.18
CO_2 MMP data at various temperatures for different oils reported in Holm and Josendal[53] and Yellig and Metcalfe[54], respectively, which is at conditions in which CO_2 is in a supercritical or dense phase.

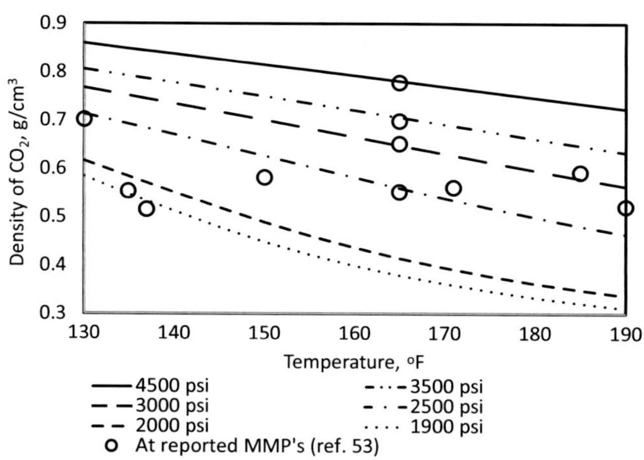

FIGURE 5.19
CO_2 densities vs. temperature in the dense phase region at selected isobars. The open circles represent CO_2 densities reported by Holm and Josendal[53] at MMP-temperature conditions shown in Figure 5.18. Points not lying exactly on the isobars or in between represent the nearest pressure or a midpoint between two pressures; temperatures are exact.

no liquids forming in the pipeline, thus eliminating the pigging requirements. For a simplified case using some assumptions, the benefits of carrying more mass can be readily realized. Note that the density will vary to some extent with distance in the case of a long-distance pipeline as pressure and temperature conditions change; however, considering Figure 5.4 if we use an average CO_2 density of 0.6 g/cm^3 and 0.15 g/cm^3 in the dense and single-phase gaseous regions, respectively, then that basically means four times (0.6/0.15 = 4) more CO_2 mass per unit volume in the dense phase.

In order to compare the frictional pressure drop in a hypothetical pipeline transporting CO_2 under various conditions, we consider the phase diagram of CO_2 as shown in Figure 5.1. Accordingly, the independently transported phase of CO_2 for comparison is gaseous or vapor phase, liquid phase, and the dense or supercritical phase, i.e., the vapor and liquid phases are those that are along the vapor pressure curve while the dense phase is a pseudo extension of the vapor pressure curve. Therefore, the densities and viscosities of the phases are at those respective conditions of temperature and pressure. In order to simplify the comparison of the frictional pressure drop we assume that these respective properties of CO_2 represent average pipeline transport conditions and remain constant throughout. Furthermore, we make the following assumptions: (1) pipeline has a fixed inside diameter of 24 in.; (2) pipeline is horizontal; (3) no phase change occurs during flow; (4) relative roughness factor is 0.0001; (5) regardless of the state or phase, CO_2 flows at a fixed velocity of 50 ft/s; and (6) given the constant or fixed CO_2 properties the frictional pressure drop can be calculated from the simple Darcy–Weisbach equation (with fanning friction factor).[55] The frictional pressure drop per unit length of the pipeline is as follows: $\frac{\Delta P}{L} = 2 f_f \frac{\rho V^2}{D}$, where ρ is the fluid density, V is the flow velocity, and D is the inside diameter of the pipeline. The Reynolds number ($\text{Re} = DV\rho/\mu$) and relative pipe roughness factor (ε/D) dependent fanning friction factor are calculated from Chen's equation (Equation 5.1).[56] Any consistent set of units can be used to calculate the dimensionless Re, e.g., diameter in meters, velocity in m/s, density in kg/m³, and viscosity in Ns/m². Using the same set of units for density, velocity and diameter will result in $\Delta P/L$ in N/m²/m, which can be converted to psi/ft:

$$\frac{1}{\sqrt{f_f}} = -4\log_{10}\left\{\frac{\varepsilon/D}{3.7065} - \frac{5.0452}{\text{Re}}\log_{10}\left[\frac{(\varepsilon/D)^{1.1098}}{2.8257} + \left(\frac{7.149}{\text{Re}}\right)^{0.8981}\right]\right\} \tag{5.1}$$

Figure 5.20 shows a comparison of the pressure drop for the three different phases as a function of ratio of density and viscosity. With the inside diameter and the flow velocity assumed constant, theoretically the variable or the ratio of density and viscosity will influence the Reynolds number, which in turn will affect the friction factor. However, given the relatively low viscosity values the Reynolds numbers for all three phases are obviously very high, and given the functional form of Chen's equation even orders of magnitude differences in the (high) Reynolds number do not seem to affect the friction factor, i.e., it remains fairly constant. Therefore, the main control on the frictional pressure drop, with other variables remaining same, is the fluid density. In the case of dense phase CO_2, a combination of nearly constant liquid-like (but somewhat lower) densities and relatively lower frictional factors result in

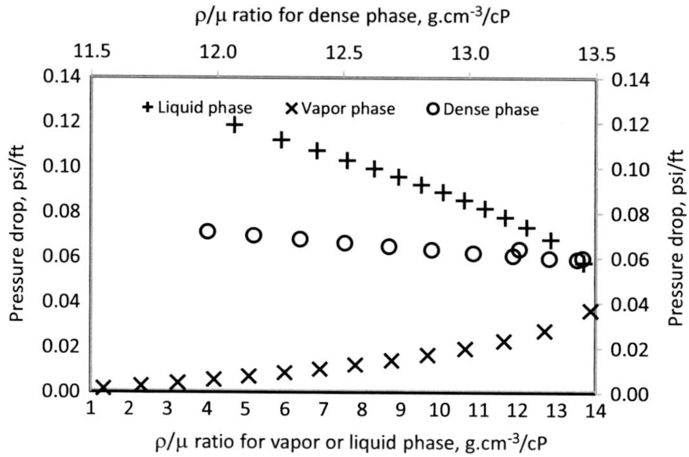

FIGURE 5.20
Comparison of frictional pressure drop for CO_2 transport in vapor, liquid, and dense phases, respectively, for an artificial scenario considering various assumptions.

pressure drop lower than that of the liquid phase CO_2. As seen in Figure 5.20, given the lower vapor phase and higher liquid phase densities of CO_2, the corresponding frictional pressure drops are lower and higher, respectively, and show an order-of-magnitude variation, while the dense phase pressure drop lies in between the two. Note that the trend seen in Figure 5.20 is somewhat analogous to Figure 5.4. The ratio of density and viscosity (considered from the Reynolds number standpoint) for the vapor and liquid phases also shows an order-of-magnitude variation. The dense phase on the other hand, given its characteristics described earlier, hardly shows any variation in the density and viscosity ratio and thus produces a frictional pressure drop that is almost constant over the entire range. Although, the frictional pressure drop hierarchy is *vapor phase* < *dense phase* < *liquid phase*; overall, from an operational standpoint the dense phase certainly offers an advantage over the vapor and liquid phases because of (1) steady or consistent frictional pressure drop and (2) the ability to carry more mass or moles.

5.6 Successful Applications—Case Studies

5.6.1 Dense Phase CO_2

From the petroleum industry perspective, perhaps, in principle, the most significant dense phase application is the CO_2—EOR. However, all reservoirs are not suitable for CO_2—EOR and thus should be screened based on factors such as reservoir geology, MMP, oil gravity, and viscosity in order to identify the most likely candidates.[57] Verma[57] presented a comprehensive survey of CO_2-based EOR projects focused on the US since that is where most of the projects are. The summary data of the 123 projects that include the temperature and pressure (given as depth) range of application mostly indicates that the state of CO_2 is dense. The success of (dense phase) CO_2—EOR projects is also evident from the historical EOR-associated oil production data that show a continued increase. In 2012, CO_2—EOR accounted for roughly 43% (same as thermal) in comparison to the meager 14% based on nitrogen and hydrocarbon gas combined. One of the most successful and time-tested applications of (dense phase) CO_2—EOR is in the Scurry Area Canyon Reef Operators Committee, also known as SACROC, which is one of the largest and oldest oil fields in the US that started CO_2 flooding in 1972.[58] Based on the production data presented by Ghahafarokhi et al.,[58] the cumulative oil production since the start of CO_2 injection until 2013 corresponds to ~33% of the 2,800 MMSTB original oil in place (OOIP). Another example of a successful (dense phase) CO_2—EOR (and sequestration) project is the Weyburn unit in Saskatchewan, Canada, which has been undergoing CO_2 injection since 2000. The CO_2 for injection is sourced via a coal gasification plant in Beulah, North Dakota. The OOIP for the Weyburn unit is ~1,400 MMSTB.[59,60] EOR oil associated with CO_2 amounting to 25% of the total daily production from the Weyburn unit and an estimated increase in production by 130 million barrels, which is 10% of the OOIP and prolonging the life of the field by 25 years have also been reported.[59] Other sources report an incremental recovery factor of 12%–20% of OOIP due to CO_2 flooding.[60] Arcia et al.[61] have also suggested the recovery of immobile retrograde condensate by cycling dense phase CO_2 in a field in Eastern Venezuela and report a condensate recovery in excess of 65% based on laboratory studies. The field implementation status is, however, unknown.

5.6.2 Dense Phase or Supercritical Water (SCW)

Water has a critical pressure and temperature of 3,200 psia and 705°F, respectively. At pressure and temperature conditions above the critical point, obviously water is super critical. Super critical water (SCW) possesses extraordinary properties in that the high temperature, somewhat closer to the range in fluid catalytic cracking, results in thermal cracking of heavy hydrocarbons. A patent on using supercritical fluids to refine hydrocarbons (water in particular) was granted in November 2014.[62] The SCW thermally splits the higher molecular weight hydrocarbons into lighter, more volatile hydrocarbons of higher value. An acronym called Supercritical Water Extraction and Refining (SCWER) resulted from this concept.[63] A number of applications that report the successful use of SCW for heavy hydrocarbon upgrading are briefly reviewed next. Watanabe et al.[64] studied the Canada oil sand bitumen upgrading in a batch reactor in what they refer to as "high-density water" at ~842°F. The oil sand bitumen sample they used was synthetic in nature in that it was made up of 10 wt% hexane precipitated asphaltenes and the rest were hexane maltenes (non-asphaltenic fraction). Although they did not specify the pressure, given the higher density of 0.1 and 0.2 g/cm^3, it certainly exceeds the 3,200 psia. Their key finding was the complete destruction of asphaltene within 30 min. Somewhat similar experiments were conducted by Kozhevnikova et al.,[65] but specifically on only the asphaltenes portion extracted from heavy Tatar oil. Their experiments also were conducted in a batch reactor or autoclave at temperature and pressure a little bit above the water critical point. These experiments demonstrated the cracking of asphaltenes by SCW yielding up to 30% liquid, ~4% gas, composed of alkanes and aromatics, and ~65% carbonized solid residue, respectively. Golmohammadi et al.[66] experimentally studied the cracking of petroleum residue obtained from the vacuum distillation unit with and without using various nanocatalysts in a batch mode SCW reactor. Their feedstock (vacuum residue) was principally somewhat similar to that of Watanabe et al., containing 79% maltenes and 21% asphaltenes. In all cases, the ashphaltene breakdown was observed and all the studied cases demonstrated the formation of lighter upgraded hydrocarbon liquids (varying quality).

5.6.3 The General Prominence of Supercritical Fluids

Based on the foregoing, the importance of supercritical fluids in industrial applications is clearly evident, which is also supported by the fact that a dedicated journal called *The Journal of Supercritical Fluids* has been in existence since 1988. This journal regularly documents numerous successful applications of supercritical fluids. Additionally, Taylor[67] covers almost all aspects pertaining to supercritical fluid extraction.

References

1. Robinson, D.B. and Peng, D.-Y. 1978. *The Characterization of the Heptanes and Heavier Fractions for the GPA Peng-Robinson Programs.* Research report/Gas Processors Association, RR-28. Tulsa, OK: Gas Processors Association.
2. Hankinson, R.W. and Schmidt, T.W. 1982. Phase behavior and dense phase design concepts for application to the supercritical fluid pipeline system. In *European Petroleum Conference.* Society of Petroleum Engineers. doi:10.2118/12567-MS.
3. McCain, W.D. 1990. *Properties of Petroleum Fluids* (2nd Edition). Tulsa, OK: Pennwell Books.
4. Lohrenz, J., Bray, B.G., and Clark, C.R. 1964. Calculating viscosities of reservoir fluids from their compositions. *Journal of Petroleum Technology* 16(10): 1171–1176. doi:10.2118/915-PA
5. Fenghour, A., Wakeham, W.A., and Vesovic, V. 1998. The viscosity of carbon dioxide. *Journal of Physical and Chemical Reference Data* 27(1).

6. Ungerer, P., Faissat, B., Leibovici, C., Zhou, H., Behar, E., Moracchini, G., and Courcy, J.P. 1995. High pressure-high temperature reservoir fluids: Investigation of synthetic condensate gases containing a solid hydrocarbon. *Fluid Phase Equilibria* 111(2): 287–311. doi:10.1016/0378-3812(95)02771-6.
7. DeBruijn, G., Skeates, C., Greenaway, R., Harrison, D., Parris, M., Simon, J., Mueller, F., et al. 2016. High-pressure, high-temperature technologies. Accessed July 19. www.slb.com/~/media/Files/resources/oilfield_review/ors08/aut08/high_pressure_high_temperature.pdf.
8. Zaostrovski, A. 2016. Kristin HPHT gas-condensate field. Accessed July 20. http://iptibm1.ipt.ntnu.no/~jsg/undervisning/naturgass/lysark/Lysark Zaostrovski2010.pdf.
9. McDermott, J.R. and Martin, B.L. 1992. Completion design for deep, sour norphlet gas wells offshore mobile, Alabama. In *SPE Annual Technical Conference and Exhibition*. Society of Petroleum Engineers. doi:10.2118/24772-MS.
10. Machado, J.J.B. and de Loos, T.W. 2004. Liquid–vapour and solid–fluid equilibria for the system methane + triacontane at high temperature and high pressure. *Fluid Phase Equilibria* 222–223: 261–267. doi:10.1016/j.fluid.2004.06.003.
11. Stamataki, S.K., Magoulas, K.G., Boukouvalas, C.J., and Tassios, D.P. 1998. Correlation and prediction of phase equilibria and volumetric behaviour of hyperbaric binary fluids. *Revue de l'Institut Français du Pétrole* 53(1): 59–69. doi:10.2516/ogst:1998009.
12. Daridon, J.L., Lagourette, B., Gaubert, J.F., Xans, P., and Montel, F. 1996. Ultrasonic velocity in a hyperbaric reservoir fluid. *Ultrasonics* 34(2): 447–449. doi:10.1016/0041-624X(96)00020-0.
13. Cañas-Marín, W.A., Ortiz-Arango, J.D., Guerrero-Aconcha, U.E., and Soto-Tavera, C.P. 2007. Thermodynamic derivative properties and densities for hyperbaric gas condensates: SRK equation of state predictions versus Monte Carlo data. *Fluid Phase Equilibria* 253(2): 147–154. doi:10.1016/j.fluid.2007.02.004.
14. Mankiewicz, P.J., Pottorf, R.J., Kozar, M.G., and Vrolijk, P. 2009. Gas geochemistry of the mobile Bay Jurassic Norphlet Formation: Thermal controls and implications for reservoir connectivity. *AAPG Bulletin* 93(10): 1319–1346. http://aapgbull.geoscienceworld.org/content/93/10/1319.abstract.
15. Arnold, G., Cavalero, S.R., Clifford, P.J., Goebel, E.M., Hutchinson, D., Leung, H., Paclibon, E.N., Weiss, T.D., and Grass, D.B. 2010. SS: Thunder Horse and Atlantis Deepwater Frontier Developments in the Gulf of Mexico: Thunder Horse Takes Reservoir Management to the Next Level. In *Offshore Technology Conference*. doi:10.4043/20396-MS.
16. Dindoruk, B. and Christman, P.G. 2004. PVT properties and viscosity correlations for Gulf of Mexico Oils. *SPE Reservoir Evaluation & Engineering* 7(6): 427–437. doi:10.2118/89030-PA.
17. EVIDENT Final Technical Report. 2016. Accessed July 26. http://cordis.europa.eu/publication/rcn/1931_fr.html.
18. Zéberg-Mikkelsen, C.K. 2001. *Viscosity Study of Hydrocarbon Fluids at Reservoir Conditions Modeling and Measurements*. Technical University of Denmark.
19. do Vale Machado, J.M.D.C.M. 2001. High pressure-high temperature viscosity modeling of hydrocarbons and synthetic mixtures present on petroleum reservoirs. Universidade do Porto.
20. Bell, K., Burgass, R., Danesh, A., Gozalpour, F., Todd, A., and Tohidi. B. 2000. EVIDENT Report 5. Project Report.

21. Pedersen, K.S., Christensen, P.L., and Shaikh, J.A. 2015. *Phase Behavior of Petroleum Reservoir Fluids* (2nd edition). Boca Raton, FL: CRC Press, Taylor & Francis Group.
22. Soave, G. 1972. Equilibrium constants from a modified Redlich-Kwong equation of state. *Chemical Engineering Science* 27(6): 1197–1203. doi:10.1016/0009-2509(72)80096-4.
23. Alani, G.H. and Kennedy, H.T. 1960. Volumes of liquid hydrocarbons at high temperatures and pressures. *Petroleum Transactions, AIME*. Society of Petroleum Engineers.
24. Dandekar, A.Y. 1994. Interfacial tension and viscosity of petroleum reservoir fluids. Heriot-Watt University.
25. Gozalpour, F., Danesh, A., Todd, A.C., and Tohidi, B. 2001. Viscosity and density data of two North Sea gas condensate samples at temperatures to 423 K and pressures to 140 MPa. *Journal of Chemical & Engineering Data* 46(5): 1305–1308. doi:10.1021/je010053b.
26. Whitson, C.H. 1983. Characterizing hydrocarbon plus fractions. *Society of Petroleum Engineers Journal* 23(4): 683–694. doi:10.2118/12233-PA.
27. Bon, J. 2016. Measurements for dense phase hydrocarbons. *Workshop Presentation*. Penang, Malaysia: Society of Petroleum Engineers.
28. Moshfeghian, M. 2016. Transportation of CO_2 in dense phase, Tip of the Month, 2012. Accessed August 1. www.jmcampbell.com/tip-of-the-month/2012/01/transportation-of-co2-in-dense-phase/.
29. Eldevik, F. 2008. Safe pipeline transmission of CO_2. *Pipeline & Gas Journal* 235 (11).
30. Chapoy, A., Burgass, R., Tohidi, B., Austell, J.M., and Eickhoff, C. 2011. Effect of common impurities on the phase behavior of carbon-dioxide-rich systems: Minimizing the risk of hydrate formation and two-phase flow. *SPE Journal* 16(4): 921–930. doi:10.2118/123778-PA.
31. Kreysa, G. and Schutze, M. 2006. *Corrosion Handbook, Corrosive Agents and Their Interaction with Materials, Volume 5, Carbonic Acid, Chlorine Dioxide, Seawater* (2nd Edition). New York: John Wiley & Sons, Inc.
32. Sloan, E.D. and Society of Petroleum Engineers (U.S.). 2000. *Hydrate Engineering*. Monograph/SPE; Henry L. Doherty series, 2973142, vol. 21. Richardson, TX: SPE.
33. Sloan, E.D. and Koh, C.A. 2008. *Clathrate Hydrates of Natural Gases* (3rd edition). Chemical Industries, vol. 119. Boca Raton, FL: CRC Press.
34. Falabella, B.J. 1975. A study of natural gas hydrates. University of Massachusetts.
35. Larson, S.D. 1955. Phase studies of the two-component carbon dioxide-water system, involving the carbon dioxide hydrate. University of Illinois.
36. Miller, S.L. and Smythe, W.D. 1970. Carbon dioxide clathrate in the martian ice cap. *Science* 170(3957): 531–533. http://science.sciencemag.org/content/170/3957/531.abstract.
37. Ng, H.-J. and Robinson, D.B. 1983. *Equilibrium Phase Composition and Hydrating Conditions in Systems Containing Methanol, Light Hydrocarbons, Carbon Dioxide, and Hydrogen Sulfide: Project 825*. Research report/Gas Processors Association, 66. Tulsa, OK: Gas Processors Association.
38. Nakano, S., Moritoki, M., and Ohgaki, K. 1998. High-pressure phase equilibrium and Raman microprobe spectroscopic studies on the CO_2 hydrate system. *Journal of Chemical & Engineering Data* 43(5): 807–810. doi:10.1021/je9800555.

39. Ohgaki, K., Makihara, Y., and Takano, K. 1993. Formation of CO_2 hydrate in pure and sea waters. *Journal of Chemical Engineering of Japan* 26(5): 558–564. doi:10.1252/jcej.26.558.
40. Katz, D.L., and Lee, R.L. 1990. *Natural Gas Engineering: Production and Storage.* McGraw-Hill Chemical Engineering Series. New York: McGraw-Hill.
41. Whitson, C.H., Brulé, M.R., and Society of Petroleum Engineers of AIME. 2000. *Phase Behavior.* SPE Monograph, Henry L. Doherty Series, 32892, vol. 20. Richardson, TX: Henry L. Doherty Memorial Fund of AIME, Society of Petroleum Engineers.
42. Deaton, W.M. and E.M. Frost. 1948. *Gas Hydrates and Their Relation to the Operation of Natural-Gas Pipe Lines*, vol. 8. Monograph (United States. Bureau of Mines). New York: Printed by the American Gas Association.
43. Makogon, T.Y. and Sloan Jr., E.D. 1994. Phase equilibrium for methane hydrate from 190 to 262 K. *Journal of Chemical & Engineering Data* 39(2): 351–353. doi:10.1021/je00014a035.
44. Kobayashi, R. and Katz, D.L. 1949. Methane hydrate at high pressure. *Journal of Petroleum Technology* 1(3): 66–70. doi:10.2118/949066–G.
45. McLeod, H.O. and Campbell, J.M. 1961. Natural gas hydrates at pressures to 10,000 psia. *Journal of Petroleum Technology* 13(6): 590–594. doi:10.2118/1566-G-PA.
46. Marshall, D.R., Saito, S., and Kobayashi, R. 1964. Hydrates at high pressures: Part I. methane-water, argon-water, and nitrogen-water systems. *AIChE Journal* 10(2): 202–205. doi:10.1002/aic.690100214.
47. Wilcox, W.I., Carson, D.B., and Katz, D.L. 1941. Natural gas hydrates. *Industrial & Engineering Chemistry* 33(5): 662–665. doi:10.1021/ie50377a027.
48. Bishnoi, P.R. and Dholabhai, P.D. 1999. Equilibrium conditions for hydrate formation for a ternary mixture of methane, propane and carbon dioxide, and a natural gas mixture in the presence of electrolytes and methanol. *Fluid Phase Equilibria* 158–160: 821–827. doi:10.1016/S0378-3812(99)00103-X.
49. Ng, H.-J. and Robinson, D.B. 1976. The measurement and prediction of hydrate formation in liquid hydrocarbon-water systems. *Industrial & Engineering Chemistry Fundamentals* 15(4): 293–298. doi:10.1021/i160060a012.
50. Whorton, L.P., Brownscombe, E.R., and Dyes, A.B. 1952. Method for producing oil by means of carbon dioxide. GEN. Google Patents. www.google.com/patents/US2623596.
51. Van Orsdol, F. 2016. Dense phase fluid measurement. Accessed August 20. www.crt-services.com/pdf/papers11/Dense Phase Fluid Measurement.pdf.
52. Holm, L.W. and Josendal, V.A. 1974. Mechanisms of oil displacement by carbon dioxide. *Journal of Petroleum Technology* 26(12): 1427–1438. doi:10.2118/4736-PA.
53. Holm, L.W. and Josendal, V.A. 1982. Effect of oil composition on miscible-type displacement by carbon dioxide. *Society of Petroleum Engineers Journal* 22(1): 87–98. doi:10.2118/8814-PA.
54. Yellig, W.F. and Metcalfe, R.S. 1980. Determination and prediction of CO_2 minimum miscibility pressures (Includes Associated Paper 8876). *Journal of Petroleum Technology* 32(1): 160–168. doi:10.2118/7477-PA.
55. Bird, B., Stewart, W., and Lightfoot, E. 2006. *Transport Phenomena*, (Revised 2nd Edition). Hoboken, NJ: John Wiley & Sons, Inc. citeulike-article-id:3279154.
56. Chen, N.H. 1979. An explicit equation for friction factor in pipe. *Industrial & Engineering Chemistry Fundamentals* 18(3): 296–297. doi:10.1021/i160071a019.

57. Verma, M.K. 2015. Fundamentals of carbon dioxide-enhanced oil recovery (CO_2-EOR)—a supporting document of the assessment methodology for hydrocarbon recovery using CO_2-EOR associated with carbon sequestration: U.S. Geological Survey Open-File Report 2015–1071, 19 p.
58. Ghahfarokhi, R.B., Pennell, S., Matson, M., and Linroth, M. 2016. Overview of CO_2 injection and WAG sensitivity in SACROC. In *SPE Improved Oil Recovery Conference*. Society of Petroleum Engineers. doi:10.2118/179569-MS.
59. The Weyburn CO_2 Injection Project. 2016. Accessed November 9. www.springer.com/cda/content/document/cda_downloaddocument/9783642253874-c2.pdf?SGWID=0-0-45-1292488-p174265368.
60. McBurney, C. and Baker, R. 2016. Field performance of the Weyburn CO_2 miscible flood and factors affecting CO_2 performance. Accessed November 9. www.co2conference.net/wp-content/uploads/2012/12/2.4-Baker_Weyburn_2011-CO2FloodingConf.pdf.
61. Einstein, M.A., Castillo, Y.C.G., and Gil, J.C. 2007. A novel improved condensate-Recovery method by cyclic supercritical CO_2 injection. In *Latin American & Caribbean Petroleum Engineering Conference*. Society of Petroleum Engineers. doi:10.2118/107283-MS.
62. Yarbro, S.L. 2014. Using supercritical fluids to refine hydrocarbons. GEN, issued November 25, 2014. www.google.com/patents/US8894846.
63. Yarbro, S.L. 2012. SNT Ventures. Personal Communication.
64. Watanabe, M., Kato, S., Ishizeki, S., Inomata, H., and Smith Jr., R.L. 2010. Heavy oil upgrading in the presence of high density water: Basic study. *The Journal of Supercritical Fluids* 53(1–3): 48–52. doi:10.1016/j.supflu.2009.11.013.
65. Kozhevnikov, I.V., Nuzhdin, A.L., and Martyanov, O.N. 2010. Transformation of petroleum asphaltenes in supercritical water. *The Journal of Supercritical Fluids* 55(1): 217–222. doi:10.1016/j.supflu.2010.08.009.
66. Golmohammadi, M., Ahmadi, S.J., and Towfighi, J. 2016 Catalytic cracking of heavy petroleum residue in supercritical water: Study on the effect of different metal oxide nanoparticles. *The Journal of Supercritical Fluids* 113: 136–143. doi:10.1016/j.supflu.2016.03.023.
67. Taylor, L.T. 1996. *Supercritical Fluid Extraction*. Techniques in Analytical Chemistry Series. New York: Wiley.

6

Special Characterization for EOR Processes

6.1 Introduction

There are several objectives associated with a fluid characterization program, ranging from input to production and surface facilities (including flow assurance) to hydrocarbon in-place calculations and input to analytical and/or numerical field development planning operations.

Direct use of laboratory and field observations is discussed in Chapters 3 and 4, along with the quality-checking process of the empirical measurements. This chapter discusses the processes involved in the characterization of reservoir fluids for EOR processes, particularly those involving miscible displacements.

Traditionally, EOS is used to represent the behavior of hydrocarbons when pressure, temperature, and fluid composition change. EOSs are used to plan, quantify and guide the impact of field development operations on hydrocarbon recovery, particularly when an EOR agent is injected into the reservoir.

6.2 EOS Recap

There are several EOSs available in the industry, such as van der Waals, Soave Redlich–Kwong (SRK), and Peng–Robinson (PR) EOSs, among others. Several modifications have been proposed to these EOSs along the years and are often used on a case-by-case basis; that is, several EOSs are tested against empirical data and the final selection is made in terms of robustness and predictive power. Derivation, strengths, and caveats of each EOS are well documented in the literature and are easily accessible. Thus, we will concentrate on the practical aspects of the EOS use rather than their derivations. A generalized form of EOS is as follows:

$$P = \frac{RT}{v-b} - \frac{a}{v(v+b)+b(v-b)}$$

TABLE 6.1
EOS Variables

	PR	SRK
K	$1-B$	
M	$A-B(2+3B)$	$A-B(1+B)$
N	$AB-B^2-B^3$	AB
A	$\Omega a\alpha\left(\frac{P_r}{T_r^2}\right)$	$\Omega a\alpha\left(\frac{P_r}{T_r^2}\right)$
B	$\Omega b\left(\frac{P_r}{T_r^2}\right)$	$\Omega b\left(\frac{P_r}{T_r^2}\right)$
Ω_a	0.457235	0.42748
Ω_b	0.077796	0.08664

and

$$Pv = ZRT$$

$$F(Z) = Z^3 - KZ^2 + mz - n = 0$$

where v, R, T, P, T_r, and P_r are the molar volume, universal gas constant, temperature, pressure, reduced temperature, and reduced pressure, respectively. Table 6.1 shows the EOR variable definitions for PR and SRK.

Overall, PR EOSs are better in predicting liquid densities and perform well for gas condensate systems, while SRK overperforms in polar systems. EOSs need, of course, to be calibrated with measured data if they are to be used for any analytical/numerical predictions. Reservoir process (depletion, miscible/immiscible injection, etc.) should be used to define the calibration strategy and expected accuracy of the EOS. It should be clear that predictive power of an EOS is limited to the processes that were used to calibrate it, and the existence of an EOS does not itself qualify it to characterize all reservoir processes, an error, all too common in the industry.

Chapter 4 describes in detail the type and quality of experiments necessary to characterize different hydrocarbon fluids and recovery processes. All of these (augmented with field observations) need to be considered during the validation and eventual calibration of the EOS, including asphaltene and other component precipitation.

6.3 EOS Role on Numerical Simulation

Before describing the EOS calibration, it is important to understand its implications on numerical models. The diffusivity equation for compositional models, which is independent of the commercial simulator, is based on the finite-difference approach for component m:

$$\Delta\left[X_m\xi_L T_L\left(\Delta P_L-\gamma_L\Delta D\right)\right]+\Delta\left[Y_m\xi_V T_V\left(\Delta P_V-\gamma_V\Delta D\right)\right]$$

$$+X_m\xi_L q_L+Y_m\xi_V q_V=\frac{VR}{\Delta t}\left[\Delta t\left[\varphi Z_m\left(\xi_L S_L+\xi_V S_V\right)\right]\right]$$

Expanded on the x-direction:

$$\left(X_m\xi_L T_{Lx}\right)_{i+1/2}\left(P_{Li+1}^{n+1}-P_{Li}^{n+1}\right)-\left(X_m\xi_L T_{Lx}\gamma_L\right)_{i+1/2}\left(D_{i+1}-D_i\right)$$

$$-\left(X_m\xi_L T_{Lx}\right)_{i-1/2}\left(P_{Li}^{n+1}-P_{Li-1}^{n+1}\right)+\left(X_m\xi_L T_{Lx}\gamma_L\right)_{i-1/2}\left(D_i-D_{i-1}\right)$$

$$+\Delta\left[Y_m\xi_V T_V\left(\Delta P_V-\gamma_V\Delta D\right)\right]+X_m\xi_L q_L+Y_m\xi_V q_V=\frac{VR}{\Delta t}\left[\Delta t\left[\phi Z_m\left(\xi_L S_L+\xi_V S_V\right)\right]\right]$$

where

X_m = liquid fraction of component m, fraction
Y_m = vapor fraction of component m, fraction
ξ_L = molar density of the liquid phase, lb mol/ft^3
ξ_V = molar density of the Vapor phase, lb mol/ft^3
T_L = transmissibility of the liquid phase, md-ft/cP
T_V = transmissibility of the Vapor phase, md-ft/cP
P_L = pressure of the Liquid phase, psia
P_V = pressure of the Vapor phase, psia
γ_L = specific gravity of the Liquid phase, psia/ft
γ_V = specific gravity of the Vapor phase, psia/ft
q_L = source term for the liquid phase, ft^3/day
q_V = source term for the Vapor phase, ft^3/day
D = distance to datum, ft
$S_{L,V}$ = liquid and Vapor saturations, fraction
K_x = permeability in the x-direction, md
K_r = relative permeability, fraction
μ = viscosity, cP
B_L, B_V = liquid and vapor relative volumes, rft^3/Scf

The EOS is used to predict the molar composition, density, specific gravity, relative volumes, and bubble-/dew-point, fugacities, among other parameters

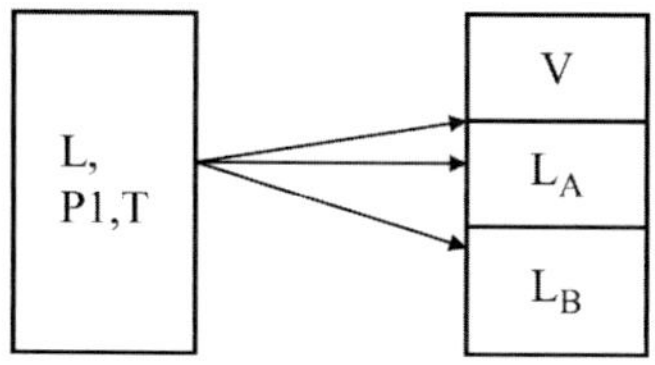

FIGURE 6.1
Three-phase flash calculation.

that affect directly the predictive fluid flow in the reservoir and thus require a higher degree of accuracy than the normal "black oil" approach where table lookups of the PVT properties are used.

Viscosity estimations are not done within the EOS, rather they use different correlations that are component based but are independent from the main EOS parameters and therefore are often calibrated separately.

When dealing with injection of an EOR agent, particularly CO_2, it is possible that more than one liquid phase is present in the reservoir and while EOS can predict such differences (see Figure 6.1), the two liquid phases are often not considered during numerical simulation as they require additional formulations for the extra liquid phase.

6.4 EOS Calibration

Several authors have proposed different workflows for EOS calibration, particularly for fluids that are highly volatile and close to the critical point. Ideally prior to any calibration exercises, the PVT/field measurements need to be validated (see Chapter 4) and their representativeness asserted. All samples that have been viewed as representative need to be included from the start on any EOS calibration (reservoirs with compositional gradient, for example) as the EOS needs to be able to reproduce their behavior under different conditions of composition, pressure, and temperature. Calibration process often involves the lumping or splitting of components, the latter being nonunique and thus requiring—for consistency—to be performed with all the samples at the same time.

EOS calibration implies changes to component specific properties—often to the plus heavy fraction, and as such it is important to understand and define uncertainty ranges (between which each property may be changed during any regression) for each component. It follows that changes of the properties of heavier fractions (C7+, C11+, etc.) are the most suitable than those of the well-characterized components such as C1.

As discussed in the previous section, majority of EOS follow a cubic equation and thus any calibration process follows the mathematical principles of regression, variable dependence, and of course optimization. Thus, mathematically several scenarios may have a high regression coefficient while providing unphysical results. Engineers should approach the calibration process with care ensuring that the new EOS is both stable and robust.

With the new fluid characterization advances, it is not uncommon to characterize fluids beyond the C7+ fraction, which was common up until recently. This new characterization does provide an opportunity to minimize the potential errors included while trying to characterize a heavy fraction without fully understanding the make of the fluid (it follows that no two C7+ fractions have similar properties) as well as limiting the degrees of freedom during the calibration process.

In the instances where the characterization stopped at C7+/C11+ and when the molar composition dominates the sample (see Figure 6.2), splitting may be necessary in order to better predict saturation pressures and critical point. Several methods may be used to perform the splitting, either using constant composition (thereby dividing the heavy fraction into several ones with the same mole fraction) or using Whitson's[1] method where the heavy fraction is divided using a three-parameter probability density function, such that a monotonical distribution is obtained. Splitting is heavily dependent on the overall composition of the sample/samples used for the calibration and it is often inside the calibration loop. Given the different approaches to splitting, all samples should undergo the same treatment to ensure consistency.

One of the main uncertainties on any fluid characterization is the properties of the heavy fraction, particularly in cases where the compositional analysis stopped at a small carbon component (i.e., C7+). Literature estimates the average error on estimation of molecular weight of heavy fractions to be in the order of 10%–20%[2–4] and thus has been extensively used as a tuning parameter. As with any experimental information, uncertainty ranges are both a function of the experimental procedure and laboratory and of course vintage of the measurement. Ideally, these ranges should be defined

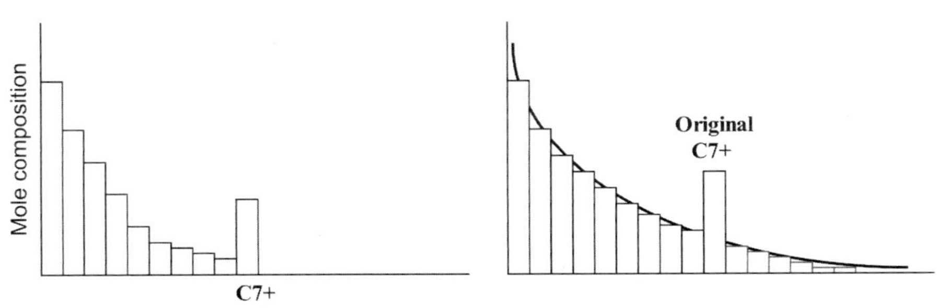

FIGURE 6.2
Example of component splitting by Whitson.

by the laboratory when the analysis was performed and it falls onto the engineer performing the calibration to define and, most importantly, justify error ranges before any changes are made. It follows that when the molecular weight of any component is changed so does the molar composition of the mixture, which in turn affects the critical point of the fluid and the saturation locus.[5] Molecular weight and density correlations are often used to estimate critical properties of the heavy fractions (necessary for the EOS) and are discussed in detail in Chapter 11.

Changes to molecular weight of the heavy fraction (along with splitting) are often performed first during the calibration exercise, giving an extra weight to the calibration of the saturation locus. In some instances, however, further changes to the critical properties of the heavy fraction/s may be required, including critical pressure and temperature. These changes are likely to affect the critical point estimation (often reducing it) and therefore saturation, liquid dropout, and saturated solution gas and/or oil.

Another strategy used for calibration is changes on the coefficients Ω_a and/or Ω_b which control the molecule attractive forces and volumes, thus allowing for similar changes as T_c and/or P_c would inflict with a limit impact on the critical point.

Acentric factors (usually named ω)—introduced by Pitzer[6]—account for the non-sphericity (acentricity) of the fluid molecules (deviation from ideal gas theory) and thus are often used to further improve on the saturation pressure, fluid dropout, and swelling following changes on critical properties of the heavy fractions.

Binary interaction coefficients may also be used to improve on the saturation locus. They represent the interaction of molecules between different pair of components and account for the deviation from the ideal gas situation where molecules/particles are assumed to be independent from each other. Small modifications on BIC, however, have the potential to substantially change the saturation locus, yielding unphysical values at lower temperatures. Thus, phase envelope should always be calculated before any changes on BIC are permanently implemented.

After the saturation pressures, swelling, oil/gas content are calibrated, volume shifts (introduced by Peneloux[7]) may be used to correct for liquid densities without significantly impacting the equilibrium conditions, thus allowing for improving on API and fluid gradients without compromising on the saturation locus.

Liquid and/or gas viscosities as such are not part of the EOS and it is often a good practice to exclude them from the EOS calibration. Viscosities should be matched alone following one of the component-based correlations where more degrees of freedom are granted given the nature of the initial estimations. It follows that changes to the component viscosity coefficients should not affect the EOS performance in any event.

A calibrated EOS should be able to reproduce, with a reasonable degree of accuracy, the results of field and laboratory observations, including reservoir and separator conditions Saturation locus should be smooth (convergence at high and low temperatures), flash, swelling and stripping calculations stable while preserving the characterization integrity of each of the components that make the hydrocarbon fluid. It is important to note that while some calibration exercises may in fact reproduce the laboratory experiments, they may fail when tested under dynamic conditions, making it paramount to test the performance of the EOS at reservoir conditions as part of the calibration process.

Use of wireline fluid characterization for PVT analysis has become a robust alternative for fluid characterization, particularly in fields with complex fluid distribution. Chapter 4 details the use of downhole fluid analysis (DFA) on quality checking of PVT samples and aiding on reservoir characterization. However, with the recent advances on DFA and delumping techniques, direct use of downhole fluid results on predicting (albeit on a less rigorous manner) fluid behavior thru EOS is well underway.[8] Early identification and validation of fluid complexity during sampling, as well as rapid (yet consistent) estimation of PVT properties (such as volumetric factors, solution gas ratios, API, etc.) are among the key advantages of these techniques, which may also be combined with the in situ evaluation of miscible processes as well as condensate reservoir monitoring.

6.4.1 Slimtube Calibration

Once the EOS has been calibrated, and its finite-difference performance tested and validated (see Section 6.5.1), some special experiments such as slimtube MMP should be introduced to further tune the EOS behavior under dynamic miscible conditions. As mentioned previously, an EOS calibrated with a set of CCE, DL, and CVD experiments is not necessarily able to reproduce fluid behavior under injection of an EOR agent (e.g., CO_2). It is, therefore, extremely important to test—and fine tune whenever necessary—the EOS behavior under such conditions to ensure consistency and validity of any numerical prediction.

As described in Chapter 4, slimtube experiments are used to estimate minimum miscibility pressure, doing so by performing a series of injection tests where both composition of the injection agent and pressure are varied. It follows that for the test to be representative reservoir oil and injection agent should reflect the current conditions of the reservoir (or those at which the injection will start in the field).

Calibration of a slimtube test requires building a 1D numerical model with the dimensions and properties of the laboratory experiment (cross-sectional area, length, porosity, permeability). Boundary conditions of injection rates, production, temperature, pressure, and composition should be directly translated from the experimental data so that the numerical model experiences the same process as the laboratory. Saturation functions (capillary forces may be ignored given their low magnitude) are often reported by the laboratory and correspond to the "packaging material" used inside the slimtube. It is often a good practice to add a small cell to either side of the slimtube (behind the injector/producer) to minimize numerical end effects.

Engineers are encouraged to work closely with the laboratory to understand the experimental setup and measurement process as they may vary for different providers: dead volumes (in the equipment) and other potential endpoint effects (albeit not expected for these displacements) should be considered while calibrating the experiment as well as measurement accuracy. Initial volumes and compositions should be verified before the calibration process starts using laboratory results, and any discrepancies should be addressed to ensure consistency.

Calibration of the slimtube does require to reproduce the laboratory injection process while looking at the effectiveness of the displacement (recovery) and other boundary conditions. This process may require modifications of the EOS calibration—following the same philosophy described above, so that for a given pressure and injection agent the recovery curves from the laboratory are reproduced. Figure 6.3 shows an example of the numerical calibration.

The calibration process should be updated if the original EOS components are grouped (see Section 6.4) to ensure consistency. It is also recommended that upon calibration the newly slimtube EOS must be tested using mechanistic models (as discussed in Section 6.5.1).

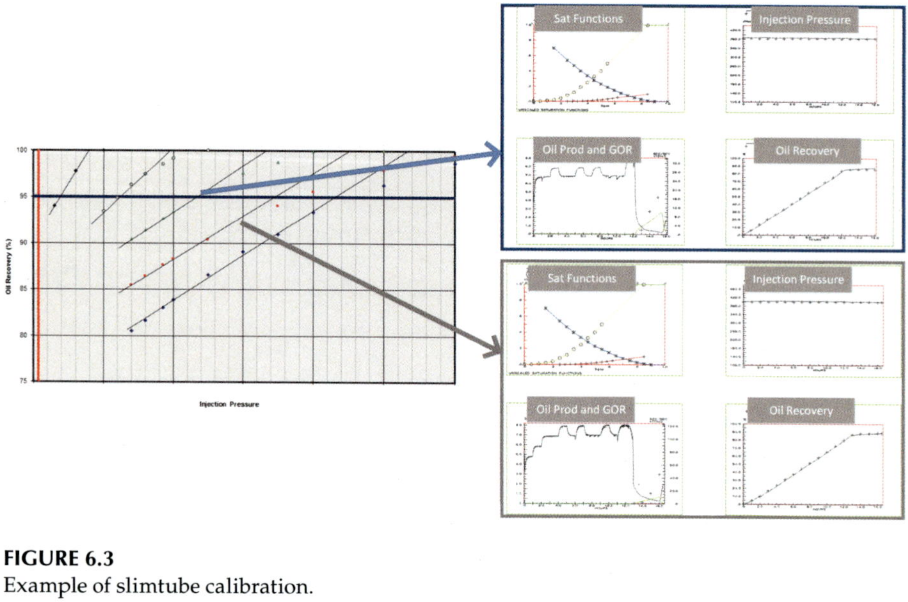

FIGURE 6.3
Example of slimtube calibration.

6.5 EOS Grouping

Grouping of single carbon number (SCN) into multiple carbon number (MCN) has been studied by several authors[1,9–11] in order to reduce finite-difference numerical simulator running times (exponentially increased with the number of components present on an EOS) while preserving the integrity and most importantly the predictive capabilities of the calibrated EOS. It is common to group SCN with components that are likely to behave on a similar fashion (close carbon number), thus requiring limited calibration of the new EOS. N_2 and CO_2 are often paired with C1 and C2, respectively, given the similar behavior (vaporizing/condensing) that each pair has when in contact with the reservoir oil. Pedersen et al.[2] proposed a slightly different approach where C1 and non-hydrocarbon components are left as pure comments and the intermediates are grouped into pseudo-components; heavier fractions are similarly reduced. Grouping strategy should be consistent with the current and future field operations and existing monitoring and surveillance systems with common sense used to further determine what components may not be grouped. If CO_2 injection is planned (either pure or with some contaminants) CO_2 and its other contaminants should not be grouped, and so on. Mixing rules are often applied to estimate the critical properties of the MCN based on the SCNs that were grouped.[12]

As components are grouped, it is expected that the critical point changes (usually reduced) as shown in Figure 6.4. If the critical point reduces

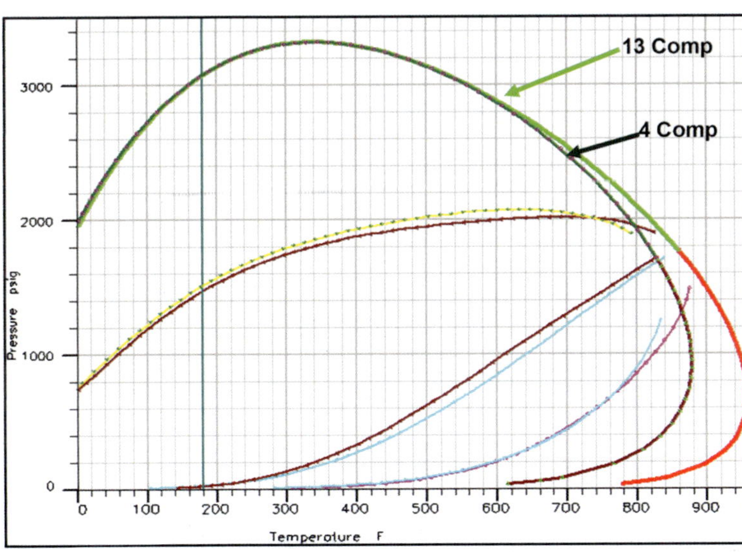

FIGURE 6.4
Phase envelope comparing different component grouping.

significantly, then it will change the predictive phase of the reservoir if reservoir temperatures are higher (or lower) than the newly predicted critical point. Thus, any changes on the grouping need to ensure consistency on the phase envelope.

Premises of EOS calibration still applies when it comes to grouping; that is, similar (if not better) behavior and stability are expected from the new "grouped" EOS and as such grouping and further calibration become part of an iterative process similar to that described in the previous section.

6.5.1 Asphaltenes

Asphaltenes pose a challenge for many reservoirs and, due to their nature, traditionally have been considered challenging for characterization purposes. Several asphaltene models have been proposed, building on improved fluid characterization techniques and reservoir surveillance. Yen–Mullins[13] proposed one of such models stating that the molecular form of asphaltenes is dominant at low concentrations (lighter oils), and as the concentration increases nanoaggregates form and ultimately, for the higher concentrations nanoaggregates form clusters. It follows that asphaltene concentrations (either mole weight or mole fraction) impacts the precipitation, flocculation, and deposition of asphaltene.

It is important to understand the reservoir process and its effects on fluid flow before embarking on the asphaltene precipitation characterization. Some of the major components are as follows: changes in porosity and permeability (due to precipitation of asphaltene molecules (typical treated as "solids"), changes on viscosity due to flocculation (from nanoaggregates to clusters), and potential impact on wettability due to compositional changes. There are several precipitation models, such as liquid solubility, colloidal, and micellization, among others. Each of these models addressed above constitute its own set of challenges. In practice, most commercial simulators have incorporated some version/s of them for asphaltene precipitation prediction. The engineers are required to fully understand the implementation process and data requirements (similar to the traditional EOS calibration) before starting the calibration process.

Results of laboratory analysis are used to validate and furbish the necessary input to the asphaltene precipitation models. Overall asphaltenes may exist[13] as a molecule dissolved in the hydrocarbon oil—as the concentration increases molecules floc onto clusters which may then deposit as solid asphaltene, these three "stages" are often used on commercial simulators to track the concentration and volume of asphaltene (molecule, floc, solid). There is an intermediate stage between molecule and floc where the asphaltene has precipitated and not yet aggregated with other molecules. There are three reversible processes which may occur: precipitation-dissolution: determining the asphaltene that has left the oil phase and may exist as an intermediate molecule–floc—reverse process is dissolution; flocculation–dissociation, where the asphaltene molecules are converted into a floc—reverse process is dissociation; and deposition–entrainment, where asphaltene flocs deposit as a solid component.

6.6 EOS Dynamic Testing

Upon successfully calibrating an EOS, the next step, prior to utilizing it for field development planning is to test its validity and stability under dynamic conditions. As mentioned in the previous sections, the calibration process, however rigorous, does not translate onto a stable, usable EOS in finite-difference numerical simulation. It should be noted that while EOS are fairly well defined, the calibration/optimization codes may differ, making it important to understand the relation between the package used to calibrate the EOS and the numerical simulator used for development planning; it is our experience that often-different approaches are taken on both software packages even if they are from the same provider.

It is highly recommended that simple models 1D or 2D can be used to test the performance and validity of the calibrated EOS under different scenarios that cover the expected changes during both historical and future performance of the reservoir. These tests are particularly important if compositional changes (such as dry gas expansion, solvent injection and/or compositional gradients) are forecasted. Furthermore, these tests should allow the EOS to cross the saturation pressure envelope (when numerical stability is often low) to minimize material balance and performance issues.

Figure 6.5 shows such an example of the expected behavior of different grouping strategies, with 13 and 4 components; other than fluid in-place and run performance, it was clear that the four-component case under dry gas injection had very similar results in terms of recovery and gas as did the 13-component version of the EOS. This process was designed to emulate a dry gas injection on a gas cap reservoir, and will need to be modified to reflect each reservoir's processes. Reservoir heterogeneity representation, while important, is not as crucial as reproducing the expected fluid frontal velocities, compositions of the reservoir fluid and injection agent as well as the pressure and temperature of the system. 1D/2D/3D mechanistic models need to be designed so that are as representative of the reservoir process for the success of this exercise. Figure 6.6 shows the oil saturation distribution of another 1D model under gas injection. Both 13- and 4-component EOS yielded similar results and thus were considered to be adequate for the characterization.

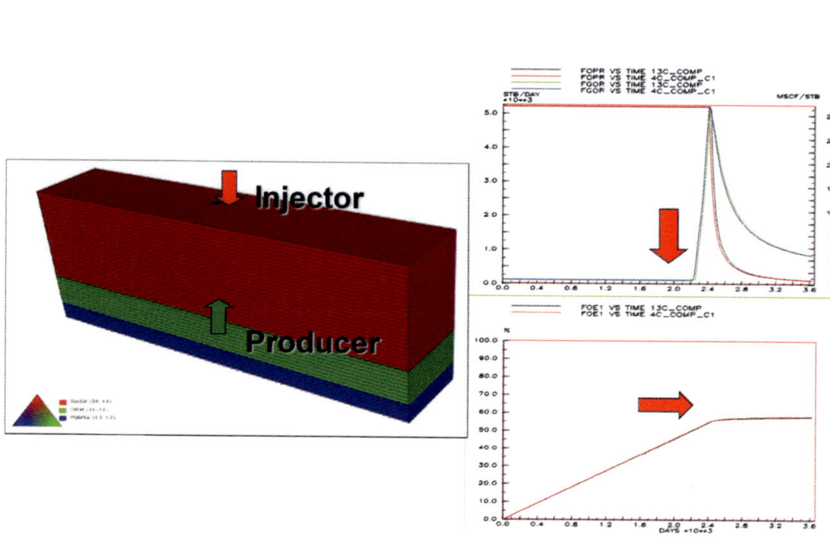

FIGURE 6.5
Example of use of 1D models to test dynamic behavior of different EOS grouping.

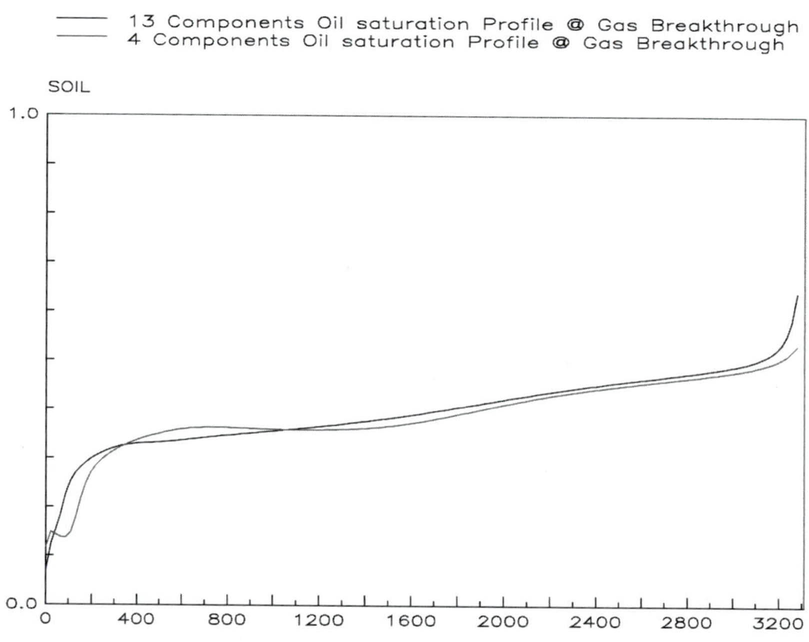

FIGURE 6.6
Oil saturation distribution on a 1D mechanistic model comparing different EOS component grouping.

6.6.1 Finite-Difference Considerations

It is important to understand the role of the calibrated EOS on the numerical simulation, not only looking at accuracy to reproduce laboratory and field data but also understanding effects on mixing and dispersion once it is used on the finite-difference grid. Several authors have investigated the effect of scale on the performance of near-miscible and miscible displacements,[14,15] all highlighting the importance of velocity (function of grid size) and reservoir heterogeneity on the accurate representation of the front advancement at miscible, partially miscible, and immiscible conditions. Figure 6.7 shows the effect of scale on a first contact miscible flood with low frontal velocities—akin to those inside the reservoir. Effect of dispersion, mixing, and of course component velocities is evident when grid resolution is as small as 1 ft.

Unfortunately, appreciation of the role that velocity plays in the convective and dispersive displacements is traditionally limited, and is often ignored. Figure 6.8 shows the effect of scale on a cross-sectional model where different finite-difference grid sizes were tested at both miscible and near-miscible conditions as explained by Moreno et al.[14] Recoveries upon circulation of 1 PV were significantly different as grid size is coarsened, as all the oil within any single cell is assumed to be contacted—and displaced—by the gaseous EOR agent. While this contacting assumption (akin to frontal smearing) applies irrespective of the actual single-cell dimensions, the actual loss of

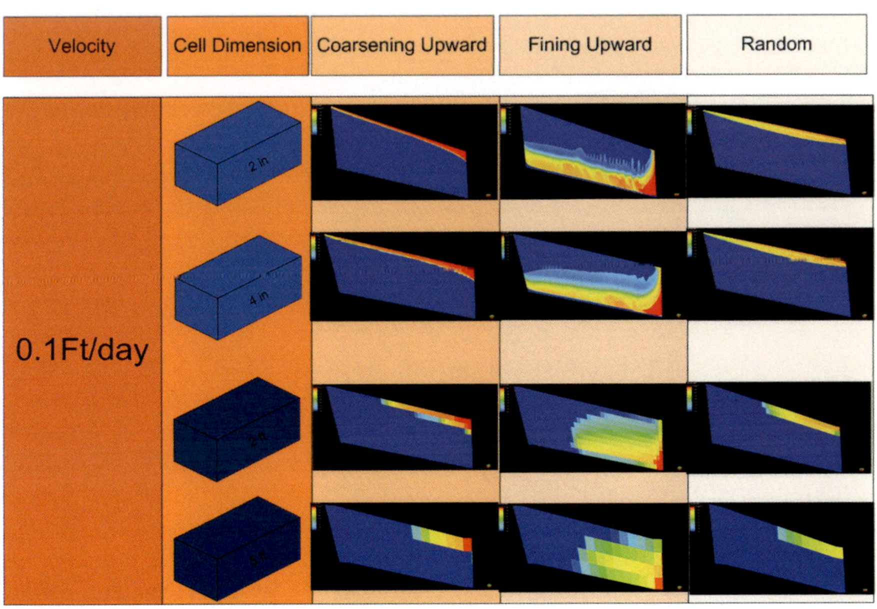

FIGURE 6.7
Displacement efficiency for 0.1 ft/day FCM front advancement for different property distributions.[14]

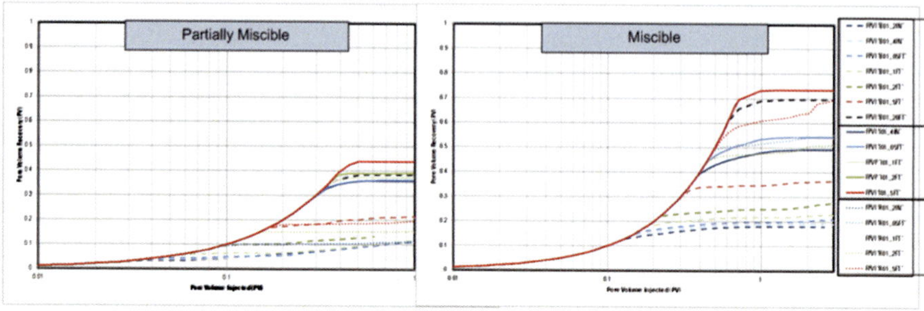

FIGURE 6.8
Effects of scale up on different rock quality distribution: near miscible (left) and FCM (right) (frontal velocity 0.1 ft/day). For different model sizes (2in. in blue to 20ft in black) and two different property distributions (*B*, which is the coarsening upward sequence and *R* to the random distribution).

grid discretization upon increasing the cell sizes through upscaling (beyond the physical limits) results on very optimistic estimations.

The issue of upscaling, of course, cannot be addressed with the compositional description alone. It does require an understanding of the dispersive and diffusive elements of the frontal flow. Dispersive flow and component mixing may be "scaled" by modifying fugacity coefficients and effectively controlling the velocity at which single components change from one phase to another. It follows that these changes need to be tested thru a similar process as described in this section to ensure that both address the dispersive flow upscaling and still maintain integrity of the predictive power of the EOS. Such changes also require a greater knowledge of the numerical simulator code to ensure that the desired effect is achieved.

One consideration which often is not foreseen when dealing with EOR injection is that of suitability and plausibility of monitoring (and of course control) systems to aid on the day-to-day operations and to guide the EOR strategy. Traditionally production and injection are monitored on a regular basis, and in some cases so is composition of the produced fluids. It is obvious that any numerical uncertainty (upscaling/dispersion) does affect the ability to use these numerical models to predict and more importantly modify the EOR strategy when field response does not resemble the expected behavior. Application of in situ (DFA) wireline compositional estimations on fluid characterization is discussed in Chapter 4 and it follows that these techniques are highly valuable to track compositional changes across the reservoir not only on EOR flooding projects but also on condensates. Through casing saturation estimations and inter-well resistivity imaging are also common practices to determine the orientation (by taking measurements on different wells), magnitude (looking at saturation changes behind the front), and overall efficiency of the frontal displacement (cross-well).

Several authors have[16,17] discussed the importance of looking at the EOR surveillance and modeling together not only to determine the appropriate monitoring tool but also to ascertain the likelihood of a successful measurement but matching expected physical/thermodynamical changes to tool resolution. As such, it does become important that relevant components are modeled explicitly in line with the expected tool requirements.

References

1. Whitson, C.H. 1983. Characterizing hydrocarbon plus fractions. *Society of Petroleum Engineers Journal* 23(4): 683–693.
2. Pedersen, K.S., Thomassen, P., and Fredenslund, A. 1989. *Properties of Oils and Natural Gases Contributions in Petroleum Geology and Engineering*, vol. 5. Houston, TX: Gulf publishing Co.
3. Thomassen, P., Pedersen, K.S., and Fredenslund, A. 1986. *Adjustment of C7+ Molecular Weights in the Characterization of Petroleum Mixtures Containing Heavy Hydrocarbons*. Dallas, TX: SPE.
4. Guo, T.M. and Du, L. 1989. *New Three-Parameter Cubic Equation of State for Reservoir Fluids: Part 3—Application to Gas Condensate*. Dallas, TX: SPE.
5. Rafael, A., Zurita, A., and McCain Jr., W.D.: An efficient tuning strategy to calibrate cubic EOS for compositional simulation. SPE 77382 presented at the 2002 SPE *Annual Technical Conference and Exhibition, San Antonio, TX*, September 29–October 2.
6. Pitzer, K.S. 1955. The volumetric and thermodynamics properties of fluids. *Journal of the American Chemical Society* 77(13): 3427–3433.
7. Peneloux, A. and Rauzy, E. 1982. A consistent correction for Redlich-Kwong-Soave volumes. *Fluid Phase Equilibria* 8: 7–23.
8. Julian, Y.Z., Zhang, D., Dubost, F., Dong, C., Mullins, O.C., O'Keffe, M., and Betancourt, S.S. EOS-Based downhole fluid characterization. SPE 114702 paper, presented at the 2008 *SPE Asia Pacific Oil and Gas Conference and Exhibition in Perth*, October 20–22.
9. Wu, R.S. and Fish, R.M. 1989. C7+ characterization for fluid properties predictions. *JCPT* 28(4): 112–117.
10. Chorn, L.G. and Mansoori, A. 1989. *C7+ Fraction Characterization*. New York: Taylor and Francis.
11. Neau, E. and Jaubert, J.N. 1993. Characterization of heavy oils. *Industrial & Engineering Chemistry Research* 32: 1196–1203.
12. Leibovici, C.F. 1993. A consistent procedure for estimation of properties associated to lumped systems. *Fluid Phase Equilibria* 87: 189–197.
13. Mullins, O.C., Andrews, A.B., Pomerantz, A.E., Dong, C., Zuo, J.Y., Pfeiffer, T., Latifzai, A.S., Elshahawi, H., Barré, L., and Larter, S. Impact of asphaltene nanoscience on understanding oilfield reservoirs SPE 146649 presented at the *Annual Technical Conference and Exhibition in Denver*, October 2011.

14. Moreno, J., Flew, S., and Gurpinar, O. EOR: Challenges of translating fine scale displacements into full field models-part 2. SPE 165298-MS presented at the 2013 *SPE Enhanced Oil Recovery Conference, Kuala Lumpur Malaysia,* July 2–4.
15. Garmeh, G. and Johns, R. Upscaling of miscible floods in heterogeneous reservoirs considering reservoir mixing. Paper SPE 124000 presented at the 2009 *SPE Annual Technical Conference and Exhibition, New Orleans,* October 4–7.
16. Altundas, Y.B., Chugunov, N., and Ramakrishnan, T.S. On the importance of accurate fluid substitution and models for seismic monitoring of CO_2. Paper presented at the 2013 *SEG Annual Meeting in Houston.*
17. Kuchuk, F., Zhan, L., Mark Ma, S., Al-shirani, A.M., Ramakrishnan, T.S., Altundas, B., Zeybek, M., de Loubens, R., and Chugunov, N. Determination of in-situ two-phase flow properties through downhole fluid movement monitoring. SPE 116068 presented at the 2008 *Annual Technical Conference and Exhibition, Denver,* September 21–24.
18. Robinson, D.B. and Peng, D.-Y. 1978. *The Characterization of the Heptanes and Heavier Fractions for the GPA Peng-Robinson Programs.* Research report/Gas Processors Association, RR-28. Tulsa, OK: Gas Processors Association.
19. McCain, W.D. 1990. *Properties of Petroleum Fluids* (2nd Edition). Tulsa, OK: Pennwell Books.
20. Shariati, A., Peters, C.J., and Moshfeghian, M. 1999. A systematic approach to characterize gas condensates and light petroleum fractions. *Fluid Phase Equilibria* 165: 165–179.
21. Manafi, H., Mansoori, G.A., and Ghotbi, S. 1999. Phase behaviour prediction of petroleum fluids with minimum characterization data. *Journal of Petroleum Science and Engineering* 22: 67–93.
22. Moreno, J., Flew, S., and Gurpinar, O. Immiscible EOR: Challenges of translating fine scale displacements into full field models. SPE 143568-PP presented at the 2011 *SPE Enhanced Oil Recovery Conference, Kuala Lumpur,* July 19–21.

7

Compositional, Fluid Property, and Phase Behavior Characteristics of Unconventional Reservoir Fluids

7.1 Significance of Unconventional Reservoir Fluids

According to the US Energy Information Administration (EIA),[1] conventional oil and natural gas is broadly defined as "Crude oil and natural gas that is produced by a well drilled into a geologic formation in which the reservoir and fluid characteristics permit the oil and natural gas to readily flow to the wellbore." Based on this definition, EIA uses an "umbrella" term for unconventional oil and natural gas that is produced by other methods that do not meet the criteria for conventional production. On the other hand, according to the US Department of Energy, as quoted by Gordon,[2] "unconventional oils have yet to be strictly defined." However, from the authors' perspective, we have attempted to expand the differentiation between the conventional and unconventional oil and gas using some specifics as follows. We define conventional systems as those possessing the following average or typical primary characteristics, and thus based on these criteria any system that does not fit this profile is termed as unconventional as it pertains to this chapter and elsewhere in the book.

- The porous medium or the reservoir rock is typical sandstone having average properties such as 20%–25% porosity; 250–500 mD absolute or intrinsic permeability; is water-wet; and has connate water saturations in the range of 20%–25%. In case of carbonates porosities as high as 30%–35%; at least two orders of magnitude lower permeability; is oil-wet; and has relatively high connate water saturations (up to 40%) owing to high capillary pressures rather than the wetting characteristics.
- The hydrocarbon fluids accumulated in the above types of rocks are those that can be classified as black oil, volatile oil, retrograde gases or gas condensates, wet gases, and dry gases, as characterized by McCain.[3]

- Typical reservoir temperatures and pressures do not exceed 300°F and 7,500 psi.[4]
- Can be drilled, completed, and produced by typical vertical or horizontal wells that are simply perforated.
- Follow the typical fundamental reservoir engineering concepts or principles; for example, see Craft and Hawkins.[5]
- Numerical simulation by finite difference type methods is relatively straightforward; for example, see Ertekin et al.[6]
- Relatively easy to recover by primary, secondary, or tertiary recovery methods.

Usman and Meehan,[7] however, make a valid point in that, what is unconventional today may become conventional in the future as technology matures. As a matter of fact, that already happens to be the case in certain oil and gas producing regions of the world as far as heavy oil and shale-based oil and gas production is concerned, i.e., the extraction of these resources is becoming routine. Especially over the last two decades or so, due to the dwindling conventional resources, the petroleum industry has focused its efforts on the exploitation of atypical or unconventional or technologically challenged oil and gas resources, which has been stimulated by vast advances in technology. In the following paragraph, in order to highlight the prominence of these unconventional oil and gas resources, we present the specifics on what these are and point out the relevant statistics on the resource base.

Unconventional gas resources typically include (1) coal bed methane (CBM gas), which is methane in coal seams; (2) tight sands gas (hydrocarbon gas in tight ultralow-permeability formations); (3) shale gas (gas in very low permeability shales); and (4) methane hydrates (methane trapped in crystal structure of water). Unconventional oil resources typically include (1) heavy oil (high-viscosity and high-density or low API oil); (2) shale oil (kerogen); (3) bitumen; and (4) tar sands (containing bitumen which has extremely high viscosities). Something that is not explicitly included in the class of unconventional gases are reservoir fluids that contain unusually large or disproportionately high fraction of non-hydrocarbon components such as carbon dioxide (CO_2) and even mercury (Hg), which pose peculiar challenges and issues in almost every aspect of dealing with these unique reservoirs. Statistics pertaining to resource and reserve estimates of conventional vs. unconventional oil and gas, based on various sources, was presented by Dandekar.[8] Although the data[8] are somewhat dated, resources and reserves do not dramatically change over a relatively short time span, and therefore, even if the most conservative estimates for technical and economic recovery of the unconventional resources are considered, it still represents a very substantial future fossil energy portfolio that dwarfs the conventional gas and oil reserves.

In summary, limited volumes of conventional oils (in particular) and gases are available with a continual decline and are relatively easy to extract using the already proven or matured technology at a low cost. However, on the other hand relatively abundant resources of unconventional oils and gases have been identified, but their extraction requires advanced technology, which may be in a transitional state, in addition to higher cost. A resource pyramid is sometimes used to illustrate this in which the top represents the conventionals with coordinates as (low volumes, easy to produce) and the base represents the unconventionals with coordinates as (very high volumes, difficult to produce). These, difficult to produce oil and gas resources, are thus well positioned to be the future of the petroleum industry, and therefore, now is the appropriate time to examine the various facets of these resources, such as their compositional, fluid properties, and phase behavior characteristics for better understanding, so that at least some barriers in the development of new technologies may be overcome.

7.2 Reservoir Fluids Containing Unusually Large Fractions of Non-hydrocarbon Components

7.2.1 Gaseous Non-hydrocarbons

According to some of the statistics,[9] more than 40% of the world's conventional gas reserves are estimated to be in reservoirs that contain unusually large fractions (as high as 70%) of hydrogen sulfide (H_2S) and CO_2, both termed as acid gases due to their propensity to form acids when they come in contact with water. Obviously, the presence of such unusual amounts of acid gases poses significant challenges to drilling, cementing, completion, production, and well interventions, as well as their disposal after processing the raw gas streams.[9] Given this importance, in 2014 the Society of Petroleum Engineers (SPE) organized a forum[9] dedicated to brainstorm technology development ideas on high CO_2 and H_2S gas fields' development.

In order to portray the prominence of reservoir fluids containing excessive amounts of gaseous non-hydrocarbon components, as much as to the extent that some are actually treated as CO_2 reservoirs than hydrocarbons reservoirs, compositional data from selected published sources were compiled. The compositional data basically fall into two categories, namely those that report somewhat unusually high N_2 as well as CO_2 and no H_2S (the balance being hydrocarbons) and those that report high H_2S as well as relatively high CO_2 and a somewhat low N_2 content (of the order of ~2 mol%) with the balance being hydrocarbons. Reservoir gases and oils containing H_2S in excess of 4 part per million by volume and 1% are classified or labeled as sour gases and sour crudes, respectively.[10] The reviewed compositional data of the two aforementioned categories are presented in Figure 7.1a, b in the form of a ternary plot in which each of the corners represents 100 mol% of CO_2, N_2 and CH_4^+ (all the hydrocarbons), and 100 mol% of CO_2, H_2S and CH_4^+, respectively. Note that given the relatively small N_2mol% (~2 mol%) in the second category it is not explicitly shown in the plot (Figure 7.1b) and has been added to CO_2mol%. The data in both the plots include gas and oil reservoirs.

7.2.2 Mercury

Mercury (Hg) exists in three different forms, namely, elemental or pure Hg, as organic mercury compounds and inorganic mercury compounds. Elemental Hg has some rather unusual properties, for example, a molecular (atomic) weight of 200.6, critical pressure and temperature in excess of 24,000 psi and 2,700°F, respectively, and an acentric factor of –0.21. Mercury deposits in the form of elemental Hg and several Hg compounds that naturally occur in petroleum reservoir fluids are often associated with geological plate boundaries, fold belts, and areas with volcanic or hydrothermal activity.[22,23] Wilhelm and McArthur[24] refer to Hg as a "naturally occurring contaminant" in petroleum reservoir fluids, which is freely distributed or partitioned (note though that always in a sub-critical state given its high P_c and T_c compared to typical temperature and pressure conditions encountered) in the upstream, midstream, and downstream sectors as well as in the consumption systems. The concentration of Hg in liquid (crude oil) and gaseous (natural gases) hydrocarbons is highly dependent on the geologic location and varies by several orders of magnitude; the approximate values quoted between 0.01 parts per billion and 10 parts per million by weight.[22] Less dense fluids such as gases mostly contain elemental Hg due to its volatility, whereas denser fluids such as condensates and crudes may contain the elemental Hg as well as organic Hg, generalized as $(C_XH_Y)_2Hg$.[24]

Wilhelm and McArthur[24] presented a table containing the estimated levels of mercury concentrations in natural gas and condensate from most geologic locations in the world, which was later expanded by Sainal et al.[25] A review of these tables clearly indicates that relatively higher concentrations of Hg both in the gas and condensate phases appear to be a phenomenon in the South East Asian locations, while the lowest concentrations are seen in the US Gulf of Mexico and the overthrust belt (Wyoming, Utah). As far as heavier hydrocarbon liquids or crude oils are concerned, Wilhelm and Bloom[22] report a Hg concentration range of 1.2 to ~3,000 ng/g as mean values of the lowest and highest halves, respectively. Details on the specificity or geographic origins of these crudes are not available; however, a data point of geographic reference for comparison is the oilfield from Argentina,[26] which reports the total Hg content from approximately 1,500 to 8,000 parts per billion by weight (ppbw; note that 1 ppbw = 1 ng/g). It is assumed that the concentrations referred to in these tables are for elemental Hg, the gas volumes are in standard conditions (for natural gas Hg concentrations) and

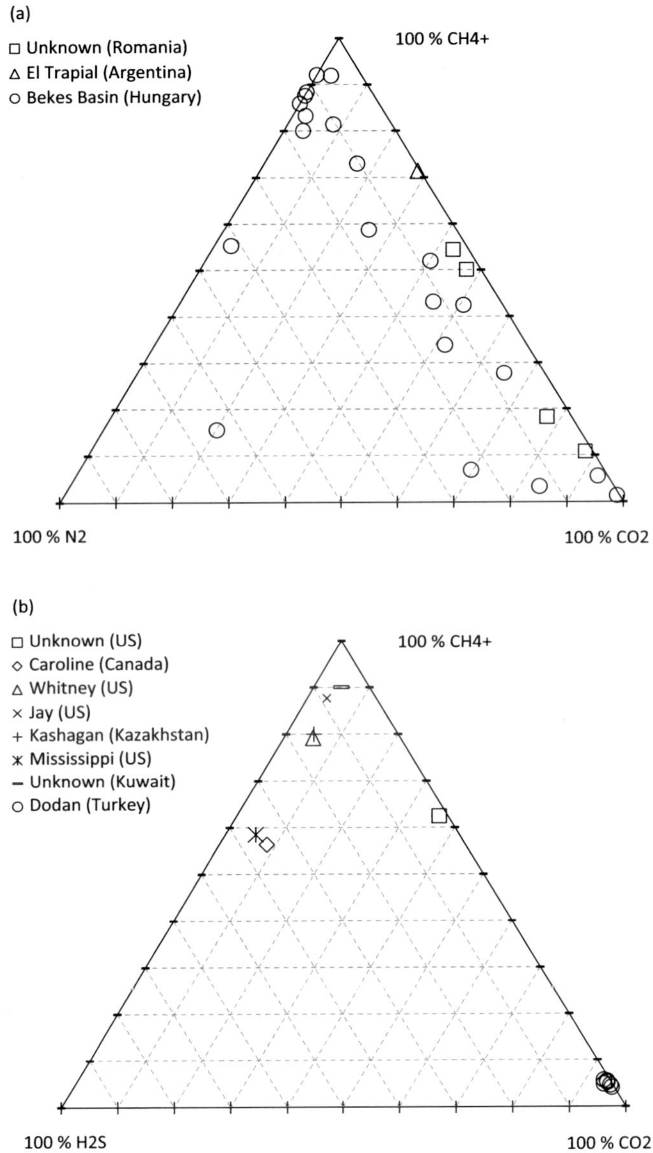

FIGURE 7.1
(a) Molar compositions of petroleum reservoir fluids from selected fields containing unusually high CO_2 and N_2. The ternary plot is constructed based on data from three literature sources. Field names, locations, and sources corresponding to the legend are listed as follows: unknown (Romania),[11] El Trapial (Argentina),[12] and Bekes Basin (Hungary).[13] (b) Molar compositions of sour petroleum reservoir fluids from selected fields. The ternary plot is constructed based on data from various literature sources. Field names, locations, and sources corresponding to the legend are listed as follows: unknown (US),[14] Caroline (Canada),[15] Whitney (US),[16] Jay (US),[17] Kashagan (Kazakhstan),[18] Mississippi (US),[19] Unknown (Kuwait),[20] and Dodan (Turkey).[21]

the masses of liquids, i.e., the condensate and the crude oil represent stock tank conditions. In order to consider these Hg ranges from a common molar composition perspective (typical from an equations of state (EOS) modeling and phase behavior standpoint), we have adopted the following approach to present the aforementioned data ranges in an alternate fashion.

For natural gases, 1 lb-mol of gas occupies 379.6 standard cubic feet (scf) volume.[8] For example, 1 μg/m³ equals to $\frac{1\times 10^{-6}\times 0.0022}{35.3147} = 6.229\times 10^{-11}$ lbm Hg/ft^3 of gas. Using the 200.6 lbm/lbm-mol atomic mass of Hg and the standard condition gas volume of 379.6 scf/lbm-mol, 100 μg/m³ equals $6.229\times 10^{-11}\times \frac{379.6}{200.6} = 1.179\times 10^{-10}$ lbm-mol Hg/lbm-mol gas. Alternatively, this means 1.179×10^{-10} lbm-mol of Hg in 1 lbm-mol of gas resulting in Hg mol fraction of $\frac{1.179\times 10^{-10}}{1.179\times 10^{-10}+1} \cong 1.179\times 10^{-10}$, which appears to be insignificant. A molecular weight of 150 and 225 lbm/lbm-mol for the stock tank condensate and stock tank oil, respectively, is reasonable to convert the Hg concentration in liquids in terms of a molar ratio. Therefore, 1 μg/kg $= 1.0\times 10^{-9}$ lbm Hg/lbm liquid, which is $1.0\times 10^{-9}\times \frac{150}{200.6} = 7.48\times 10^{-10}$ lbm-mol Hg/lbm-mol condensate and $1.0\times 10^{-9}\times \frac{225}{200.6} = 1.12\times 10^{-9}$ lbm-mol Hg/lbm-mol oil. Similar to the gas calculations shown previously, this means the mole fraction of Hg in the condensate and oil $\cong 7.48\times 10^{-10}$ and 1.12×10^{-9}, respectively. All the data ranges reported in the second column of Table 7.1 are converted in this manner.

7.2.2.1 Impact of Association of Hg with Petroleum Reservoir Fluids

As far as the interaction of elemental Hg with petroleum reservoir fluids is concerned, first and foremost, one needs to look back at what were the standard practices decades ago that involved the use of Hg as a hydraulic or displacement fluid in sampling cylinders and various PVT systems. Although the use of elemental Hg has been virtually eliminated in various operations, perhaps some systems are still in existence that use Hg. Elemental Hg is something we consider as a "liquid piston," which obviously offers a significant advantage of minimizing or even zeroing the various dead volumes, especially in a PVT system. In some cases where a floating piston is used (e.g., sampling bottles), Hg is not in direct contact with the hydrocarbons because the floating piston acts as a barrier between the two; however, there were/are some in which Hg is directly in contact with the hydrocarbons. Let us examine the impacts of these. In a sampling operation when a separator liquid sample is collected in a sample cylinder that uses Hg (and no floating piston), then it is quite possible that the Hg that actually "originated" from the

TABLE 7.1

Elemental Hg Mole Fraction Range in Natural Gases, Liquid Condensates, and Crude Oils of Different Geographic Origins

Geographic Origin	Natural Gases (×10⁻⁷)	Liquid Condensates (×10⁻⁵)	Crude Oils (×10⁻⁵)
Europe	0.118–0.177[24]	–	–
South America	0.059–0.141[24]	0.004–0.007[24]	0.112–0.598[26]
Gulf of Thailand	0.118–0.472[24]	0.030–0.090[24]	–
Africa	0.094–0.118[24]	0.037–0.075[24]	–
Gulf of Mexico (USA)	0.00002–0.00047[24]	–	–
Overthrust Belt (USA)	0.006–0.018[24]	0.0001–0.0004[24]	–
North Africa	0.059–0.094[24]	0.001–0.004[24]	–
Malaysia	0.001–0.236[25]	0.001–0.007[25]	–
Indonesia	0.236–0.354[25]	0.001–0.037[25]	–
Unknown	–	–	0.0001–0.224[22]

Table is constructed based on the estimates (in different units converted into mole fractions, see text for description) presented by Wilhelm and Bloom[22]; Wilhelm and McArthur[24]; Sainal et al.[25] and Salvá and Gallup.[26]

reservoir would drop down due to cooling[23] into the liquid phase and blend with the Hg used as a hydraulic fluid, i.e., the former remaining as unaccounted for in further analyses, thus introducing an uncertainty. Similarly, when it comes to PVT systems, Hg may be in direct contact throughout with the tested fluid (oil, gas condensate, etc., in CCE, DL, CVD tests); i.e., with the oil ready to undergo various CCE or DL pressure reduction steps, but when one considers a CVD Hg is initially in contact with the gas phase (above the retrograde dew point pressure) and then with the retrograde condensate phase (below the dew point), owing to gravity differences. Such close interactions or associations of elemental Hg with the hydrocarbon fluid phases are bound to influence the volumetric results, fluid properties due to mutual solubility (although small) effects; however, to the best of our knowledge, no corrections to PVT data or these interactions have been discussed or reported at least in the open literature. One of the exceptions, though, is an Hg volatility correction applied to the pressure measurement of compressed gases when Hg is used as a confining or pressure transmission fluid in a gauge[27]; conceptually such corrections could be potentially tailored to the aforementioned PVT and sampling type operations.

In principle, it is possible to carry out equilibrium flash calculations on reservoir fluids that contain Hg to determine the partitioning of Hg into, for example, a separator gas, separator liquid, stock tank gas, and stock tank liquid or in other words in the overall productions systems, just the way it is carried out for usual components such as methane, ethane, etc. This is also supported by Wilhelm et al.,[28] who state that computational models can be used to predict the Hg concentrations (we mean mole fractions) in the separated vapor and liquid streams. Note though that only elemental Hg is considered in this type of modeling.[27] However, referring to Table 7.1, although the originally reported values are estimates and assumptions have been made in arriving at the Hg mole fractions, even if an order of magnitude higher values are considered, these mole fractions appear to be rather insignificant at least from the coarse or bulk level (as opposed to a finer molecular level) standpoint of exerting any influence on the phase behavior, equilibrium flash, partitioning and fluid properties modeling traditionally carried out using EOS models in commercial PVT simulators. An alternate view that we have is given such insignificant mole fractions, the existing EOS or thermodynamic models are simply not capable of "capturing" the Hg partitioning or for that matter solubility in equilibrium vapor and liquid phases. Additionally, the unusual critical properties of Hg are certainly a "misfit" or an outlier in the reservoir fluid composition "ensemble." The option of Hg as a component in most commercial PVT simulators does not even explicitly exist; although in most cases the user may be able to define "own component," i.e., Hg with its own properties and mole fraction in the reservoir fluid composition. Exceptions, however, do exist, which we have discussed in the ensuing paragraph. Another issue is the instability of most Hg species, i.e., their propensity to undergo changes in their physical states and chemical compositions[23] as temperature and pressure conditions continuously alter through the production system. The various chemical forms of Hg exhibit significantly different chemical and physical behavior.[22] Note that this is generally not the case when dealing with typical constituents (the hydrocarbon as well as non-hydrocarbon components such as methane through the plus fraction, CO_2, N_2, H_2S) in a reservoir fluid stream, be it single-phase vapor or liquid or these splitting into equilibrium vapor and liquid; i.e., they remain stable throughout and do not undergo any chemical changes or breakdowns at any stage of oilfield production systems. This also poses a particular challenge in the phase behavior modeling of reservoir fluids containing Hg. Therefore, model prediction accuracy and reliability needs to be established by comparing with empirical (we interpret that as actual measured) data.[28]

The general lack of systematic and controlled experimental data especially the saturation pressure, compositional, and fluid properties, etc., for systems containing Hg; i.e., the type that is fairly routinely measured on reservoir fluids against which EOS models are tested, is the primary reason for the insufficiency of adequate thermodynamic models. Although a large amount of experimental data on properties of Hg (e.g., vapor pressure, density) as well as solubility of Hg in pure hydrocarbons exist, the open literature does not contain any data on hydrocarbon mixtures, which is really what is needed for developing reliable models for the type of applications of importance in the upstream sector. Time, cost, toxicity of Hg, and the difficulty in anticipating the wide range of process conditions may be some of the reasons for lack of data. In two of the recent publications,[29,30] good predictive capability of

EOS models is demonstrated; however, this is for Hg vapor pressure, density, diffusion coefficients, and solubility in pure hydrocarbons (normal pentane and above). Khalifa and Lue[30] rightly state that predicting the solubility of Hg in liquids and gases (interpreted as pure systems as they have done) gives an indication of inter-phase Hg pathways or mass transfer. Therefore, it is possible that these EOS models[29,30] may be extended to the type of predictive capability required in the upstream operations but that appears to be somewhat stifled by the lack of systematic data we alluded to earlier, although some progress appears to have been made in this regard.[30]

Polishuk et al.[29] state that several companies have developed simulators for modeling the phase equilibria of Hg and its compounds in oil and gas systems that is based on largely proprietary underlying assumptions and the development methodology, which also includes the use of certain binary interaction parameters. Even much of the experimental data on Hg partitioning in multicomponent systems are kept as classified and not published in the open literature. For example, see Gompos[31], in which model predictions for multicomponent systems are compared with confidential experimental data and the information presented very implicitly. Lam et al.[32] implemented a Unified Thermodynamics with a single fluid model, named as Multiflash™ Cubic Plus Association (MF-CPA) that can handle all the fluid as well as solid phases (hydrates, waxes, and asphaltenes) and Hg partitioning. In their modeling (case) studies, they demonstrated the superiority of their approach over the traditional EOS models that are ubiquitous in every PVT simulator. The former partitioned the Hg in hydrocarbon vapor, condensate as well as the aqueous phase, while the latter erroneously partitioned all Hg to the aqueous phase. The Hg partitioning as predicted by the Multiflash™ vs. time for a 10-year period can thus be construed as a "forecast," presumably based on some type of history match or tuning of the model to observed data. Nevertheless, what is actually necessary is a (unified) platform such as this that is capable of accepting input in the form of feed molar composition, including Hg, which is flashed to certain processing conditions, and the thermodynamic model would then predict the partitioning of Hg in the newly formed equilibrium phases. Wilhelm et al.[28] have stated that it is feasible to make better predictions on the presence of Hg in appreciable quantities in reservoirs and geologists, based on analogs, can predict the likely range of reservoir Hg concentrations with reasonable accuracy. The Hg concentration so predicted or perhaps based on collected in situ fluid samples can thus be the part of the reservoir fluid composition ensemble for Hg partitioning predictions. A flash calculation (case study) example for a natural gas type mixture containing water and Hg at a fixed temperature and pressure is shown in the user guide for Multiflash™ for Windows.[33] Note that the Hg mole fraction in this example is $\sim 3.5 \times 10^{-9}$, which is of similar order of magnitude as shown in Table 7.1, is shown to split or partition into the equilibrium vapor, liquid phases, and the aqueous phase. However, the relative precision of Hg of such near zero mole fractions, as compared to other components such as methane, etc., into similarly small values into the newly formed phases can only be verified by comparison with controlled experimental data gathered from laboratory scale experiments to determine the solubility of Hg in well-defined hydrocarbon fluids. Although data collected from field measurements can fill some of these gaps and are valuable, the systems may not be in full equilibrium, fluctuate with time and thus subject to considerable uncertainties.[34] As an example, the Hg partitioning case study that is shown in the Multiflash™ user guide is plotted in Figure 7.2 in terms of equilibrium ratios or partitioning coefficients $\left(K_i = \dfrac{Y_i}{X_i}; \text{where } i = 1 \text{ to } N \text{ components}\right)$ to show the Hg distribution (along with other components) in the vapor and liquid phases. As seen in the figure, a relatively higher value of the partitioning coefficient indicates a reasonably higher presence of Hg in the vapor phase, which is attributed to its volatility. Similar levels of elemental Hg equilibrium ratios based on a number of field observations also appear in Edmonds et al.[34]

7.3 Formation of Solid CO_2 and Effect of CO_2 on Paraffin Wax

In Figure 7.1a, b, examples of reservoir fluids containing very high concentration of CO_2 were presented. Gas fields containing such high CO_2 concentrations are termed as low quality and many remain as stranded or are

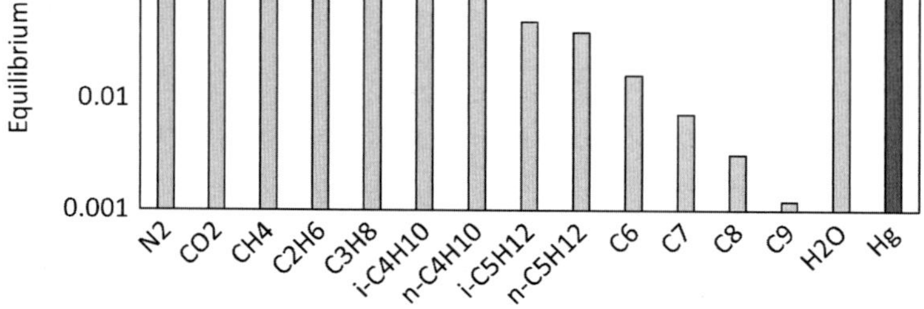

FIGURE 7.2
Equilibrium ratios of various components and Hg based on the flash calculations carried out on a natural gas type mixture containing ~3.5 × 10^{-9} mole fraction Hg. Plot constructed from tabulated data presented in the Multiflash™ user guide.[33]

considered poor from a commercial standpoint, a case in point is Malaysia where more than 13+ trillion standard cubic feet (TSCF) of gas reserves remain undeveloped.[35] Despite the high CO_2 content such fields are in need of development due to increasing demands. Obviously, in order to get a saleable product, i.e., a sweetened gas or oil, the CO_2 must be removed and its concentration reduced to acceptable specifications. Cryogenic-based separation[36] or removal of CO_2 has been identified as one of the most suitable technologies for such high CO_2 gas streams since traditional methods may not be sustainable. However, the use of low temperatures in itself may give rise to the dropout of solid CO_2 or dry ice or CO_2 freeze out leading to blocked equipment and other operational issues.[37] The formation of solid CO_2 has been attributed[38] to its relatively high triple point (–70.2°F), i.e., when all three phases are in equilibrium. The plugging problems caused by solid CO_2 formation are akin to gas hydrates in natural gas production wells. Another potential area where solid CO_2 may form is due to the rapid depressurization of natural gas streams containing high CO_2 when significant Joule–Thomson cooling may occur. Therefore, vapor–liquid–solid equilibria (VLSE), typically at very low temperatures, of systems comprised of CO_2 and the usual hydrocarbons components assumes great importance in the cryogenic processing. Of particular importance are the safe operating conditions that may be determined from such type of equilibria calculations.

Although binary systems comprised of CO_2 and various alkanes have been experimentally studied extensively as well as successfully modeled,[39] measured data on multicomponent systems, needed for verifying the efficacy of thermodynamic models, are quite limiting. The most amount of experimental data on the VLSE that is published in the literature dating back to the 1950s are on the CO_2–CH_4 binary system; perhaps because CH_4 being the lightest alkane is the primary constituent in a natural gas stream. The phase behavior of such systems in which solid CO_2 may form has been discussed in more detail by Hlavinka et al.[40]; however, some of the fundamentals of the VLSE which extend beyond the typical phase envelopes/behavior known and/or of importance to reservoir engineers is presented in Figure 7.3 to demonstrate the importance of the solid CO_2 formation. At least, for completeness, Figure 7.3 shows the interplay of phase envelopes/behavior in conjunction with the formation of the solid CO_2 phase; a more thorough discussion on this topic is beyond the scope of this book. Note that the lower end of the phase envelope in the cryogenic region is generally of no significance to reservoir engineers from a typical reservoir fluid phase behavior perspective because reservoir temperatures are usually much higher; however, in this case when dealing with solid CO_2, this region is of importance from a processing/separation standpoint. Before describing the total phase behavior of a binary system made up of CO_2 and CH_4, let us first consider Gibb's phase rule. The degrees of freedom, F, for a binary system are $F = 4 - P$, which means that two (bivariant) and one

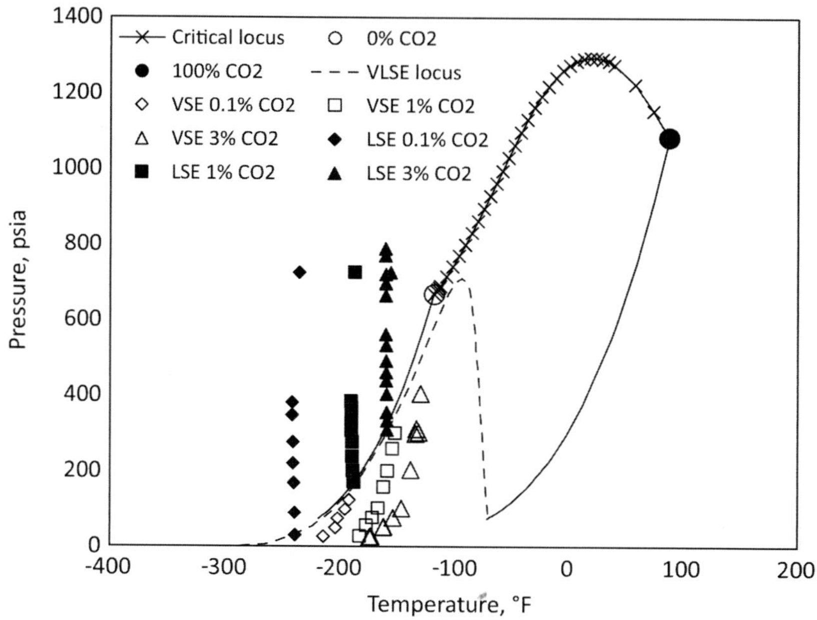

FIGURE 7.3
Composite phase behavior for a CH_4–CO_2 binary system. The curves ending with open and solid circles represent the vapor pressure curves of pure CH_4 and CO_2, respectively. The critical locus represents the collection of critical points of all the phase envelopes of binary systems of CH_4 and CO_2. The VLSE locus (dashed line) is PT values digitized from the curve presented in Davis et al.[41] The other open and solid symbols represent vapor–solid and liquid–solid equilibria, i.e., VSE and LSE, respectively, for binaries having overall CO_2 of 0.1, 1 and 3 mol%. The experimental VSE and LSE data are from Agarwal and Laverman,[42] ZareNezhad,[43] and Pikaar,[44] whereas the 3% LSE data (except at 725 psia) are predictions from Hlavinka et al.[40]

(univariant) degrees of freedom are required for the binary to be defined fully for VLE and VLSE, respectively. For VLE what this means is regardless of the overall composition of the binary, the equilibrium vapor and liquid phase compositions in the two-phase region will always be the same because temperature and pressure are the only two independent variables required to be fixed. In principle, similar to the phase envelope, the VLSE region also is depicted by a locus of PT points. However, it should be noted that although the placement of phase envelopes will depend on the overall composition, the PT position of VLSE phase envelope or locus is fixed because the system is univariant,[41] i.e., if the temperature is fixed the pressure as well as the vapor and liquid phase compositions also are fixed[40] (see the three and two-phase envelope intersections in Figure 7.3). The unique VLSE locus will, however, not be the case in multicomponent mixtures because the degrees of freedom will be more than one. Let us now examine some specifics of Figure 7.3. As far as the VLE region for this binary is concerned, it is bounded by the vapor pressure curves of the lightest and the heaviest components, i.e., CH_4 and CO_2, respectively, in this case, which means that all phase envelopes belonging to various binaries represented by certain mol% of either CH_4 or CO_2 will be within these two vapor pressure curves and obviously the respective critical points will form the critical locus that starts and ends with CH_4 and CO_2 critical pressure and temperature. The VLSE locus shown in Figure 7.3 represents the equilibria between vapor–liquid–solid phases. The open symbols enclosed by the VLSE curve denote the equilibria between vapor and solid for the stated CO_2mol% in the overall mixture, i.e., 0.1, 1, etc., which are also referred to as "frost" points. The solid symbols on the left of the CH_4 vapor pressure and the VLSE curve (some lying on it) denote the equilibria between liquid and solid for the stated overall CO_2mol%. Since the LSE points are on the left of the CH_4 vapor pressure curve and the critical locus the state of any binary would be in a liquid phase (from a conventional phase behavior perspective), and thus with the cryogenic level temperatures resulting in an equilibria between liquid and solid (along the line called as "freezing" marked by these respective points). This means that pressure–temperature points on the right indicate a 100% liquid phase state. As far as the VSE points are concerned, two possibilities exist, i.e., (1) if one were to generate a phase envelope for a mixture containing only 0.1% CO_2 then it would be so narrow (almost like the CH_4 vapor pressure curve) that VSE data points would denote a single phase vapor, thus simply representing the VSE or (2) if the system were to be in a two-phase then the liquid created would be frozen,[40] i.e., a vapor–solid equilibria. Also, note that by virtue of the phase rule, for a binary system, regardless of the overall composition, the two-phase (liquid and vapor) compositions will always be the same. Various publications[36,38–40] claim good EOS prediction capabilities of these intricate phase equilibria calculations that suggest the safe operating regions as far as staying away from solid CO_2 or dry ice is concerned.

CO_2 given its superior solvent properties has the capability to extract heavier hydrocarbon components (precipitating as paraffin wax) up to C_{30}[45] and thus has a positive effect on the wax appearance temperature (WAT), and is considered as a flow improver.[45] Hosseinipour et al.[45] reported 4°C–9°C reduction in the WAT's of Malaysian and Sudanese oils. Pan et al.[46] also concluded that the experimental and predicted results provide evidence of significant decrease in WAT (referred to as cloud point temperature or CPT; an alternative term) on dissolving CO_2 (as well as light hydrocarbons) in crude oils. The CPT's at different pressures for the original oil (oil 4) are ~2°C–2.4°C higher than the one containing 30% CO_2, again indicating a positive effect on the CPT or WAT; although the impact of adding same % of lighter alkanes (methane, propane, and pentane) is much more pronounced. Modeling results of Pedrosa[47] also demonstrate the lowering of WAT due to CO_2; however, there is a certain threshold concentration of CO_2 up to which the beneficial effects are observed.

7.4 Compositional Characteristics of Shale Gas and Shale Oil vs. Conventional Reservoir Fluids

According to the US EIA[48] statistics, United States is the top commercial producer of shale gas and shale (referred to as tight oil) oil (not the same as oil shales). In fact, the split between the daily non-shale and shale-based gas and oil production appears to be nearly 50–50, thus indicating the significance of the latter. In the US, the Marcellus region has contributed the most as far as shale gas is concerned, whereas a large portion of shale oil has originated from the Eagle Ford and the Bakken regions[48], respectively. Therefore, we consider strictly the compositional characteristics of the fluids produced from these shale plays and compare them with the "benchmark" or conventional or some typical reservoir fluids; the reader should refer to other publications as far as the shale gas/oil vs. conventional reservoir fluids' geological aspects or characteristics are concerned.

Figure 7.4 compares the molar composition of gases from the three shale plays, namely Antrim, Barnett, and New Albany with gases from three conventional systems, namely Mobile Bay, Hugoton, and Dog Lake. Note that multiple compositions are available for the shale plays; however, only one is shown for each to do a quick comparison. As seen in Figure 7.4, nothing unusual is observed as far as gases from shale plays are concerned. As a

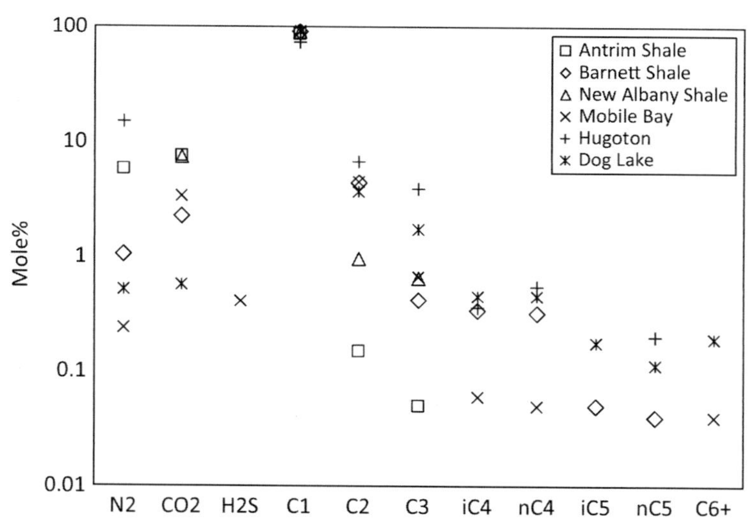

FIGURE 7.4
Comparison of molar compositions of gases from shale and non-shale plays. The Antrim, Barnett, and New Albany shale data are from Martini et al.,[49] Hill et al.,[50] and Martini et al.,[51] respectively. The Mobile Bay data are from Mankiewicz et al.,[52] whereas the Hugoton and Dog Lake data are as reported by Sloan.[53]

matter of fact, gases from both shale and non-shale exhibit very similar compositional characteristics (see the interspersed data points), which are typical of dry gases, i.e., they all contain methane of the order of 90% and contain a plus fraction or heavy end (C_{6+}) from nil to negligible. It should be noted that it is the plus fraction, and not the well-defined components, that typically imparts the "uniqueness" or diversity to every reservoir fluid, i.e., methane present in any fluid has the same properties but that is not the case with a plus fraction, other than just labeling it as such.

Next we compare the compositional characteristics of oils originating from the shale plays with some benchmark or conventional oils. On a ternary diagram in Figure 7.5, the molar compositions of the oils from the shale plays (Bakken and Eagle Ford volatile oil and gas condensate) are compared with conventional fluids from the Brent reservoir (note that this is not a blend), oil and condensate from Southwest Texas (these liquids are perhaps part of the West Texas Intermediate, WTI blend). Strictly speaking, from a compositional standpoint there is nothing that stands out as

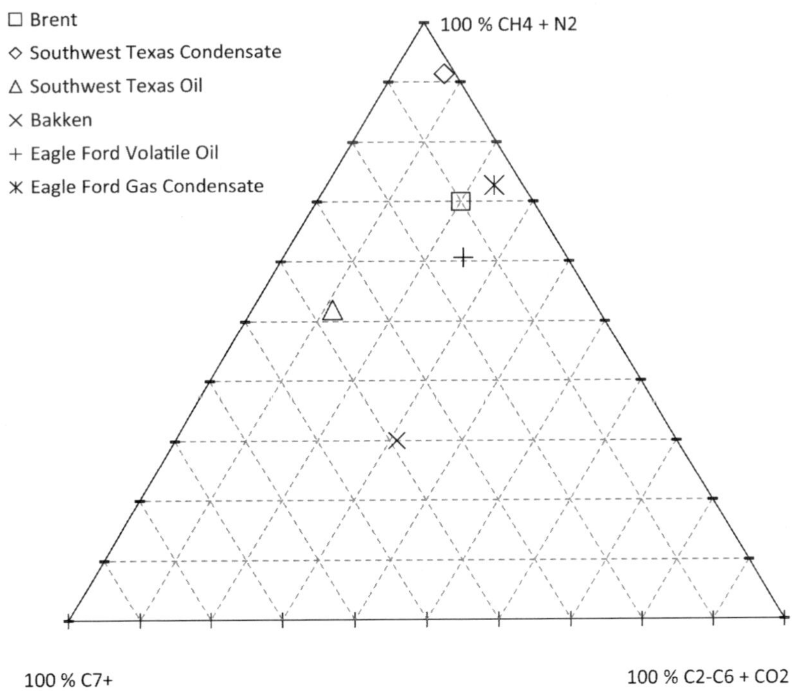

FIGURE 7.5
Comparison of molar compositions of three conventional reservoir fluids and three shale-based fluids. The ternary diagram is constructed based on the data for Brent and the Southwest Texas oil and condensate from Bath et al.[54] and Hoffmann et al.,[55] whereas Bakken and Eagle Ford data are from Adekunle and Hoffman[56] and Gherabati et al.[57]

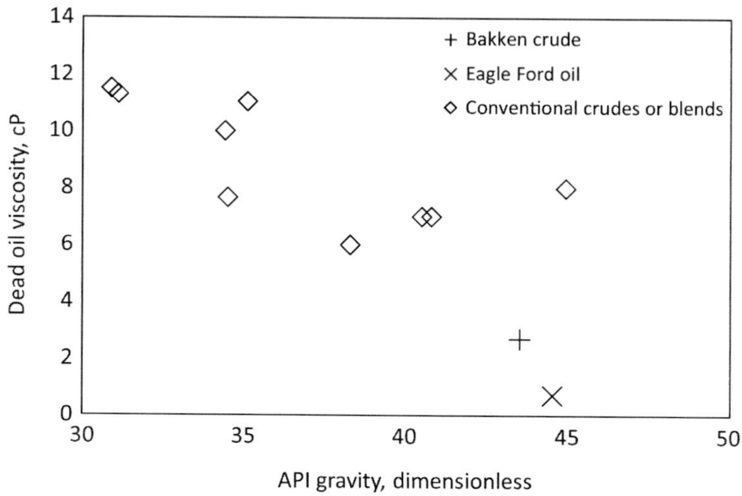

FIGURE 7.6
Comparison of API gravities and dead oil viscosities of shale-based fluids and conventional crudes or blends (Brent blend; WTI; Alaska North Slope; Danish North Sea; Forcados blend; Kirkuk blend; Murban; Oseberg blend and Tapis blend). The data for Bakken crude and Eagle Ford oil are from Intertek report[58] and Whitson and Sunjerga,[59] whereas the rest are from the Environmental Science and Technology Centre of Canada database.[60]

far the shale-based fluids are concerned. Figure 7.6 shows the API gravity [API Gravity or $°\text{API} = \dfrac{141.5}{\gamma_o} - 131.5$, where γ_o is the specific gravity of oil or oil density relative to water density at 60°F and atmospheric pressure], and viscosity comparison of the stock tank liquids originating from the Bakken, Eagle Ford, and many other conventional crudes or blends. Similar to the molar compositional distribution, the API gravity and viscosity of both the shale-based fluids also fit in the usual ranges of conventional fluids and seems to follow the low API—high viscosity or high API—low viscosity trend. As a matter of fact, both the Bakken crude and the Eagle Ford oil appear to be much lighter (high API gravity) compared to the conventional fluids.

Similar to that was mentioned earlier, the well-defined components such as methane—(even) hexane or the single carbon number C_6 (assumed as normal hexane) have the same properties such as the critical temperature, critical pressure, and acentric factor, regardless of the reservoir fluid they are present in; however, that is not the case with the heavy end (single carbon number or SCN fractions and the plus fraction or residue). Although some similarities (or are assumed as such if generalized properties are used) may exist as far SCNs (C_7, C_8, C_9, etc.) are concerned, which is certainly not the case with the plus fraction, for example, C_{10+} or C_{20+}. As a matter of fact, appreciable variations in the plus fraction properties can be observed even within the same fluid types. Therefore, even though the plus fraction is merely labeled as such, for example, C_{10+}, its properties (and mole fraction) differ from fluid to fluid and are the principle distinguishing factor in reservoir fluids. It is due to the increasing chemical complexity (paraffin–naphthene–aromatic distribution and the number of isomers) with the carbon number that generally gives rise to the differences in properties of the SCNs and the plus fraction in each reservoir fluid. Although it is quite common to use the extended reservoir fluid compositions up to a plus fraction of C_{20+} in numerical simulations, a somewhat standard practice or format used in reporting is to limit the description up to C_{7+}, along with its molecular weight and specific gravity, which are used in characterizing the plus fraction. The following balance equations govern the characterization procedure:

$$Z_{7+} = \sum_{n=7}^{N+} Z_n \tag{7.1}$$

$$Z_{7+}MW_{7+} = \sum_{n=7}^{N+} Z_n MW_n \tag{7.2}$$

$$\frac{Z_{7+}MW_{7+}}{\gamma_{7+}} = \sum_{n=7}^{N+} \frac{Z_n MW_n}{\gamma_n} \tag{7.3}$$

where Z_{7+}, MW_{7+} and γ_{7+} are the mole fraction, molecular weight, and the specific gravity of the C_{7+} fraction (typically experimental values), whereas the summations on the right-hand side signify splitting of the C_{7+} fraction into a certain number of SCNs and an extended plus fraction. For example, if the chosen extended plus fraction is 20+, then $n = C_7$ through C_{19} and $N+ = C_{20+}$. Mole fraction splitting of the C_{7+} can be achieved via any of the methods presented in Pedersen et al.,[61] following which generalized molecular weight and specific gravity can be assigned to the SCNs and then the extended plus fraction mole fraction, molecular weight, and the specific gravity determined by the above balance equations. Note that these balance equations can also be written for a C_{6+} fraction by simply replacing 7 with 6.

Therefore, given the traditional approach in reporting reservoir fluid compositions, a somewhat closer examination of the liquids produced from the unconventional and conventional systems can be achieved by comparing the C_{7+} characteristics. Figure 7.7 presents a cross plot of C_{7+} molecular weight and specific gravity. Overall, although the correlation between the two parameters is not very strong, which is to be expected, most data points follow a trend of increasing specific gravity with increasing molecular weight; higher molecular weight or specific gravity simply means a relatively heavier heavy end. In this case as well the data for the Eagle Ford condensates pretty much falls within the range of values for conventional fluids. Note that for the Eagle

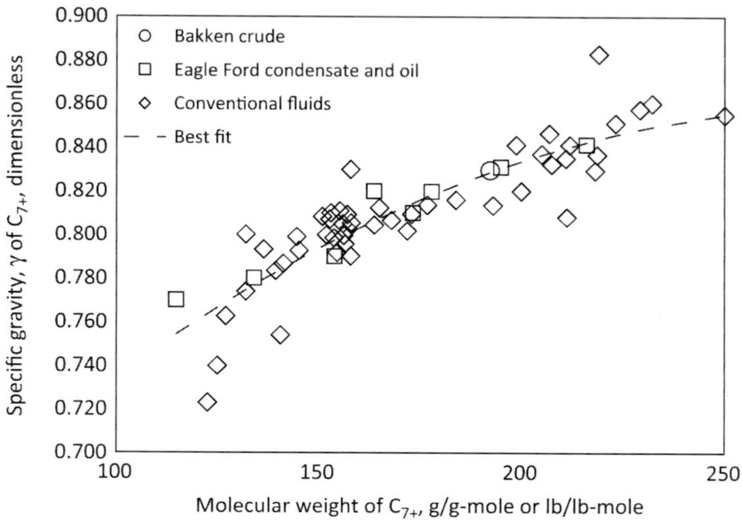

FIGURE 7.7
Crossplot of C_{7+} molecular and specific gravity of shale-based fluids and conventional fluids. The data for Eagle Ford condensate and oil are from Gherabati et al.[57] and Whitson and Sunjerga[59], respectively, whereas the data for Bakken crude are from Nojabaei.[62] The data on conventional fluids are from a large database for which the references are cited in Dandekar.[63]

Ford oil (plausible oil composition as stated by Whitson and Sunjerga[59]) and Bakken crude only molecular weight data has been reported from which the specific gravities have been estimated according to the best fit; however, this appears to correlate quite well with the rest of the data.

Taking a closer look at the SCN fraction' (sometimes referred to as True Boiling Point or TBP fractions) molecular weight and specific gravity data, Figures 7.8a, b, respectively, compares these values for the Bakken and Eagle Ford fluids with some typical fluids. As far as the Eagle Ford is concerned it appears that the data (together with the critical properties and acentric factor) is regressed because Whitson and Sunjerga[59] state "plausible usage for Eagle Ford reservoir fluids." The dead oils are typically distilled according to a prescribed boiling point range and TBP fractions are collected and labeled as C_7, C_8, C_9, etc., while the distillation residue is termed as C_{20+} or C_{30+} depending on how long the distillation is continued from a boiling range standpoint. The average properties of the collected SCN or TBP fractions are measured (shown in these two plots), which are generally used for estimating[65] the critical properties and acentric factor for use in EOS modeling. Also shown in these plots are the generalized values that are typically used as default when such measured data are unavailable; note that generalized critical properties and acentric factor values also are available for use in EOS modeling. Clearly, the comparison shown in Figure 7.8a, b also does not point to anything unusual (the only exception being the Bakken C_{13} specific gravity, for an unknown

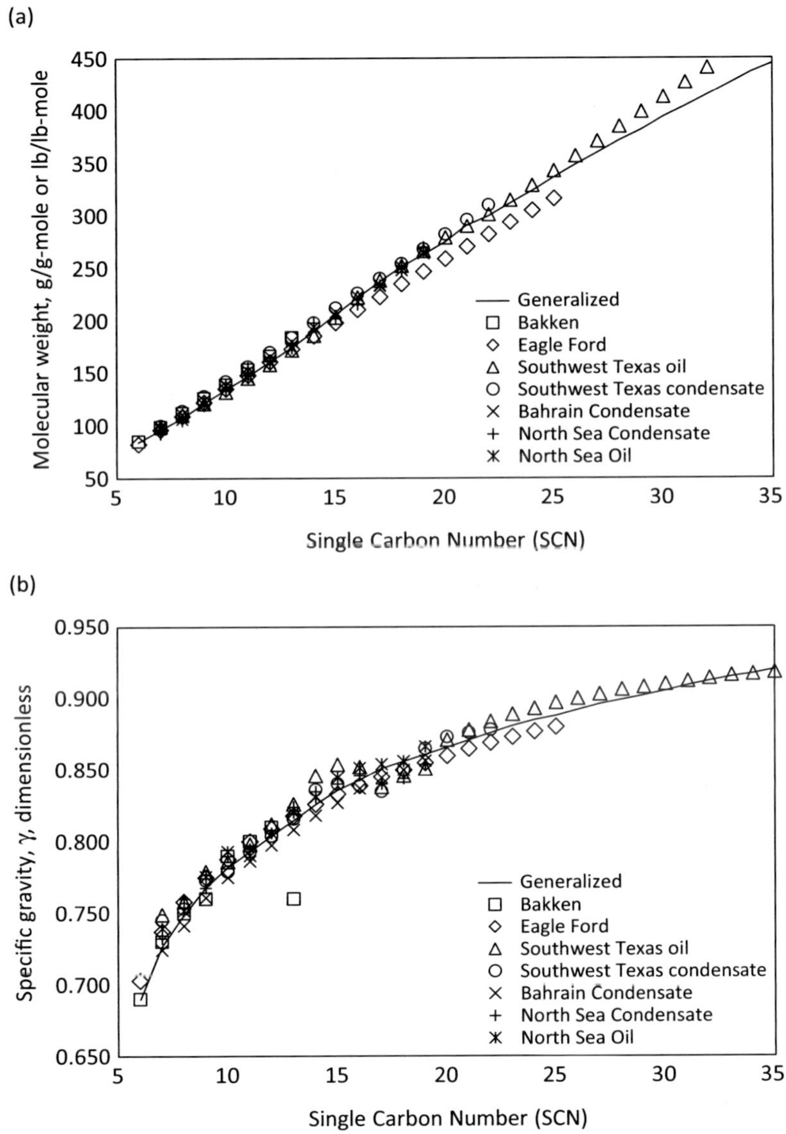

FIGURE 7.8
(a) Comparison of SCN fraction molecular weights of Bakken,[58] Eagle Ford,[59] fluids and various typical conventional fluids from Southwest Texas,[55] Bahrain,[64] and the North Sea.[61] Also shown on the plot are reference or generalized[61] values, which are used as default when actual measured data is unavailable. Note: the superscripts indicate the references from which the data has been taken to construct this plot. (b) Comparison of SCN fraction specific gravities of Bakken,[58] Eagle Ford,[59] fluids and various typical conventional fluids from Southwest Texas,[55] Bahrain,[64] and the North Sea.[61] Also shown on the plot are reference or generalized[61] values, which are used as default when actual measured data is unavailable. Note: the superscripts indicate the references from which the data have been taken to construct this plot.

reason) as far as shale-based fluids are concerned as the data seem to align quite well with the conventional fluids as well as the generalized values.

In summary, the comparative plots and corresponding discussion above provide sufficient evidence that inherently the shale-based fluids do not possess any atypical compositional characteristics, but rather are quite similar to conventional fluids. The primary difference in the way these fluids behave in situ, however, stems from the fact the shale-based fluids reside in a very low permeability (nano darcy range) porous media as opposed to orders of magnitude higher intrinsic permeability porous media from which conventional fluids are produced. It is due to such ultra-low permeabilities the shale-based fluids are produced at a much lower rate than a conventional fluid in a conventional porous media, at the same drawdown.[59]

7.5 Fluid Property and PVT Characteristics of Fluids Containing Large Proportions of Non-hydrocarbons and Shale-Based Fluids vs. Conventional Fluids

7.5.1 Fluid Phase Behavior of Acid Gases and Their Mixtures with Hydrocarbons

In this section, we present and discuss a variety of experimental data on mixtures of (1) pure acid gases, CO_2, and H_2S; (2) CO_2, H_2S, and normal alkanes; (3) high CO_2 and/or H_2S content reservoir fluids. These type of data, as presented, are of significant value in evaluating the accuracy and reliability of EOS models.

Figure 7.9 depicts the overall phase behavior of the CO_2 and H_2S binary system, which is constructed based on experimental data from three different sources ranging from the 1950s to 2001.[66–68] As seen, for the most part the data appear to be consistent with the exception of the critical points as reported by Stouffer et al.[68]—potentially due to the errors associated in interpolating/fitting[69] (see Figure 7.9 caption). Qualitatively, the overall phase behavior shows the characteristics of what is seen in a typical binary mixture. Note though, however, that the critical locus is somewhat flat because of the small difference between the critical pressure and critical temperature of CO_2 and H_2S, which is less than 250 psia and 125°F, respectively. In cases of binary mixtures made up of markedly different critical properties (one component much lighter than the other or larger differences in molecular sizes), such as, even CH_4 and nC_4H_{10}, the critical locus tends to be dome shaped, or stretched upwards in the direction of higher pressure. This happens simply because relatively higher pressure is necessary to keep the lighter component "dissolved" in the heavier component or the mixture existing in the single phase when it passes the critical locus.

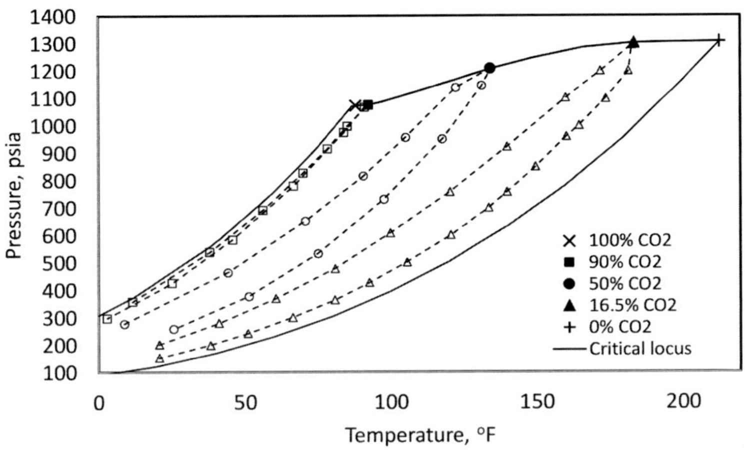

FIGURE 7.9
Overall phase behavior of the CO_2–H_2S binary system constructed from experimental data reported in three different sources. PT points along the entire critical locus (100% CO_2–100% H_2S with a spacing of ~10 mol%) is experimental data from Bierlein and Kay,[66] whereas the 16.5 mol% CO_2 (rest being H_2S) phase envelope is based on data reported by Sobocinski and Kurata.[67] The precise mol% reported for the other two-phase envelopes is 50.01 and 90.45, respectively, for which the data are from Stouffer et al.[68] Note that these two-phase envelopes are forced to merge with the data of Bierlein and Kay,[66] which actually is for slightly different mol% of 47.3 and 90.09, respectively. The critical points reported by Stouffer et al. for their mixtures overshoot the critical locus of Bierlein and Kay; attributed to probable errors given the way they were indirectly determined as discussed by Chapoy et al.[69]

For any binary system by virtue of the Gibb's phase rule, only two degrees of freedom or independent variables are necessary for the mixture to exist in two phases ($F = 2 - 2 + 2 = 2$); which are temperature and pressure. This means regardless of the overall composition of the mixture the equilibrium vapor and liquid phases in the two-phase region (within the space bounded by the two vapor pressure curves and the critical locus) will have the same mole fractions and the densities at a given PT point. Therefore, data for a certain selected CO_2–H_2S binary are presented in Figures 7.10 and 7.11, respectively. In Figure 7.10 (first and the second plot), mole fractions of CO_2 and H_2S in the equilibrium vapor and liquid phases are shown as a function of pressure at a temperature of 68°F. The data from all three sources appear to fall on the same common best-fit trendline, indicating consistency in the measurements and thus establishing the accuracy and reliability of the data. This particular compositional behavior in the two-phase region has been elucidated by Dandekar,[8] and those explanations qualitatively apply in the case of data shown in Figure 7.10 as well. Figure 7.11 shows the plot of equilibrium vapor and liquid phase densities for the CO_2 and H_2S binary system at 68°F. Bierlein and Kay[66] reported the experimental molar volumes at various pressures and

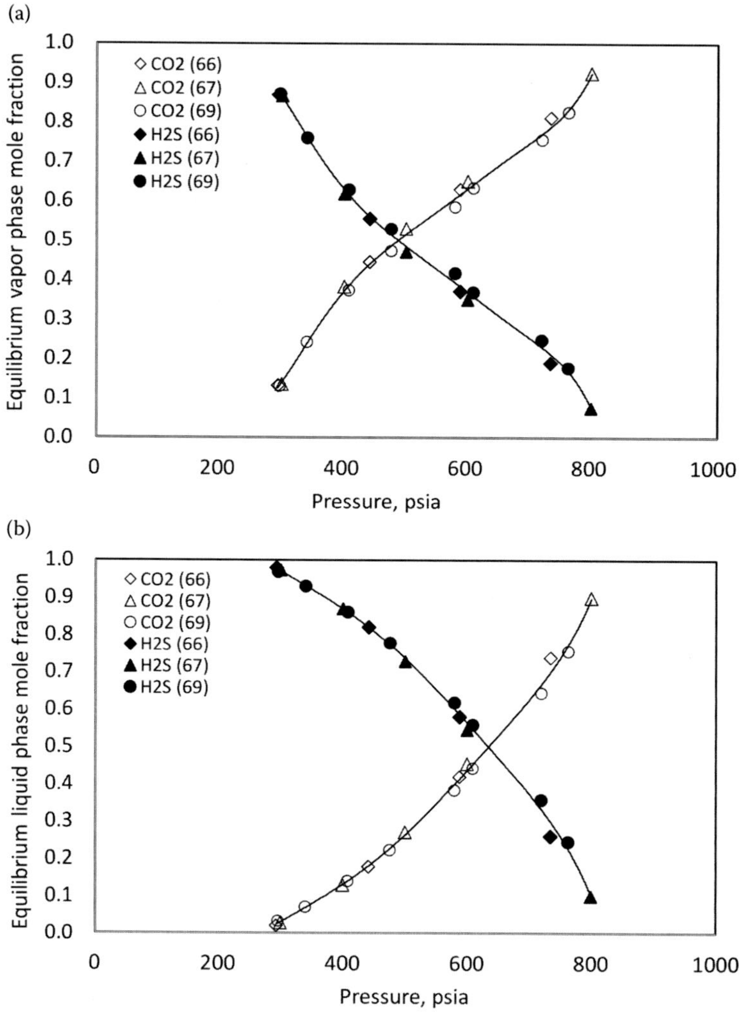

FIGURE 7.10
Mole fractions of CO_2 and H_2S in the equilibrium vapor (first plot) and liquid phases (second plot) at 68°F. Numbers in parentheses in the legend are the references from which the experimental data is taken to construct these plots.

temperatures for this system, which were converted to densities in g/cm³ using the compositional data of Bierlein and Kay[66] as well as Sobocinski and Kurata[67]: $\text{density in g/cm}^3 = \left(\dfrac{1}{\text{molar volume}}\right) \times \text{mixture molecular weight}$ calculated from compositions. The behavior as displayed in Figure 7.11 is typical of what is expected in the two-phase region, i.e., density of the

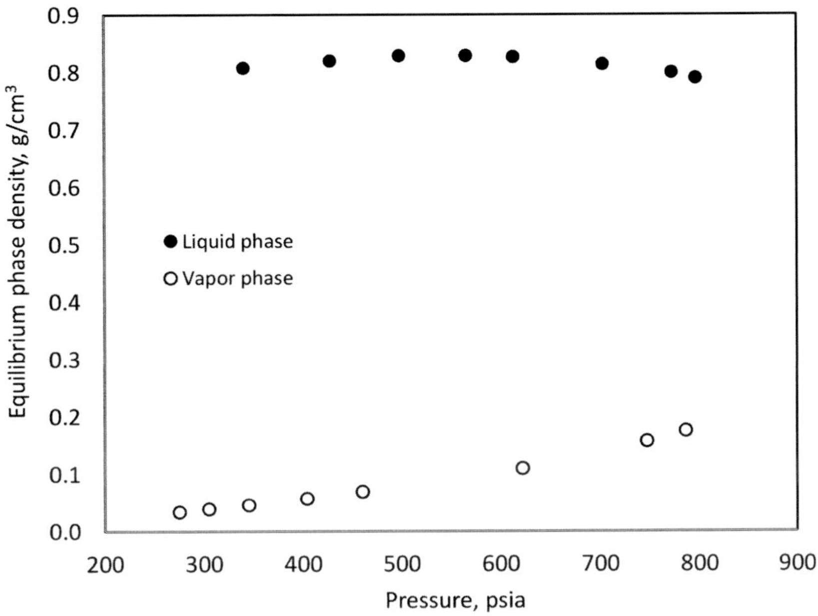

FIGURE 7.11
Densities of the equilibrium vapor and liquid phases for the CO_2 and H_2S binary system at 68°F. The plot is constructed based on the experimental data on saturated molar vapor and liquid volumes, two-phase compositional data by Bierlein and Kay[66] and the two-phase compositional data of Sobocinski and Kurata.[67]

equilibrium liquid phase increases despite the decrease in pressure (dominant compositional effect over-riding the pressure), whereas the density of the equilibrium vapor phase decreases with pressure (pressure effect dominant compared to compositions). Note though that perhaps due to the interpolated compositions, the equilibrium liquid phase densities' increase (expected) with decreasing pressure somewhat dissipates in the lower pressure region. Dandekar[8] has discussed the equilibrium vapor and liquid phase density characteristics in detail to which the reader is referred. The final data presented for the CO_2 and H_2S binary system are the compressibility factors, which are plotted vs. the ratio of pseudoreduced pressure and pseudoreduced temperature (P_{pr}/T_{pr}) in Figure 7.12. The data presented in Bierlein and Kay[66] and Stouffer et al.[68], respectively, on which Figure 7.12 is based were converted for presenting as follows. The molar volumes (V/n), given pressures, and temperatures in Bierlein and Kay[66] were directly used to compute the Z factors from the real gas equation $Z = \frac{P}{RT}\frac{V}{n}$ (using consistent set of units). However, Stouffer et al.[68] directly reported the Z factors for the tested compositions and pressures and temperatures. P_{pr} and T_{pr} were calculated from the given pressures

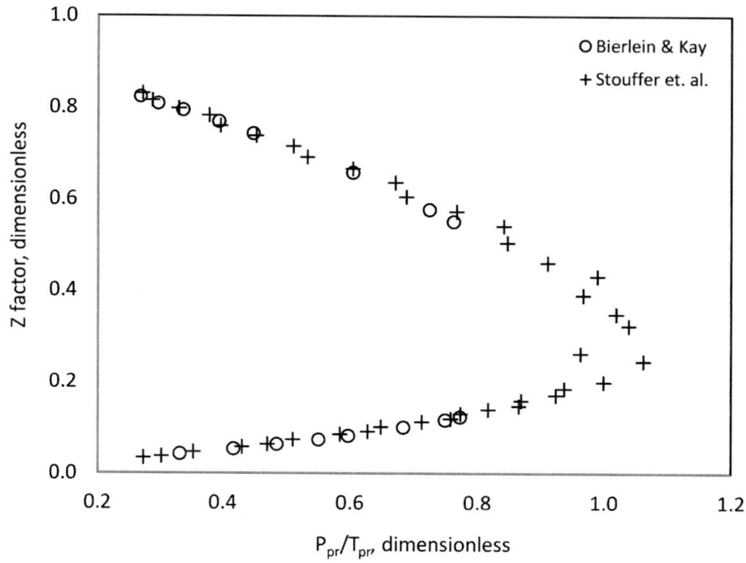

FIGURE 7.12
CO_2 and H_2S binary system compressibility (Z) factors plotted as a function of pseudoreduced pressure and pseudoreduced temperature ratio (P_{pr}/T_{pr}). The plot is constructed based on experimental data on molar volumes, compositions, given pressure and temperature, directly reported Z factors and the critical constants from Bierlein and Kay[66] and Stouffer et al.,[68] respectively.

(P) and temperatures (T), compositions (X or Y_{CO_2}, X or Y_{H_2S}), and critical constants: $\left(P_{cCO_2}, P_{cH_2S}, T_{cCO_2}\ T_{cH_2S}\right)$; $P_{pr} = \dfrac{P}{X\,or\,Y_{CO_2} \times Pc_{CO_2} + X\,\text{or}\,Y_{H_2S} \times Pc_{H_2S}}$ and $T_{pr} = \dfrac{T}{X\,or\,Y_{CO_2} \times Tc_{CO_2} + X\,\text{or}\,Y_{H_2S} \times Tc_{H_2S}}$. Note that corrections were not applied to the molar mixing rules. As seen in Figure 7.12, the data from both sources aligns very well showing the consistency.

Given the prominence of CO_2, a large amount of reliable experimental phase behavior data for the CO_2 and normal alkanes systems, especially binaries, exist—this database and the numerous associated references have been tabulated by Vitu et al.[70] As a matter of fact, the phase behavior of CO_2 and crude oil systems has also been experimentally studied in more details from the standpoint of CO_2-based EOR applications (e.g., see Gardner et al.[71]; Orr et al.[72]; Shelton and Yarborough[73]). Note though that in such applications, the oil intrinsically does not contain high CO_2 but the overall mol% of CO_2 may become as high as 60%–70% in the total mixture later during the injection process. The binary systems made up of CO_2 and alkanes (depending on the carbon number) exhibit rather intricate and somewhat non-trivial phase behavior, especially as it pertains to liquid–liquid (LL) immiscibility, which also is seen in the case of CO_2 and crude oil systems, identified[71–73] almost four decades ago. Given the availability of the relatively recently published experimental data and their analysis for the CO_2–alkane binary systems, the purpose of this discussion accompanied by the supporting figures is to illuminate the phase behavior of well-defined CO_2 binaries and then draw parallels with CO_2–crude oil systems, which are covered in detail in Chapter 9.

In phase behavior terminology, six different types of phase behaviors have been classified by Van Konynenburg and Scott[74] for binary systems, and elaborated (for generic or generalized or non-specific systems) in detail by Sadus.[75] The reason binary systems are categorized accordingly because they (in particular the critical locus) are bounded by the vapor pressure curves of the lightest and the heaviest component, respectively. Out of these six phase behaviors, the simplest phase behavior is Type I in which the critical locus begins with the critical point of the lightest component and ends with the critical point of the heaviest component and no other facets exist. Type II also is very similar to Type I; however, these binaries exhibit an LL critical line which rapidly rises to high pressures into an area outside the critical locus, after intersecting the vapor pressure curve of the lightest component, from what is known as "upper critical end point" (UCEP).[75] The UCEP appears on the top of the three-phase liquid–liquid–vapor (LLV) equilibria line inside the critical locus. Sadus[75] notes that solid phase formation at very low temperatures often obscures the difference between Types I and II.

From the standpoint of a generalized description of Type III, the critical locus or the LV line from the critical point of the heaviest component usually extends part of the way to the critical point of the lightest component, but then veers off sharply to high pressures.[75] The UCEP in the case of Type III also appears on the top of the three-phase LLV equilibria line (within the two-phase region); however, unlike Type II, in this case there is continuity between the critical point of the lightest component and the UCEP, i.e., the UCEP is the end of the discontinuous critical locus. Depending on the position of the main critical locus, various subclasses of Type III also exist. Often "VV immiscibility" is observed when the main critical locus extends to temperatures greater than the critical temperature of the heaviest component. The three-phase LLV region can also be located either above or below the vapor pressure curve of the lightest component.

There are a number of similarities between Types IV and V; both have LLV lines placed within the two-phase region. There is a typical round shape critical locus with discontinuities in the vicinity of vapor pressure curve of the lightest component, and both have the UCEP and lower critical end point (LCEP). The critical locus from the critical point of the heaviest component ends at the LCEP, whereas the critical locus from the lightest component terminates at the UCEP. In the case of Type IV systems, there is also an LL critical line at low temperatures that are somewhat higher than the critical temperature of the lightest component, which begins with the "lower" UCEP (top of the LLV line within the two-phase region) and extends into the high pressure regions—note that this particular phenomenon is absent in Type V systems.[75]

Van Konynenburg and Scott,[74] Fall and Luks,[76] and Vitu et al.[70] have grouped the various CO_2–normal alkane binaries into different phase behavior types, which are presented in the form of a table (see Table 7.2). With the exception of *n*-Tridecane (n-$C_{13}H_{28}$), all other alkanes group into a certain phase behavior type. According to Van Konynenburg and Scott,[74] the CO_2–*n*-Tridecane binary represents a Type II to Type III and it presents a Type IV behavior, which is alternatively stated by Vitu et al.[70] as "the transitions from Type II to Type IV and from Type IV to Type III which experimentally occur when the carbon atom number of the alkane lies, respectively, between 12 and 13 and between 13 and 14," i.e., although the carbon number between 12 and 13 or 13 and 14 is hypothetical (but can be considered from the molecular weight standpoint if the pseudo binary concept is applied); from a phase

TABLE 7.2

Grouping of Various CO_2–Normal Alkane Binary Systems and CO_2–Normal Alkane Mixtures According to Phase Behavior Types

n-Alkane or Their Mixtures	Molecular Weight(s), lbm/lbm-mol	Phase Behavior Type
Methane through *n*-Hexane	16 (CH_4)–86 (n-C_6H_{14})	Type I
n-Heptane through *n*-Dodecane	100 (n-C_7H_{16})–170 (n-$C_{12}H_{26}$)	Type II
n-Tridecane	184	Type II to Type III transition; presenting a Type IV behavior (see corresponding text)
n-Tetradecane through *n*-Docosane and higher	198 (n-$C_{14}H_{30}$)–310 (n-$C_{22}H_{46}$), and higher	Type III
Methane and *n*-Butane[77]	25.4–25.9	Type I
n-Butane, *n*-Pentane, *n*-Hexane, and *n*-Heptane[78]	67.9–79.1	Type I; also stated as such by the authors based on the critical locus measurements
n-Decane and *n*-Undecane[79]	146.7	Type II Based on molecular weight
n-Undecane and *n*-Dodecane[79]	163.9	Type II Based on molecular weight

The superscripts for normal alkane mixtures represent the literature references.

behavior standpoint, *n*-Tridecane is bracketed as a Type IV system. Using a pseudo-binary concept, grouping is also presented for mixtures of normal alkanes, i.e., CO_2 is treated as a discrete component, whereas rest of the components are recombined as one pseudo component for which the molecular weights are calculated to place them according to the well-defined CO_2–normal alkane binary system phase behavior types.

Figure 7.13 shows the critical loci for the various Type I CO_2–normal alkane binaries. Obviously, CH_4 being lighter than CO_2, the vapor pressure curve is placed on the left, whereas C_2H_6 being closer to CO_2, but slightly heavier, is on the right. With increasing dissimilarity between CO_2 and the normal alkane, the critical locus stretches upward as well as spreads on the right. Note that Vitu et al.[70] show a predicted LLV, UCEP, and the LL region for the CO_2–CH_4 binary (actually making it a Type II system); however, Cismondi et al.[81] have stated that the UCEP is "virtual" because of the interference of the solid–fluid transition, given the very low temperatures. The CO_2–C_2H_6 binary is unique among the collection shown in Figure 7.13, i.e., the system has two critical pressures at all temperatures between ~62°F and the critical temperatures of CO_2 and C_2H_6, which are very close: a peculiarity that has been ascribed to azeotropy.

A plot similar to Figure 7.13 is shown in Figure 7.14 treating the multicomponent mixtures containing CO_2 as pseudo binaries. Two mixtures M1 and

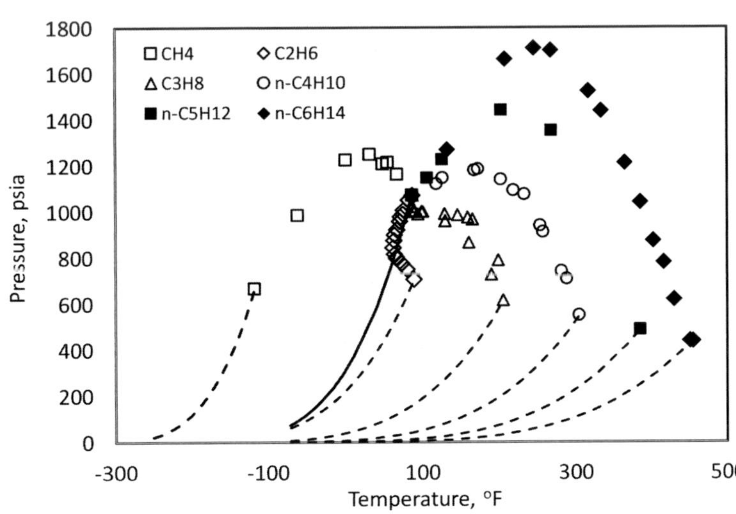

FIGURE 7.13
Critical loci of CO_2 normal alkane binary systems of Type I. Markers are experimental data, solid curve is CO_2 vapor pressure, and broken curves from left to right are vapor pressures of methane through *n*-hexane. Note that the markers are continuity points on the critical locus formed by the respective binaries. CO_2–CH_4; CO_2–C_3H_8; CO_2–*n*-C_4H_{10}, and CO_2–*n*-C_5H_{12} data are from various sources listed in Hicks and Young,[80] whereas CO_2–C_2H_6 and CO_2–*n*-hexane data are from sources listed in Vitu et al.[70] and Cismondi et al., respectively.[81]

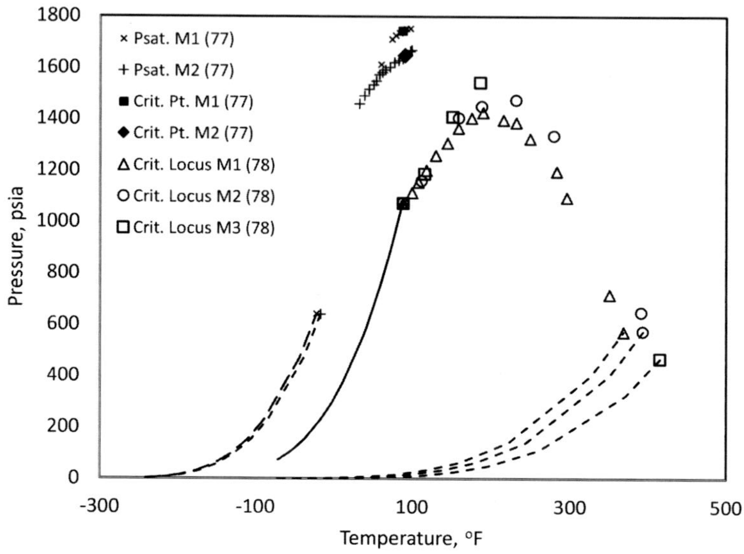

FIGURE 7.14
Critical loci of CO_2 and five different pseudo binary systems of Type I. Numbers in parentheses in the legend are the references from which the pertinent experimental (all markers) data is taken to construct these plots. Normalized mole fractions of M1[77] are $CH_4 = 0.78$ and $nC_4H_{10} = 0.22$; M2[77] are $CH_4 = 0.76$ and $nC_4H_{10} = 0.24$. Normalized mole fractions of the quaternary mixtures M1[78] are $nC_4H_{10} = 0.489$, $nC_5H_{12} = 0.349$, $nC_6H_{14} = 0.129$, $nC_7H_{16} = 0.033$; M2[78] are $nC_4H_{10} = 0.35$, $nC_5H_{12} = 0.40$, $nC_6H_{14} = 0.15$, $nC_7H_{16} = 0.10$ and M3[78] are 0.25 each. Broken curves from left to right are pseudo vapor pressure curves of M1–M2,[77] M1–M3[78] (in that order), whereas the solid curve is vapor pressure of CO_2.

M2 (Pan et al.[77]) are made up of methane and *n*-butane for which both the bubble and dew point and interpolated critical points were reported. Using the normalized mole fractions of CH_4 and n-C_4H_{10}, pseudo critical temperature, pressure, and acentric factor were calculated from which the pseudo vapor pressures were estimated using the Lee–Kesler[82] equations in order to represent the system as a pseudo binary. A somewhat similar approach was adopted for the three quaternary mixtures (M1–M3), composed of n-C_4H_{10}–n-C_7H_{16}, that were tested by Qiu et al.[78] Their measurements included the critical points for all three mixtures with CO_2, as well as the critical points of the hydrocarbon portions of M1 and M2. Again, using the normalized mole fractions of n-C_4H_{10}–n-C_7H_{16}, the pseudo acentric factors were calculated for all three mixtures and pseudo-critical temperature and pressure for M3; these values were then employed to compute the pseudo-vapor pressure curves to represent the system as a pseudo binary. As seen in Figure 7.14, at least mixtures M1–M3 appear to superimpose or fit well (critical loci as well as the pseudo-vapor pressures) within the various heavier true binaries shown in Figure 7.13, i.e., displaying the typical characteristics of Type I systems (also refer to molecular weight cut-offs shown in Table 7.2).

Shariati et al.[79] tested several multicomponent systems containing CO_2 in which the CO_2 mole fraction was nearly constant at 0.25. Two of the systems they tested were made up of n-$C_{10}H_{22}$ and n-$C_{11}H_{24}$ (denoted as M1) and $nC_{11}H_{24}$ and $nC_{12}H_{26}$ (denoted as M2), respectively. Their experimental data included several bubble-point measurements for these systems. As per the molecular weight classification proposed above in Table 7.2, these systems would fall into Type II category of phase behavior. Although other experimental data were unavailable for these systems, a composite plot has been prepared, as shown in Figure 7.15. Again, using the pseudo binary concept the pseudo vapor pressures of the two mixtures were calculated by the Lee–Kesler equations. Clearly, M2 being heavier than M1 has its vapor pressure curve on the far right. Similar observations can be made for the predicted phase envelopes of M1 and M2 (placed on the far right with high cricondentherm and critical temperature). Note though that since M2 contains slightly higher mole fraction of CO_2, the low-temperature end of the bubble-point curve is placed a little on the left of M1 curve and somewhat closer to the vapor pressure of CO_2. Also, given the somewhat similar overall compositional characteristics, the bubble-point curves for both systems on the left of the cricondenbar almost overlap each other resulting in essentially same

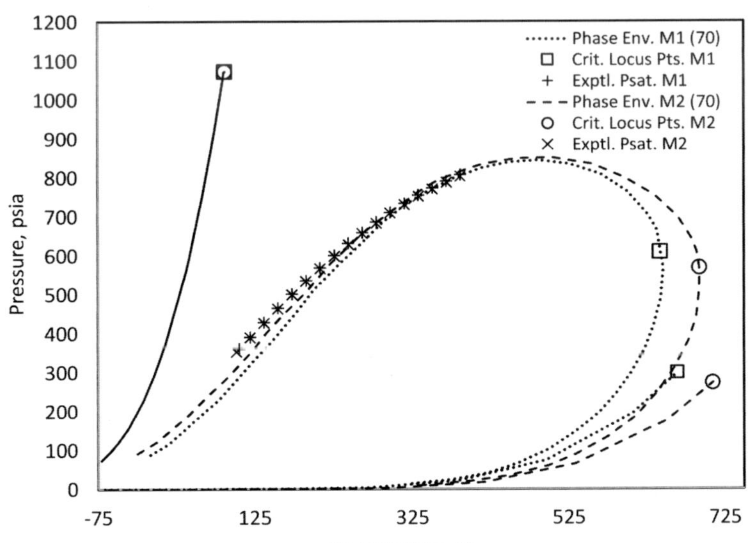

FIGURE 7.15
Pseudo critical locus points of CO_2 and two different pseudo binary systems of Type II. Plus and Cross markers represent the experimental bubble points of two of the mixtures tested by Shariati et al.[79] Normalized mole fractions of M1[79] are $nC_{10}H_{22} = 0.68$ and $nC_{11}H_{24} = 0.32$; M2[79] are $nC_{11}H_{24} = 0.45$ and $nC_{12}H_{26} = 0.55$. Dashed curves on the extreme right ending with a square and circle are pseudo vapor pressures of M1 and M2, respectively, whereas the solid curve is that of CO_2. The phase envelopes of M1 and M2 (including the critical point) are digitized EOS prediction values from Vitu et al.[70]

bubble-point pressures, which is more prominent as far as the experimental data are concerned.

Figure 7.16 shows the critical loci for the various Type I H_2S–normal alkane binaries. Obviously, both CH_4 and C_2H_6 being lighter than H_2S, the vapor pressure curve is placed on the left of H_2S. Usually, the increasing critical temperature is accompanied by a corresponding decrease in the critical pressure; however, H_2S is an exception; for example, the critical temperature is slightly higher than that of C_3H_8 but on the contrary the critical pressure is nearly double that of C_3H_8, placing the C_3H_8 vapor pressure curve on the right of H_2S. Additionally, given the nearly similar critical temperatures of H_2S and propane but doubled critical pressures, the critical locus tends to be somewhat vertical with an inward curvature; note that something quite similar is also the case with a CO_2–C_2H_6 system (see Figure 7.13). With increasing difference between the critical temperature of H_2S and the binary pair, the critical locus begins to take on a dome shape (such as in H_2S–nC_4H_{10}), and eventually displaying the typical characteristics as seen in the H_2S and nC_5H_{12} and $nC_{10}H_{22}$ binaries. As far as the H_2S–CH_4 binary system is concerned, between a certain overall mole fraction of H_2S (0.458–0.110), at least based on the mixtures studied by Kohn and Kurata,[84] the critical points are not "visible" and the iso-vols tend to creep upward into relatively higher

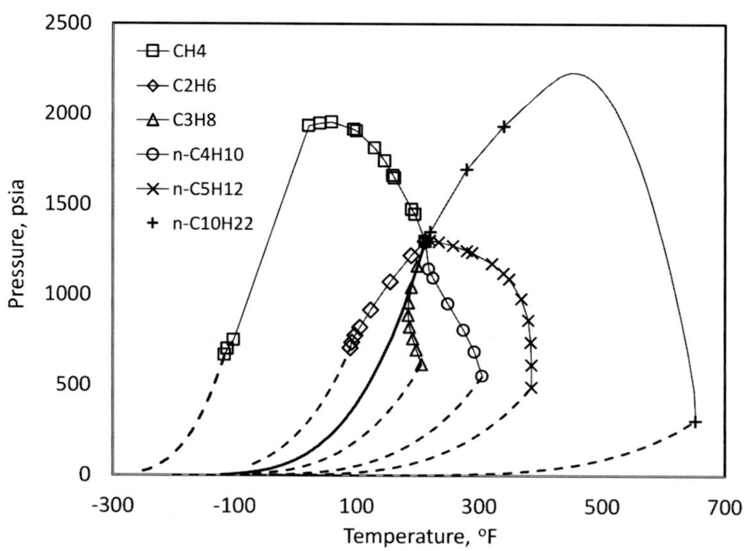

FIGURE 7.16
Critical loci of H_2S normal alkane binary systems of Type I. Markers are experimental data, solid curve is H_2S vapor pressure and broken curves from left to right are vapor pressures of methane through *n*-decane. Note that the markers are continuity points on the critical locus formed by the respective binaries. H_2S–CH_4[83,84]; H_2S–C_2H_6[85]; H_2S–C_3H_8[86]; H_2S–n-C_4H_{10}[87]; H_2S–n-C_5H_{12}[88] and H_2S–n-$C_{10}H_{22}$[89] data is from various sources which are identified by the respective superscripts.

pressures, thus leaving the phase envelopes somewhat open ended in the low-temperature region. This behavior (not shown in Figure 7.16, but basically the gap in the markers in the H_2S–CH_4 binary critical locus) is seen in ~ –40°F to –100°F range, which is not of much significance as far as reservoir engineering applications are concerned but may be of importance in cryogenic processing of natural gases. Similar to the CO_2–hydrocarbon systems, two liquid phases also are observed in the case of H_2S. For example, Kohn and Kurata[84] report on the presence of an H_2S-rich and methane-rich liquid phase, which appears to be mostly confined to very low (negative) temperatures unlike CO_2–hydrocarbon systems.

Nichita et al.[90] presented multiphase equilibria calculation for two synthetic hydrocarbon systems containing appreciable amounts of H_2S, namely a quaternary and a six-component mixture comprised of ~42 and 11 mol% H_2S, respectively. Both these mixtures contained CO_2 as well of roughly an order of magnitude lower mol%. At least based on the multiphase flash calculations, they reported the formation of two liquid phases (H_2S rich and hydrocarbon rich) in equilibrium. Again, somewhat similar to Kohn and Kurata,[84] these two liquid phases exist in equilibrium at very low temperatures (–101°F to –144°F, respectively) and relatively low pressures (~218–616 psia), which contrasts with the behavior exhibited by CO_2 hydrocarbon mixtures, occurring at temperatures that are of practical significance from an EOR standpoint. In order to compare the equilibrium ratios or partitioning potential of H_2S with CO_2 into the two equilibrium liquid phases, a crossplot based on the predicted values of Nichita et al.[90] for the two synthetic systems is presented in Figure 7.17. A comparison of CO_2 and H_2S equilibrium ratios

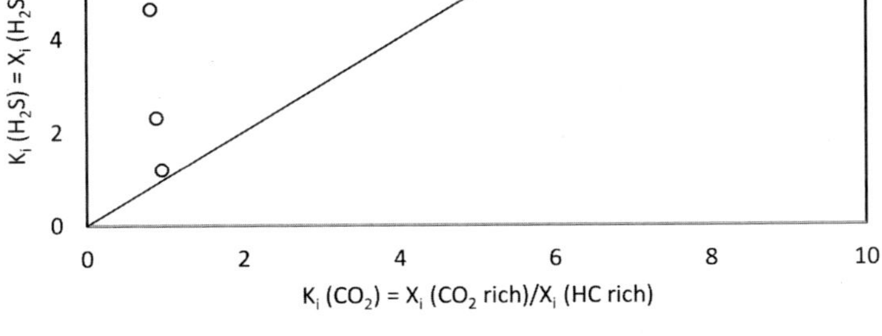

FIGURE 7.17
Comparison of equilibrium ratios of H_2S and CO_2 in the H_2S and CO_2 rich and hydrocarbon rich (HC) phases for two synthetic systems. The K_i values are calculated and the crossplot constructed based on the three-phase flash calculations reported by Nichita et al.[90]

shown in Figure 7.17 indicates a nearly equal partitioning (K_i values close to one) between the CO_2 and hydrocarbon rich liquid phases, respectively. However, that is not the case with H_2S, having rather high K_i values, which means much of it is concentrated in the H_2S-rich liquid phase, thus reducing the amount of hydrocarbon components present, i.e., the phase being literally defined as such. Besides the synthetic systems, to the best of our knowledge, there are no documented studies that report on the formation of a second liquid phase for H_2S–crude oil mixtures.

In contrast with the vapor–(H_2S rich) liquid–(hydrocarbon rich) liquid equilibria, a simple vapor–liquid equilibrium, given the ethane/propane like H_2S characteristics, however, shows that H_2S partitions almost equally between the vapor and liquid phases. In order to demonstrate this near equal partitioning of H_2S, Figure 7.18 shows a crossplot of H_2S mole fractions in the vapor and liquid phases, respectively, that are in equilibrium at various temperature and pressure conditions for two crude oils and one gas condensate system. The overall H_2S concentration in one of the oils[14] and the gas condensate[91] is almost the same—34.8 and 34 mol%, respectively, whereas in the other oil[92] it is 5 mol%. Note that although from the standpoint of plotting the H_2S vapor and liquid mole fractions, the oil from Jacoby and Rzasa[92] is treated as one, in actuality the data are for three different mixtures that are blends of the same natural gas, absorber oil, and crude oil in varying proportions but with the same constant mol% of H_2S as well as CO_2. As seen in Figure 7.18, with some exceptions, the mole fractions seem to line-up around the 45° line indicating a fairly equal distribution of H_2S in the vapor and liquid phases.

Although the production of oil with high CO_2 is not a rarity given the use of CO_2 as an injection gas in EOR processes or gas floods, reservoirs fluids that inherently contain unusually high CO_2 may pose unique production challenges or even provide some interesting opportunities. In the El Trapial field in Argentina, the high CO_2 is not due to flooding but is naturally present, ranging from 45% to 75% of the composition of the associated gas.[12] Despite the high CO_2 content, the authors[12] do not report anything unusual in terms of phase behavior, but one important consequence is the relatively high gas gravity of ~1.6, which is influenced by CO_2 properties. In this case, the fluid property challenge associated with such high CO_2 content is actually associated with the accurate prediction of gas densities required in using this gas in gas lift opeartions.[12] The accurate prediction of gas densities obviously requires accurate gas deviation or compressibility (Z) factors, which are dependent on the "corrected" pseudocritical temperature and pressure of the gas mixture. This was successfully achieved by the authors[12] by incorporating the Wichert and Aziz correction[93] and the Benedict–Webb–Rubin (BWR)[94] EOS model. As it has been pointed out by the authors, high CO_2 gas lift gas application references are scarce in the literature, but the El Trapial case marks a successful use of high CO_2 gas lift contributing to 37% of the oil produced.

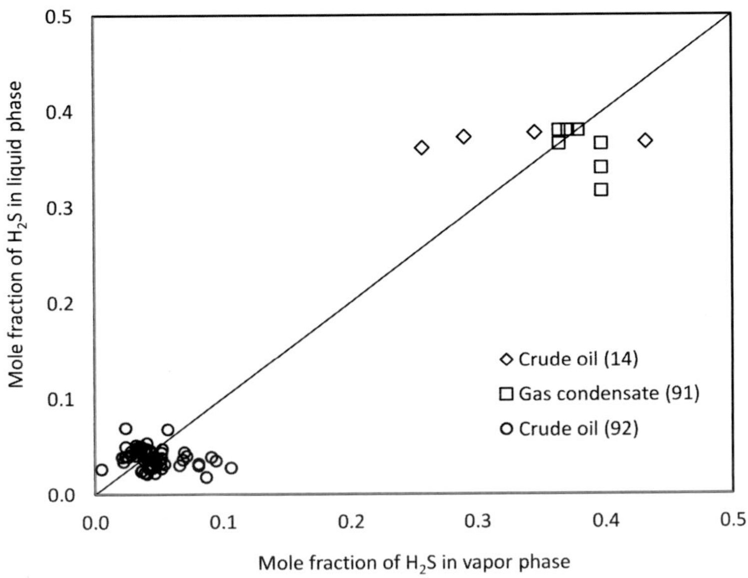

FIGURE 7.18
Partitioning of H_2S in the equilibrium vapor and liquid phases for two crude oils and a gas condensate. The plot is constructed on the basis of data reported in the references that are indicated in the legend parentheses.

As far as high CO_2 content reservoir gases are concerned, the main issue is accurate and reliable estimation of the *Z* factors because they are integral part of equations that are used in calculating the gas density, gas formation volume factor, coefficient of isothermal compressibility, and viscosity. In most cases, the *Z* factor is estimated using various correlations or EOS models (for example, see previous paragraph). These correlations or EOS models make use of pseudoreduced temperatures and pseudoreduced pressures, which obviously require pseudocritical temperatures and pseudoreduced pressures—typically calculated using simple molar mixing rules. Although these molar mixing rules work well for similar components, i.e., light hydrocarbons, they do not provide appropriate representation when gas mixtures contain significant amounts of non-hydrocarbons such as CO_2. Therefore, either corrections to the mixing rules or new correlations are required for accurate estimation of the *Z* factors. New correlations and new experimental data on *Z* factors of gases containing substantial amounts of CO_2 appear in[95–98] that the reader is referred to.

In the case of the gas condensate (Caroline field),[91] as pressure falls below the dew point a buildup of H_2S concentration and then staying almost constant in the retrograde condensate will take place up to a certain pressure. Subsequently, as pressure continues to fall, some revaporization of H_2S will occur resulting in a corresponding small increase in the H_2S mole fraction (and then remaining steady) in the vapor phase. However, given the nearly equal partitioning of H_2S between the vapor and condensate phases, the preceding mole fraction changes with pressure will not be very dramatic, but nevertheless, part of the revaporized H_2S together with what is originally present in the vapor phase will cause a slight increase in the H_2S mole fraction in the produced raw gas as pressure continues to decline. This discussion can be visually understood by looking at the data points that do not place precisely on the 45° line in Figure 7.18 as well as the field average H_2S concentration plot shown by Seto and Beliveau.[91] The higher H_2S concentration in the overall composition of the oil studied by Vagtborg[14] suggests that H_2S somewhat controls the phase behavior in the two-phase region; it lowers the equilibrium ratios of methane because H_2S along with similar components (ethane and propane) will also be present in the vapor phase in relatively significant amounts, thereby reducing the mole fraction of methane. The nearly equal partitioning of H_2S as well as ethane and propane in the vapor and liquid phases and the relatively lower equilibrium ratio of methane, in high H_2S content oils will perhaps keep the oil phase, below the bubble-point pressures, "lighter," which may be of benefit from a production standpoint.

7.5.2 Fluid Phase Behavior and Properties of Shale Reservoirs

Given the fact that the shale-based fluids reside and also flow in very low nano darcy permeability range rocks, their in situ fluid phase behavior with changing pressure deserves a separate discussion, because of what is typically termed as "confinement (low permeability)," which is negligible in conventional porous media and absent in bulk. For example, a fluid of given overall composition exhibiting a saturation pressure at the reservoir temperature in bulk (in a PVT cell) may not be the same under confinement. Similarly, the properties such as density, viscosity or those of reservoir engineering significance may differ with and without confinement. Therefore, an understanding of why these differences occur and their magnitude as a function of the level of confinement is of obvious importance given the prominence of unconventionals. The theoretical aspects and the mechanism of confinement effect on fluid phase behavior and properties are covered in details in Chapter 8; however, the magnitude of confinement and its corresponding effect is presented in this subsection.

Over the recent years, numerous publications have appeared in the literature on fluid phase behavior and PVT of shale reservoirs emphasizing the effect of confinement. Some of which are reviewed and graphically presented as a collection for ready reference and/or correlation. One of the first attempts to experimentally study the effect of porous media on phase behavior was made by Sigmund et al.[99] They measured the dew- and bubble-point pressures of methane–*n*-butane and methane–*n*-pentane binary systems in bulk (unpacked) and in the porous media (packed with 30–40 U.S. mesh glass bead packing). Their experimentation did not show any major differences in the dew- and bubble-point pressures given the relatively larger pore sizes created by the packing. However, their theoretical analysis indicated differences in the dew- and bubble-point pressures at high surface curvatures; mean surface curvature is given by $J = \left(\frac{1}{r_1} + \frac{1}{r_2}\right)$, where r_1 and r_2 are the principal radii of curvature, which means lower values of the pore sizes. Danesh[100] has stated that surface forces can be significant in tight pores, affecting the phase behavior of fluids; in particular when the pores are typically less than 10 nm, which is orders of magnitude smaller than the most conventional pore sizes of 20–200 µm as suggested by Tissot and Welte.[101] In shale reservoirs, the common pore size distribution ranges between ~1 and 20 nm[102] with dominant pore diameters smaller than 2 nm.[103]

The fundamental difference between the fluid phase behavior in bulk and perhaps from practical engineering standpoint in conventionally sized pores, and in nano-sized pores is suggested to be due to the intimate interaction between the fluid and the wall of the pore (kerogen).[104] The molecular interaction being somewhat random in bulk as opposed to orderly and layered in nano pores, which can cause alterations in the thermodynamic properties (critical constants) and fluid phase behavior.[104]

In order to understand the impact of confinement on fluid properties, we first look at the critical constants for pure components as these are the building blocks of petroleum reservoir fluids. Based on the published results of Pitakbunkate et al.[104] and Singh et al.,[105] plotted in Figure 7.19a, b are the pseudoreduced critical temperature and pressure, respectively, as a function of the magnitude of confinement. Note that the values appearing in both sources are based on the Grand Canonical Monte Carlo (GCMC) simulations using graphite slabs separated by a certain variable (different pore sizes) spacing representing a kerogen pore,[104] and thus do not represent any experimental data. We define the pseudoreduced critical temperature and pressure as the ratio of T_c or P_c due to confinement and in bulk. Overall, as seen in Figure 7.19a, b, higher confinement vis-à-vis smaller pore size appears to suppress the critical temperature as well as pressure with very small values for very narrow pore sizes and vice-versa, which is logical in that if the pore size is hypothetically zeroed then there would be no matter present and thus no or zero critical temperature and pressure. At least in the case of pseudoreduced critical temperature, 10 nm seems to be a threshold where the value is one. Additionally, the pseudoreduced critical temperatures vs. the magnitude of confinement for the four components aligns and correlates reasonably well on a composite curve. However, in the case of pseudoreduced critical pressure the data groups according to the sources, i.e., methane and ethane[104] and *n*-butane and *n*-octane,[105] respectively. Although for the most part the trend is qualitatively similar, i.e., increasing pseudoreduced critical pressure with increasing pore size, the threshold confinement appears to vary with the data sources, and most notably a reversal is seen in the case of *n*-butane and

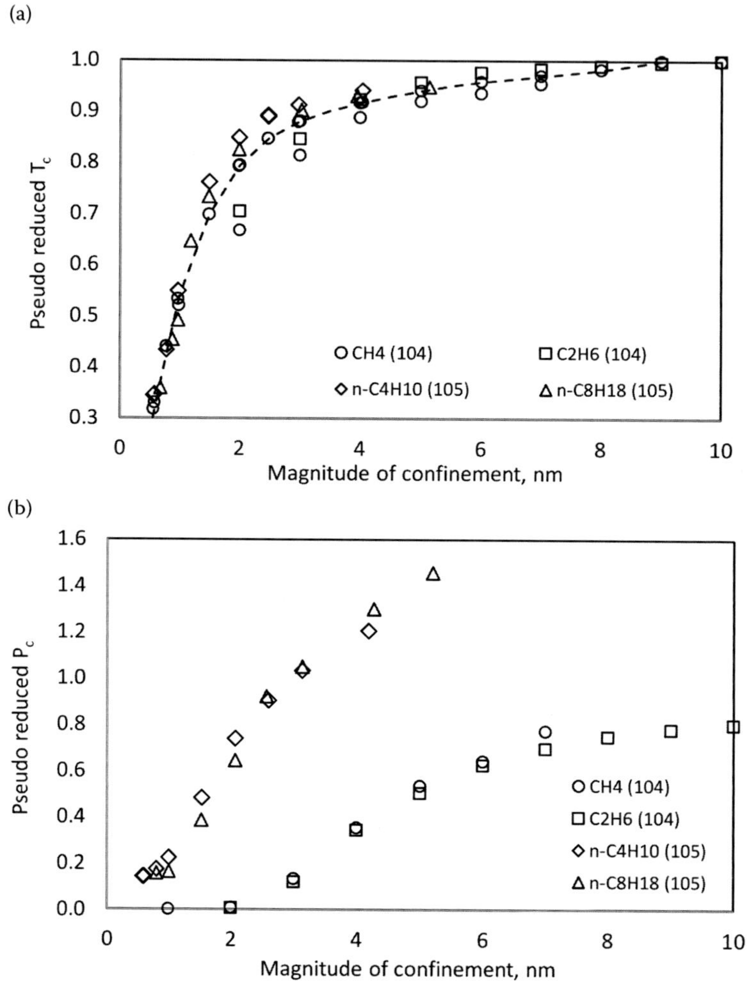

FIGURE 7.19
Pseudoreduced critical temperature (first plot) and pressure (second plot) vs. magnitude of confinement for four different pure components. Numbers in the legend parentheses are sources of data based on which these plots are constructed.

n-octane. The authors[105] have not provided any explanation for the increase in the critical pressure (or in other words values of pseudoreduced critical pressures above one) in a confinement range of ~3–5 nm.

Singh et al.[105] also presented data on the reduced orthobaric densities (along the vapor pressure curves) for both *n*-butane and *n*-octane vs. reduced temperature for two different levels of confinement, 0.8 and 4 nm, respectively, which is shown as a combined plot in Figure 7.20. Although the principle of corresponding states does not seem to apply across the different confinements, the two components do obey the principle under the same

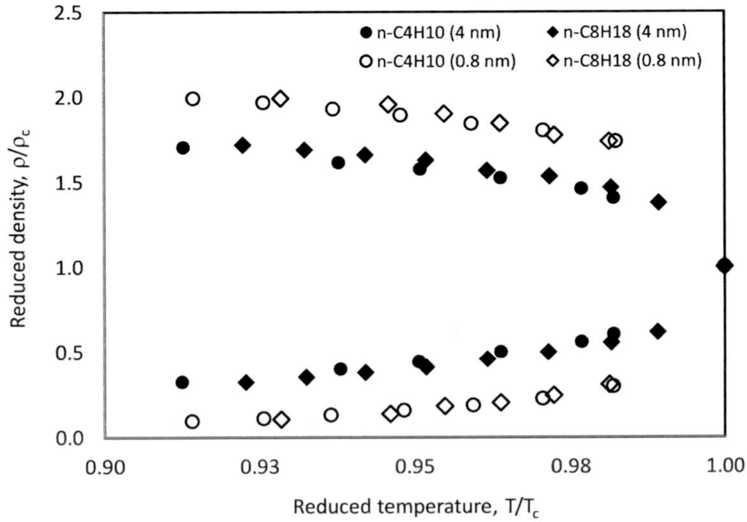

FIGURE 7.20
Corresponding states orthobaric densities as a function of reduced temperature for *n*-butane and *n*-octane in two different pore sizes. Plot is constructed based on the digitized values from Singh et al.[105]

level of confinement (data points close to each other). Pitakbunkate et al.[104] also compared methane and ethane densities as a function of pressure in bulk and under various confinements in the range of 1–10 nm at two different isotherms to represent the minimum and maximum Eagle Ford shale reservoir temperatures (both components are supercritical). All densities are higher than that in the bulk and increase with decreasing pore size, i.e., the highest values are at 1 nm confinement. The authors[104] ascribed the higher densities to the strength of the attractive potential allowing more molecules to pack the system. Finally, as far as the utility of the critical constants of pure components under confinement along with the pore size distribution (from core analyses) is concerned, these values can become the input for reservoir simulation applications.[104]

The decreasing pure component critical temperature and pressure with decreasing pore size or increasing confinement naturally translates into reduced mixture critical temperature and pressure, respectively, and as a result the phase envelope shrinks in comparison to that in the bulk. The phase envelope shrinkage in comparison to bulk is because of the fact that the bubble and dew point curves merge at the critical point. A comparison of the critical locus in bulk and under a 5-nm confinement as shown in Figure 7.21 best depicts this suppression. As shown in Pitakbunkate et al.,[104] the shrunk phase envelope under confinement conditions actually is pocketed inside the one in bulk. We show this in the form of a "subtracted" phase envelope PT points in Figure 7.22 which illustrates this effect. As seen in

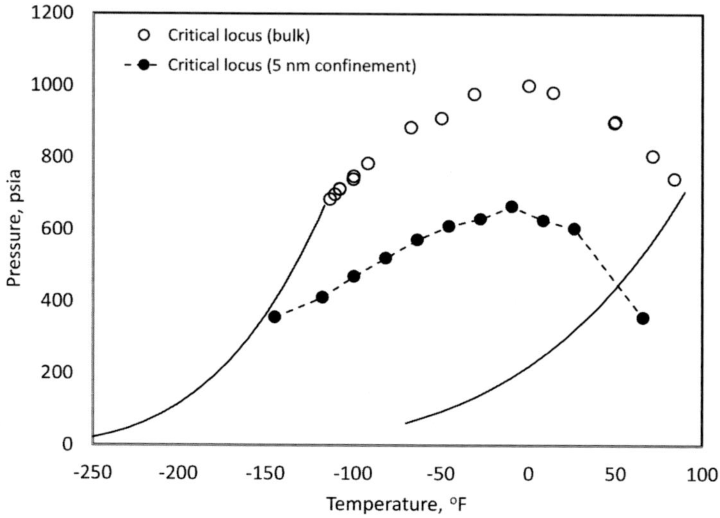

FIGURE 7.21
Comparison of methane and ethane binary system critical locus in bulk and under a 5-nm confinement. The first and the second solid curves are vapor pressures of methane and ethane, respectively. The critical locus PT points in bulk is a collection of experimental data from various sources tabulated in Hicks and Young,[80] whereas those under a 5-nm confinement are deduced from the pressure–composition plot in Pitakbunkate et al.[104]

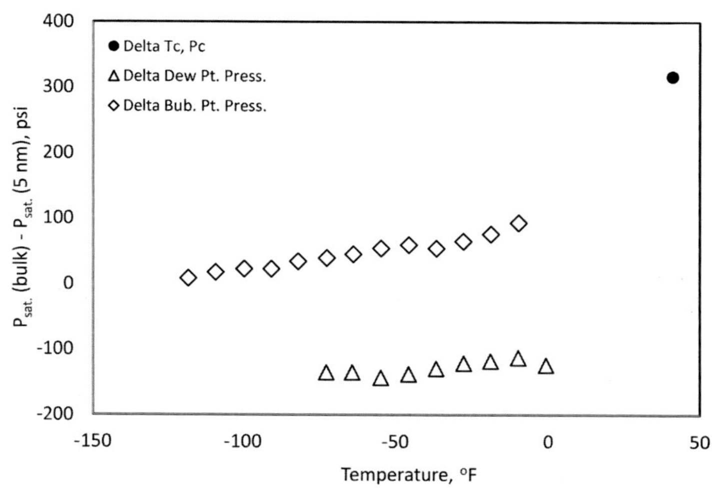

FIGURE 7.22
Subtracted phase envelope PT points for the 30.02 mol% methane and 69.98 mol% ethane binary system. The subtraction or delta or the difference between the saturation (dew or bubble) pressures is as shown by the *y*-axis title. The saturation pressures in bulk are from Bloomer et al.,[106] whereas those under the 5-nm confinement are from Pitakbunkate et al.[104] Note that the solid circle PT point represents a difference between the mixture critical temperature and pressure in bulk and under the 5-nm confinement, respectively.

Figure 7.22, the binary under confinement shows higher dew point pressures (indicated by the –ve saturation pressure difference) but mostly lower bubble-point pressures (indicated by the +ve saturation pressure difference). On studying this binary system, Pitakbunkate et al.[104] have also commented on the compositional characteristics in the two-phase region. Given the closer interaction between the pores and the heavier component (ethane in this case), the confined system as compared to bulk (or a conventional porous media for that matter) would adsorb ethane than methane.[104] This is somewhat congruent and significant with what the authors[104] have stated, i.e., in the initial stages of production, shale reservoirs tend to release lighter components first and the heavier components left behind due to adsorption. After the lighter components are completely removed the relatively heavier components will then begin to show up in the produced fluid.[104]

Nojabaei[62] also compared the phase envelopes for four different binaries made up of 70 mol% methane (fixed) and other alkanes ranging from propane to *n*-decane for a 10 nm confinement or pore size. In contrast with Pitakbunkate et al.,[104] the critical points in bulk and under confinement were shown to be the same as an anchor point and all the 10 nm pore size phase envelopes rotated slightly (implementing capillary effect) in a counter clockwise fashion thus resulting in lower bubble and higher dew point pressures; however, these differences appear to be relatively small. The same critical point in bulk and under confinement is because of the use of identical bulk critical properties used in generating the phase envelopes for the two cases; note that Pitakbunkate et al.[104] have actually used the suppressed mixture critical point in generating their 5-nm confinement phase envelope. For the methane and *n*-hexane binary, Nojabaei[62] also compared the phase envelopes in bulk and under confinement (a 20 nm pore size) for three different overall methane mol%, namely 70, 50, and 30. Qualitatively, the behavior is similar to the 10 nm pore size; however, pore size effect appears to be much more pronounced for the system with the lowest methane concentration resulting in bubble-point differences of as much as 300 psi at the lower temperatures.

Dong et al.[107] studied a quaternary mixture composed of methane, *n*-butane, *n*-pentane, and *n*-hexane in bulk and under 5- and 10-nm confinement for which they also showed higher dew point pressures as compared to bulk, i.e., a –ve saturation pressure difference as we have defined earlier. In addition to the synthetic mixture, they also showed similar differences between the bulk and confined dew point pressures for the Marcellus shale gas. As shown in Figure 7.23, the range of dew point pressure increase appears to be similar to that of Pitakbunkate et al.,[104] with the confinement effect somewhat diminishing with increasing pore size. Civan et al.[108] simulated the behavior of a shale gas/condensate fluid represented by a synthetic ternary mixture composed of methane, *n*-butane, and (supposedly) *n*-octane. Using the altered critical constants (as suggested in Singh et al.[105]) for the mixture constituents, they generated the phase envelopes under 2-, 4-, and 5-nm confinement and

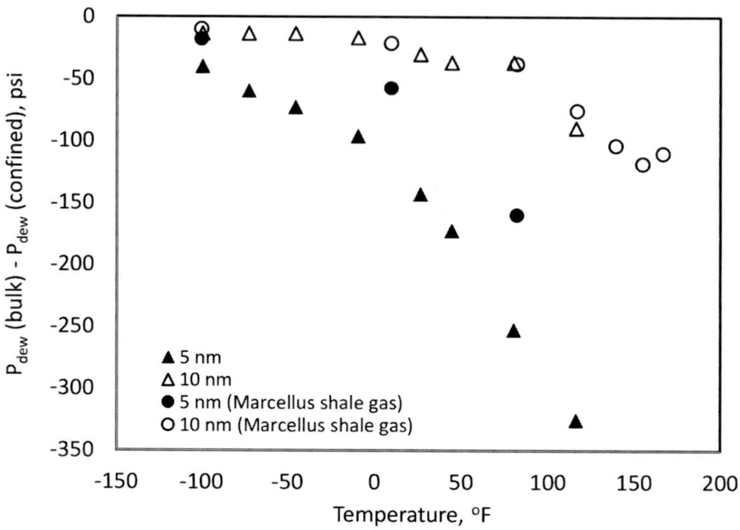

FIGURE 7.23
Differences between the dew point pressures in bulk and under -5 and 10-nm confinement for a quaternary system composed of methane, *n*-butane, *n*-pentane, and *n*-hexane, and the Marcellus shale gas. Plot constructed based on the values reported in Dong et al.[107]

compared that with the bulk fluid phase envelope. Although a suppressed or shrunk and pocketed phase envelope for the 2-nm confinement compared with bulk was observed, the low-temperature parts of the 4- and 5-nm confinement phase envelopes were above that of the one in bulk; the mixture critical temperatures under all confinements less than that in bulk but the 4- and 5-nm critical pressures higher than that in bulk. The higher critical pressures are supposedly due to the reversal in critical pressures as noted in Figure 7.19b, for *n*-butane and *n*-octane, respectively. Note that the consistent trend of higher (lower or non-retrograde) dew point pressures under confinement as opposed to bulk also is seen in the case of the synthetic mixture studied by Civan et al.[108] Something that is remarkable though is the fact that the fluid of the same overall composition under a certain confinement (2 nm in this case) completely changes its character in that it becomes either a dry or a wet gas, as compared with bulk; for example, in the system studied by Civan et al.[108] the cricondentherm is lower than the reservoir temperature of 190°F. Note that the cricondentherm in all cases of confinement are less than bulk, which does affect the (maximum) liquid dropout curve, which is less than the 4- and 5-nm confinement. Obviously, this is a manifestation of the altered critical constants of the mixture constituents, but this alteration in the phase behavior does have important consequences as far as production from the tight confined porous media is concerned.

As far as real reservoir fluids such as Bakken and Eagle Ford in tight formation are concerned, a consistent trend of suppressed bubble-point pressures under confinement compared to bulk emerges. Basically, with the same base critical properties used for both in bulk and under confinement, the fluid has a common critical point for the former and latter, with the confined phase envelope shifting in a counterclockwise direction thereby producing lower bubble-point pressures. The gap between the bulk and confined bubble-point pressures widens with decreasing temperatures as well as decreasing pore sizes and completely diminishes at the critical temperature. Essentially, if phase envelopes for a variety of nm scale pore sizes are generated and compared with the bulk phase envelope then the bubble and dew point curves of all confinements merge at the common critical point in a manner identical to the iso-vols or quality lines. Bulk phase envelopes for the Bakken[62] and the Eagle Ford[109] oils compared with the 10- and 20-nm confinement are qualitatively very similar showing a suppression in the bubble-point pressures.

Teklu et al.[110] carried out an in-depth analysis of the effect of confinement on the bubble-point pressures for the Bakken oil using (a) only the altered critical temperature and pressure of CO_2, methane, the intermediate components, and the four lumped groups; (b) only incorporating the capillary pressure correction in flash calculations; and (c) incorporating both the critical properties shift and the capillary pressure correction. They used the approach suggested by Singh et al.[105] to obtain the modified critical properties under confinement. Similar to Pitakbunkate et al.,[104] the reduction in the critical properties was found to be larger for smaller pore sizes. Teklu et al.[110] also observed the reduction in the critical properties of heavier components to be much more significant compared to the lighter one, which they attribute to the ability of the pore to accommodate fewer of the heavy component molecules. Overall, all phase envelopes under confinement shrink and varying degree of suppression in bubble-point pressures is observed in all cases. Using the phase envelopes Teklu et al.[110] also discuss the dew point temperature variation at a 1,000 psia isobar and compare the corresponding dew-point temperatures for 3 1,000-nm confinement, ranging from 400°F to 1,000°F; however, considering the Bakken reservoir temperature of 241°F, the dew points are perhaps not of major practical significance. Considering the unconfined or bulk bubble-point pressure of the Bakken oil at the reservoir temperature of 241°F, Figure 7.24 shows the bubble-point pressure difference as a function of pore radius in the range of 3–50 nm (the modal range for pore size in Bakken is from 10 to 50 nm[62]). As seen in Figure 7.24, the bubble-point pressure difference decreases with increasing pore size since the critical properties shift as well as the capillary pressure correction also decreases. Also shown in Figure 7.24 is the bubble-point pressure difference for the Eagle Ford oil at the reservoir temperature of 238°F for 10- and 20-nm confinement based on the values reported by Xiong et al.[109] Note that Xiong et al.[109] applied only the capillary pressure correction to the bubble point, which is perhaps the reason the magnitude of suppression is not the same for Bakken and Eagle Ford oils; additionally, when only the capillary pressure

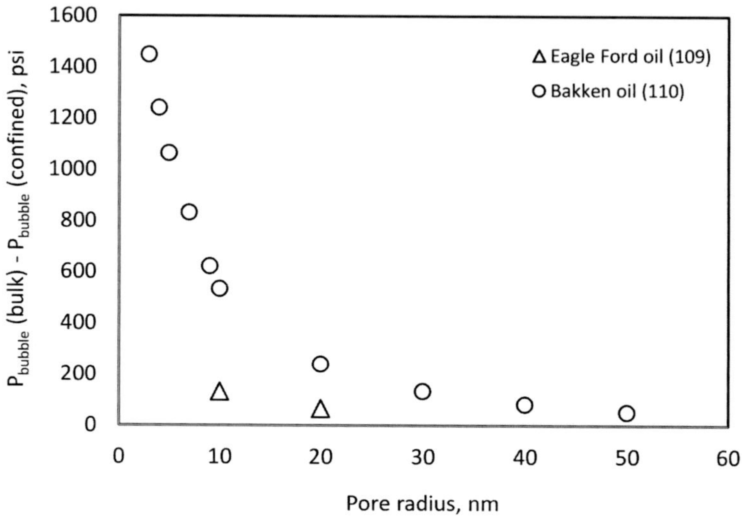

FIGURE 7.24
Differences between the bubble-point pressures in bulk and under different levels of confinements (pore radius) for Eagle Ford and Bakken oils. Numbers in the legend parentheses represent the references from which the values are taken and the plot constructed.

correction in the case of Bakken oil is considered then the degree of suppression is relatively smaller compared to the case when all three effects are considered (see above). From the usual fluid phase behavior terminology, the suppression in bubble point basically means a relatively large degree of undersaturation.[3] With regards to the bubble-point calculations under confinement, Teklu et al.[110] make a very valid and crucial point in that the corrections are purely theoretical and apply to single pore size, whereas in reality the actual or real nano porous media in which hydrocarbons reside would typically have a pore size distribution, i.e., different levels of confinement, which is also termed as heterogeneity; a case in which it will be challenging to interpret the phase behavior. The second goal of Teklu et al.'s[110] paper was to determine the effect of confinement on the minimum miscibility pressure (MMP). Based on their calculations they concluded that the shifts (reductions) in the critical temperature and pressure contribute to the reduction in MMP when compared with bulk.

Next we review the fluid properties under confinement, namely the two-phase densities, viscosities, and the typical reservoir engineering properties such as solution gas-to-oil ratio (GOR), R_s, and the oil formation volume factor, B_o, in conjunction with the dissolved gas. Given the fact that in multicomponent systems such as reservoir fluids the dissolved gas is the primary controlling parameter on the qualitative and quantitative characteristics of these properties, it is best to first consider the mechanism of gas evolution or liberation as pressure falls below the bubble point in a confined system.

Honarpour et al.[111] have provided a very appropriate explanation for the gas liberation phenomenon in tight porous media that contributes to the characteristics of the two-phase properties. As opposed to conventional reservoirs, in shale plays the rise in gas production below the (bulk) bubble-point pressure may not be the case because the gas separation or liberation has to overcome the relatively higher capillary pressures and thus due to this artifact the gas remains in a sort of "pseudo" solution (keeping the oil somewhat light) for a much longer time despite the continued decline in pressure. This will contribute to the R_s remaining fairly constant (relatively higher, see later plots) and B_o values relatively higher for a "longer time until the gas (slowly) separates out of the oil after overcoming the high capillary pressures in nanopres"[111]; in other words, the difficulty experienced by the gas to actually get liberated. In conventional reservoirs there is a slight dip in the producing GOR after reservoir pressure falls below the bubble point until the gas saturation reaches the critical value (S_{gc})[8,111]; however, in case of shale reservoirs (controlled by the pore sizes), this phenomenon may extend for a longer duration due to suppressed bubble-point pressures and moreover due to relatively lesser volume of released gas[111] (gas "held" up in the oil[111]) as well as the delay in the gas liberation. This phenomenon of the delayed gas release and lesser volumes is useful in understanding the fluid properties as a function of pressure in nano porous media.

Although in the case of pure components the densities increase with increasing confinement (smaller pore sizes), due to the packing effect as suggested by Pitakbunkate et al.,[104] quite the opposite is the case with real reservoir fluids, especially for the oil phase below the bubble-point pressures (as would be the case in a differential liberation sequence) where the densities actually decrease with increasing confinement.[62] Clearly, for the oil phase, the aforementioned phenomenon of (excess) gas held up in the oil making it somewhat lighter, i.e., higher density appears to be intuitively at play. However, once the (free) gas is actually released then it may conform to the packing effect, thereby resulting in an increase in the gas density with decreasing pore sizes. An identical oil and gas viscosity behavior is also exhibited, supposedly for the same reasons, as shown by Nojabaei.[62] Note though that Nojabaei[62] suggests that gas density and viscosity have no particular trend with pore size, which appears to be true but only in the vicinity of the bubble point, as a matter of fact the gas densities show an opposite trend in the low pressure region (see Figure 7.25). The viscosity trend is shown in Figure 7.26; note that unlike the density plot there are no single phase viscosities since this is a bubble-point system and thus there is no free gas above the bubble-point pressure. The held-up gas, delayed gas release, etc., will affect R_s and B_o as well, an obvious outcome being relatively higher values for both, which have been qualitatively described in the preceding paragraph. In order to demonstrate the impact of confinement on reservoir engineering properties, as an example, Figure 7.27 shows the R_s behavior as a function of pressure for the Bakken oil. Note that in the single-phase region,

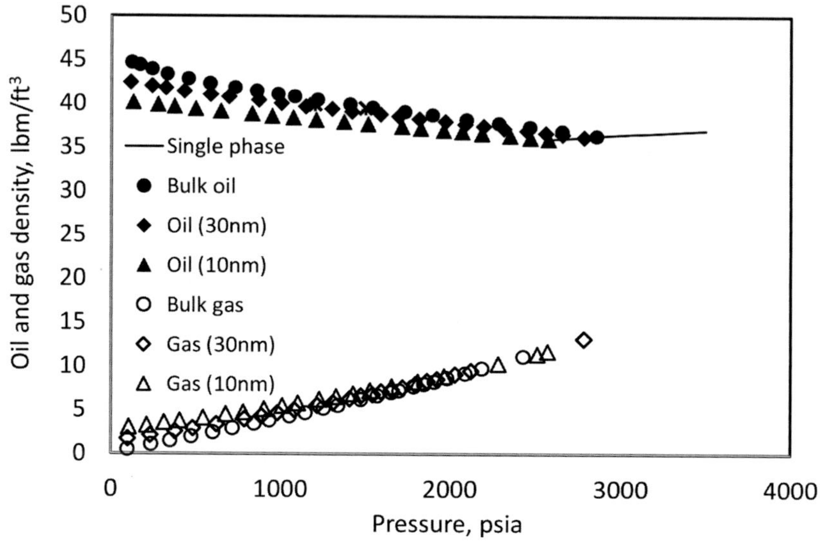

FIGURE 7.25
Oil and gas densities for the Bakken fluid in the two-phase region in bulk and under 30- and 10-nm confinements. Plot is constructed based on the separately reported values of oil and gas densities by Nojabaei.[62]

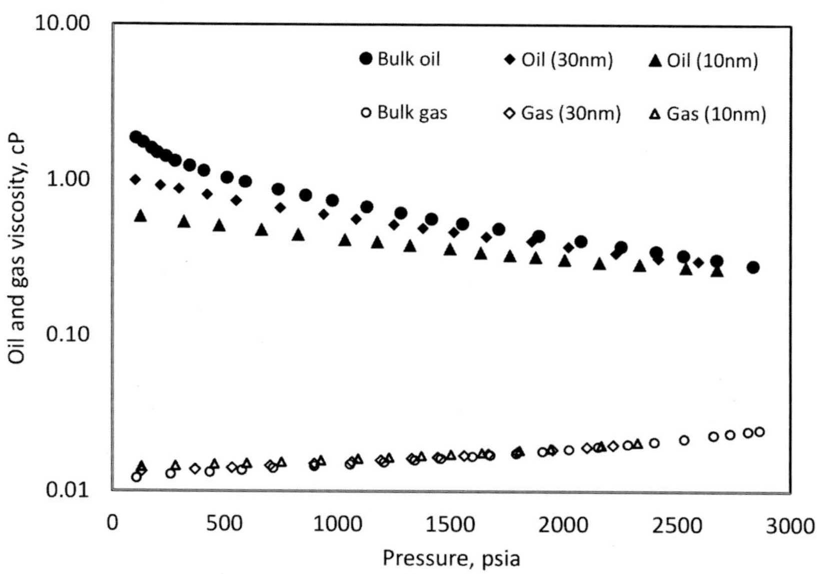

FIGURE 7.26
Oil and gas viscosities for the Bakken fluid in the two-phase region in bulk and under 30- and 10-nm confinements. Plot is constructed based on the separately reported values of oil and gas viscosities by Nojabaei.[62]

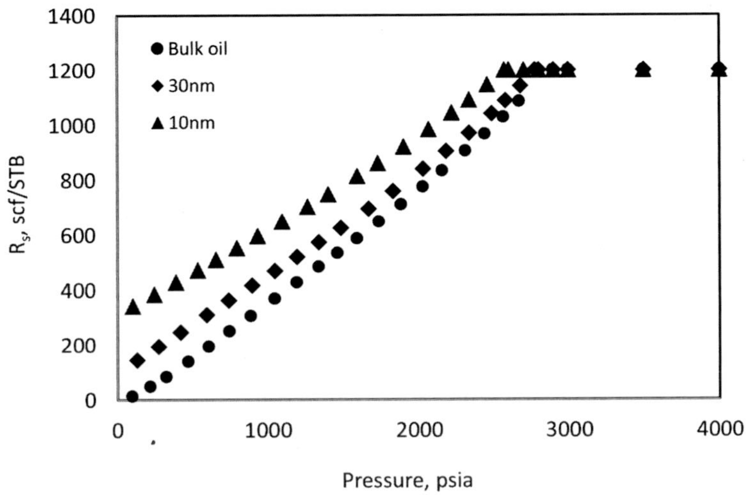

FIGURE 7.27
Solution GOR in bulk and under confinement for the Bakken fluid. Plot is constructed based on the values reported by Nojabaei.[62]

i.e., above the bubble-point pressure, oil properties are constant with pore size since capillary pressure is not defned.[62] The values plotted in Figures 7.25–7.27 (based on Nojabaei[62]) for Bakken oil supposedly represent a typical differential liberation sequence; note that these are not measured. The oil density and viscosity trends similar to Figure 7.25 and 7.26 are also indicated by Xiong et al.[109] for the Eagle Ford oil and by Du and Chu[112] for the Bakken oil (not shown here).

7.6 Measurement and EOS Modeling of Unconventional Reservoir Fluids—State of the Art

To date, there have been no systematic fluid phase behavior and property measurement techniques developed for acquiring the needed experimental data of shale-based fluids under confinement. This is mainly due to the fact that it is practically impossible or too difficult because of technology gaps to actually conduct these experiments and make nano-pore PVT measurements.[111] However, a forerunner to the challenge related to conducting fluid phase behavior and PVT measurements for shale-based fluids is the practical difficulty in capturing representative fluid samples (bottom hole and surface). In liquid-rich shales (LRS), given the very low permeability porous medium in which these fluids reside, the gas and oil rates fluctuate and continue to decline after the well is put on production.[111] Therefore, steady-state flow conditions or in other words a stable GOR, which is a requirement in collecting representative fluid samples, early in the life of a well, is never reached. Although this may happen much later in a certain time frame, and thus conducive for sampling, as shown by Honarpour et al.,[111] it may be too late from the standpoint of obtaining reliable fluid information. Related to the GOR two other unknowns are the time it will stabilize and the recombination value to be used, especially given the fluctuations in the gas and oil rates (and thus the ratio as well).[111] Honarpour et al.,[111] state that despite these issues or shortcomings, it is still a standard practice to recombine the surface separator samples in the laboratory on which fluid phase behavior and PVT measurements are carried out in bulk. It is then plausible to correct the collected bulk data for high capillary pressures in nano-sized pores.[111] The pore size corrected PVT data or calibrated EOS models then become part of reservoir simulations to determine the oil and gas recovery for shale-based reservoirs. The importance of incorporating confinement effects (via capillary pressure) are demonstrated by Nojabaei[62]; for example, the results show up to a 90% increase in recovery when capillary pressure is included in flash calculations. Chapter 8 deals with the specific aspects of incorporating (nano) porous media effects on fluid phase behavior.

Recently some authors have attempted to overcome the technology gaps and actually reported bubble-point measurements carried out in situ in nano-sized porous media using differential scanning calorimetry (DSC)[113] and lab-on-a-chip technology that enables the visualization of fluid behavior inside nanoscale channels.[114] Using DSC[113] authors report bubble-point temperature measurements on octane, decane and their binary mixtures under ~38 and 4 nm pore confinements, whereas hexane, heptane, and octane are tested using lab-on-a-chip technology for 50 nm confinement. In both cases PR EOS[115] model incorporating capillary pressure effect was used to compare the obtained experimental results. As the level of confinement increased (smaller pore size of ~4 nm), the differences between the measured and modeled values also increased, which were of the order of ~14%–19%.[113] It is quite plausible that in future such approaches may be utilized to obtain some limited data on saturation pressures of real reservoir fluids, which may in turn be used to rigorously calibrate or tune the EOS models for prediction of the required reservoir engineering properties. However, handling real reservoir fluids with multiple components, much higher pressure and temperature conditions is perhaps relatively not as trivial as is the case with the simple systems tested by the authors.[113,114]

For LRS, instead of explicitly accounting for high capillary pressures, Whitson and Sunjerga,[59] on the other hand, propose a different, but somewhat practical approach; the underlying problem statement though being the fact that "what you produce at the surface is not what you have in the reservoir." If producing streams from LRS and conventional systems are concerned then they differ in that the former produces a much leaner fluid than what is in situ or initially present, GOR significantly varies, which is attributed to substantially low permeabilities, whereas in the latter, for months or years, the characteristics of the initial fluid stream are quite similar to the in situ fluid system.[59] A special laboratory test called the multi-mixture constant composition expansion (CCE) is proposed by Whitson and Sunjerga,[59] which is designed to be experimentally simple; cost effective and most importantly cover the range of compositions from producing wellstream to possible in situ reservoir fluids and lastly provide the PVT data that guarantees the EOS ability to predict the fluid phase behavior and reservoir engineering properties for all relevant flowing reservoir compositions. The recommended CCE test is with a minimum of four mixtures that vary in terms of leanness or richness. An example included by Whitson and Sunjerga[59] procedurally illustrates this concept. For example, a well is producing at a stable OGR (inverse of GOR) of X SPbbl/MMscf (note: SPbbl is separator barrel and 1 MMscf = 1.0E[06] scf or standard cubic feet) after one month of production, at the end of which separator gas and liquid samples are collected. However, prior to the well stabilizing at X SPbbl/MMscf, if the maximum OGR of Y SPbbl/MMscf is observed then the laboratory is given the four recombinations as follows: (1) the minimum of 60 SPbbl/MMscf (leanest fluid); (2) the maximum of Y SPbbl/MMscf (richest fluid) and the two intermediates of $X+0.33(Y-X)$ and $X+0.67(Y-X)$, respectively. Any EOS model that matches the multi-mixture CCE data should be considered as reliable in describing the wide variation or range of fluids originating from the same reservoir.[59] For details on the procedure and EOS modeling, the reader is referred to the publication of Whitson and Sunjerga.

References

1. Glossary. 2016. Accessed November 22. www.eia.gov/tools/glossary/index.php?id=U.
2. Gordon, D. 2012. Understanding unconventional oil. *The Carnegie Papers*. http://carnegieendowment.org/files/unconventional_oil.pdf.
3. McCain, W.D. 1990. *Properties of Petroleum Fluids* (2nd Edition). Tulsa, OK: Pennwell Books.
4. Ungerer, P., Faissat, B., Leibovici, C., Zhou, H., Behar, E., Moracchini, G., and Courcy, J.P. 1995. High pressure-high temperature reservoir fluids: Investigation of synthetic condensate gases containing a solid hydrocarbon. *Fluid Phase Equilibria* 111(2): 287–311. doi:10.1016/0378-3812(95)02771-6.
5. Craft, B.C. and Hawkins, M.F. 1959. *Applied Petroleum Reservoir Engineering* (3rd Edition). Prentice Hall. http://books.google.com/books/about/Applied_petroleum_reservoir_engineering.html?id=t7RTAAAAMAAJ&pgis=1.
6. Ertekin, T., Abou-Kassem, J., and King, G. 2001. *Basic Applied Reservoir Simulation*. Society of Petroleum Engineers. http://books.google.com.hk/books?id=M41dAQAACAAJ.

7. Ahmed, U. and Meehan, D.N. 2016. *Unconventional Oil and Gas Resources: Exploitation and Development*. Edited by A.Y., Dandekar. CRC Press. https://books.google.com.my/books?id=n8P1CwAAQBAJ.
8. Dandekar, A.Y. 2013. *Petroleum Reservoir Rock and Fluid Properties*. Boca Raton, FL: CRC Press.
9. SPE high CO_2 and H_2S gas fields development. 2014. www.spe.org/events/14fsap/pages/about/.
10. No title. 2016. Accessed December 24. www.petrowiki.org.
11. Popp, V.V., Marinescu, M., Manoiu, D., Popp, A., and Chitu, P.M. 1998. Possibilities of energy recovery from CO_2 reservoirs. In *Proceedings of SPE Annual Technical Conference and Exhibition*, 33–41. Society of Petroleum Engineers. doi:10.2523/48925-MS.
12. Blann, J.R. and Laville, G.M. 1997. Gas lifting a major oil field in Argentina with high CO_2 content associated gas. *SPE Production & Facilities* 12(1): 41–45. doi:10.2118/30638-PA.
13. Clayton, J.L., Spencer, C.W., Koncz, I., and Szalay, A. 1990. Origin and migration of hydrocarbon gases and carbon dioxide, Békés Basin, Southeastern Hungary. *Organic Geochemistry* 15(3): 233–247. doi:10.1016/0146-6380(90)90002-H.
14. Vagtborg, H. Jr. 1954. Equilibrium vaporization ratios for a reservoir fluid containing a high concentration of hydrogen sulfide. In *Society of Petroleum Engineers*.
15. Harvey, P.A., Coppes, J.L., Webb, R.J., and Lebiadowski, G.P. 1993. Petroleum engineering nineties style: Caroline major sour gas condensate project. In *SPE Gas Technology Symposium*. Society of Petroleum Engineers. doi:10.2118/26140-MS.
16. Kilstrom, K. 1983. Whitney canyon sour gas well completion techniques. *Journal of Petroleum Technology* 35(1): 40–46. doi:0.2118/10079-PA.
17. Chambers, J.B. and Hillegeist, R.B. 1975. Producing sour oil and gas in the jay field. *Journal of Petroleum Technology* 27(6): 701–706. doi:10.2118/4998-PA.
18. Malik, Z., Charfeddine, M., Moore, S., Francia, L., Denby, P., and Kco, A. 2005. IPTC 10636 the supergiant Kashagan field : Making a sweet development out of sour crude. In *Simulation. International Petroleum Technology Conference*. doi:10.2523/IPTC-10636-MS.
19. Hamby, T.W. Jr. 1981. Development of high-pressure sour gas technology. *Journal of Petroleum Technology* 33(5): 792–798. doi:10.2118/8309-PA.
20. Al Qaoud, B.N. 2012. Challenges in sour gas handling for Kuwait Jurassic sour gas. In *SPE Middle East Unconventional Gas Conference and Exhibition*. Society of Petroleum Engineers. doi:10.2118/154452-MS.
21. Hoşgörmez, H., Yalçin, M.N., Soylu, C., and Bahtiyar, I. 2014. Origin of the hydrocarbon gases carbon dioxide and hydrogen sulfide in dodan field (SE-Turkey). *Marine and Petroleum Geology* 57(November): 433–444. doi:10.1016/j.marpetgeo.2014.05.012.
22. Wilhelm, S.M. and Bloom, N. 2000. Mercury in petroleum. *Fuel Processing Technology* 63(1): 1–27. doi:10.1016/S0378-3820(99)00068-5.
23. Boschee, P. 2015. Advancements in the removal of mercury from crude oil. *Oil and Gas Facilities* 2(2): 1–9. doi:0.2118/0413-0012-OGF.
24. Wilhelm, S.M. and McArthur, A. 1995. Removal and treatment of mercury contamination at gas processing facilities. In *Proceedings of the SPE/EPA Exploration & Production Environmental Conference*, 319–334. Society of Petroleum Engineers. doi:10.2118/29721-MS.

25. Sainal, M.R., Carigali, P., Bhd, S., Shafawi, A., and Scientific Services. 2007. Mercury removal system for upstream application: Experience in treating mercury from raw condensate SPE-106610. In *Proceedings of the E&P Environmental and Safety Conference*, 1–10. Society of Petroleum Engineers. doi:10.2118/106610-MS.
26. Salvá, C. and Gallup, D. 2010. Mercury removal process is applied to crude oil of Southern Argentina. In *SPE Latin American and Caribbean Petroleum Engineering Conference Proceedings*. Vol. 1. Society of Petroleum Engineers. doi:10.2118/138333-MS.
27. Jepson, W.B. and Rowlinson, J.S. 1955. Calculation of the correction to be applied to gas isotherms measured in the presence of mercury. *The Journal of Chemical Physics* 23(9). doi:10.1063/1.1742393.
28. Wilhelm, S., Isa, A.A.M., and Safri, S. 2008. Mercury in SE Asia produced fluids—Holistic approach to managing offshore impacts. In *Proceedings of International Petroleum Technology Conference*. Society of Petroleum Engineers. doi:10.2523/12821-MS.
29. Polishuk, I., Nakonechny, F., and Brauner, N. 2016. Predicting phase behavior of metallic mercury in liquid and compressed gaseous hydrocarbons. *Fuel* 174: 197–205. doi:10.1016/j.fuel.2016.02.002.
30. Khalifa, M. and Lue, L. 2017. A group contribution method for predicting the solubility of mercury. *Fluid Phase Equilibria* 432: 76–84. doi:10.1016/j.fluid.2016.10.025.
31. Gompos, A. 2015. *Evaluation of Mercury Distribution Models in Process Simulators*. National Technical University of Athens.
32. Lam, K., K B C Advanced Technologies Sdn, Zhang, X., Watson, M.J., and K B C Process. 2016. Unified thermodynamics and life of field in modelling complex upstream. In *Proceedings of the Offshore Technology Conference Asia*, 1–16. Offshore Technology Conference. doi:10.4043/26704-MS.
33. *Multiflash™ for Windows, User Guide*. 2007. Infochem Computer Services Ltd.
34. Edmonds, B., Moorwood, R.A.S., and Szczepanski, R. 1996. Mercury partitioning in natural gases and condensates. In *Paper Presented to the GPA European Chapter Meeting*. London.
35. Darman, N.H. and Harun, A.R. 2006. Technical Challenges and Solutions on Natural Gas Development in Malaysia. *The Petroleum Policy and Management (PPM) Project 4th Workshop of the China—Sichuan Basin Case Study Beijing*. www.ccop.or.th/ppm/document/CHWS4/CHWS4DOC08_nasir.pdf.
36. Turunawarasu, D., Norpiah, R.B.M., and Pottayil, A. 2015. A study on phase behavior and its application to cryogenic technologies for treatment of high CO_2 gas fields. In *Proceedings of the OTC Brazil. Offshore Technology Conference*. doi:10.4043/26070-MS.
37. Salihuddina, R.S., Rahmana, F.H., Isab, M.F.M. 2016. *Prediction of Dry Ice Formation in High-CO_2 Processes*. Penang, Malaysia: Society of Petroleum Engineers.
38. Ozturk, M. 2010. *Modeling Vapor-Liquid-Solid Phase Behavior in Natural Gas Systems*. Rice University.
39. De Guido, G., Langè, S., Moioli, S., and Pellegrini, L.A. 2014. Thermodynamic method for the prediction of solid CO_2 formation from multicomponent mixtures. *Process Safety and Environmental Protection* 92(1): 70–79. doi:10.1016/j.psep.2013.08.001.

40. Hlavinka, M.W., Hernandez, V.N., and McCartney, D. 2006. Proper interpretation of freezing and hydrate prediction results from process simulation. *GPA Annual Convention Proceedings*. M.W. Hlavinka, V.N. Hernandez, and D. McCartney. Proper Interpretation of Freezing and Hydrate Prediction Results from Process Simulation.
41. Davis, J.A., Rodewald, N., and Kurata, F. 1962. Solid-liquid-vapor phase behavior of the methane-carbon dioxide system. *AIChE Journal* 8(4): 537–539. doi:10.1002/aic.690080423.
42. Agarwal, G.M. and Laverman, R.J. 1974. Phase behavior of the methane carbon dioxide system in the solid-vapor region. In *Advances in Cryogenic Engineering*. Edited by K.D. Timmerhaus, 19:317–338. Boston, MA: Springer US. doi:10.1007/978-1-4613-9847-9_40.
43. ZareNezhad, B. 2006. Prediction of CO_2 freezing points for the mixtures of CO_2-CH_4 at cryogenic conditions of NGL extraction plants. *Korean Journal of Chemical Engineering* 23(5): 827–831. doi:10.1007/BF02705935.
44. Pikaar, M.J. 1959. *A Study of Phase Equilibria in Hydrocarbon-CO_2 System*. University of London.
45. Hosseinipour, A., Japper-Jaafar, A.B., and Yusup, S. 2016. The effect of CO_2 on wax appearance temperature of crude oils. *Procedia Engineering* 148: 1022–1029. doi:10.1016/j.proeng.2016.06.580.
46. Pan, H.Q., Firoozabadi, A., and Fotland, P. 1997. Pressure and composition effect on wax precipitation: Experimental data and model results. *SPE Production & Facilities* 12(4): 250–258. doi:10.2118/36740-PA.
47. Pedrosa, N. 2016. *Evaluating the Effect of High CO_2 Contents on N-Paraffin Solubility, Precipitation and Deposition*. Penang, Malaysia: Society of Petroleum Engineers.
48. United States, and Geological Survey. 2014. Today in energy. *U.S. Energy Information Administration Website*. www.eia.gov/todayinenergy/detail.php?id=19991#.
49. Martini, A.M., Walter, L.M., Ku, T.C.W., Budai, J.M., McIntosh, J.C., and Schoell, M. 2003. Microbial production and modification of gases in sedimentary basins: A geochemical case study from a Devonian shale gas play, Michigan basin. *AAPG Bulletin* 87(8): 1355–1375. doi:10.1306/031903200184.
50. Hill, R.J., Jarvie, D.M., Zumberge, J., Henry, M., and Pollastro, R.M. 2007. Oil and gas geochemistry and petroleum systems of the Fort Worth basin. *AAPG Bulletin* 91(4): 445–473. doi:10.1306/11030606014.
51. Martini, A.M., Walter, L.M., and McIntosh, J.C. 2008. Identification of microbial and thermogenic gas components from upper Devonian black shale cores, Illinois and Michigan Basins. *AAPG Bulletin* 92(3): 327–339. doi:10.1306/10180706037.
52. Mankiewicz, P.J., Pottorf, R.J., Kozar, M.G., and Vrolijk, P. 2009. Gas geochemistry of the mobile bay Jurassic Norphlet formation: Thermal controls and implications for reservoir connectivity. *AAPG Bulletin* 93(10): 1319–1346. doi:10.1306/05220908171.
53. Sloan, E.D. and Society of Petroleum Engineers (U.S.) 2000. Hydrate Engineering. Monograph/SPE; Henry L. Doherty series, vol. 21; Monograph (Society of Petroleum Engineers (U.S.)), v. 21; Monograph (Society of Petroleum Engineers (U.S.)), Henry L. Doherty series. 2973142. Richardson, TX: Society of Petroleum Engineers.
54. Bath, P.G., Fowler, W.N., and Russell, M.P. 1980. The brent field, a reservoir engineering review. In *Proceedings of the European Offshore Technology Conference and Exhibition*. Society of Petroleum Engineers. doi:10.2118/164-1980-MS.

55. Hoffmann, A.E., Crump, J.S., and Hocott, C.R. 1953. Equilibrium constants for a gas-condensate system. *SPE* 198(219-G): 1–10. doi:10.2118/219-G.
56. Adekunle, O. and Todd Hoffman, B. 2016. Experimental and analytical methods to determine minimum miscibility pressure (MMP) for Bakken formation crude oil. *Journal of Petroleum Science and Engineering* 146: 170–182. doi:10.1016/j.petrol.2016.04.013.
57. Gherabati, S.A., Browning, J., Male, F., Ikonnikova, S.A., and McDaid, G. 2016. The impact of pressure and fluid property variation on well performance of liquid-rich eagle ford shale. *Journal of Natural Gas Science and Engineering* 33: 1056–1068. doi:10.1016/j.jngse.2016.06.019.
58. No title. 2017. Accessed January 25. www.desmogblog.com/sites/beta.desmogblog.com/files/NorthernPlains Bakken Crude Oil Sample Chemical Composition.pdf.
59. Whitson, C.H. and Sunjerga, S. 2012. PVT in liquid-rich shale reservoirs. In *Proceedings of the SPE Annual Technical Conference and Exhibition*. Society of Petroleum Engineers. doi:10.2118/155499-MS.
60. No title. 2017. Accessed January 25. www.etc-cte.ec.gc.ca/databases/Oilproperties.
61. Pedersen, K.S., Christensen, P.L., and Shaikh, J.A. 2015. *Phase Behavior of Petroleum Reservoir Fluids* (2nd Edition). Boca Raton, FL: CRC Press, Taylor & Francis Group.
62. Nojabaei, B. 2015. Phase behavior and flow analysis of shale reservoirs using a compositionally-extended black-oil approach. The Pennsylvania State University.
63. Dandekar, A. 2015. Critical evaluation of empirical gas condensate correlations. *Journal of Natural Gas Science and Engineering* 27, Part 1 (November): 298–305. doi:10.1016/j.jngse.2015.08.072.
64. Al-Mahroos, F.M. and Tjoa, G.H. 1987. Analysis and phase behavior of Khuff gas system in Bahrain field. In *Middle East Oil Show*, 631–640. Manama: Society of Petroleum Engineers. doi:10.2118/15766-MS.
65. Riazi, M.R. and Daubert, T.E. 1987. Characterization parameters for petroleum fractions. *Industrial & Engineering Chemistry Research* 26(4): 755–759. doi:10.1021/ie00064a023.
66. Bierlein, J.A. and Kay, W.B. 1953. Phase-equilibrium properties of system carbon dioxide-hydrogen sulfide. *Industrial and Engineering Chemistry* 45(3): 618–624. doi:10.1021/ie50519a043.
67. Sobocinski, D.P. and Kurata, F. 1959. Heterogeneous phase equilibria of the hydrogen sulfide-carbon dioxide system. *AIChE Journal* 5(4): 545–551. doi:10.1002/aic.690050425.
68. Stouffer, C.E., Kellerman, S.J., Hall, K.R., Holste, J.C., Gammon, B.E., and Marsh, K.N. 2001. Densities of carbon dioxide + hydrogen sulfide mixtures from 220 K to 450 K at pressures up to 25 MPa. *Journal of Chemical and Engineering Data* 46(5): 1309–1318. doi:10.1021/je000182c.
69. Chapoy, A., Coquelet, C., Liu, H., Valtz, A., and Tohidi, B. 2013. Vapour-liquid equilibrium data for the hydrogen sulphide (H_2S)+carbon dioxide (CO_2) system at temperatures from 258 to 313K. *Fluid Phase Equilibria* 356: 223–228. doi:10.1016/j.fluid.2013.07.050.
70. Vitu, S., Privat, R., Jaubert, J.N., and Mutelet, F. 2008. Predicting the phase equilibria of CO_2 + hydrocarbon systems with the PPR78 model (PR EOS and K_{ij} calculated through a group contribution method). *Journal of Supercritical Fluids* 45(1): 1–26. doi:10.1016/j.supflu.2007.11.015.

71. Gardner, J.W., Orr, F.M., and Patel, P.D. 1981. The effect of phase behavior on CO_2-flood displacement efficiency. *Journal of Petroleum Technology* 33(11): 2067–2081. doi:10.2118/8367-PA.
72. Orr, F.M., Yu, A.D., and Lien, C.L. 1981. Phase behavior of CO_2 and crude oil in low-temperature reservoirs. *Society of Petroleum Engineers Journal* 21(4): 480–492. doi:10.2118/8813-PA.
73. Shelton, J.L. and Yarborough, L. 1977. Multiple phase behavior in porous media during CO_2 or rich-gas flooding. *Journal of Petroleum Technology* 29(9): 1171–1178. doi:10.2118/5827-PA.
74. Konynenburg, P.H.V. and Scott, R.L. 1980. Critical lines and phase equilibria in binary van der Waals mixtures. *Philosophical Transactions of the Royal Society A: Mathematical, Physical and Engineering Sciences* 298(1442): 495–540. doi:10.1098/rsta.1980.0266.
75. Sadus, R.J. 1994. Calculating critical transitions of fluid mixtures: Theory vs. experiment. *AIChE Journal* 40(8): 1376–1403. doi:10.1002/aic.690400810.
76. Fall, D.J. and Luks, K.D. 1985. Liquid-liquid-vapor phase equilibria of the binary system carbon dioxide + N-tridecane. *Journal of Chemical and Engineering Data* 30(3): 276–279. doi:10.1021/je00041a012.
77. Pan, H.Q., Yang, T., and Guo, T.M. 1995. Measurement and prediction of the bubble/dew point locus in the near-critical region and of the compressed fluid densities of the methane-carbon dioxide-N-butane ternary system. *Fluid Phase Equilibria* 105(2): 259–271. doi:10.1016/0378–3812(94)02620–G.
78. Qiu, D., Fu, J., Cai, J., Peng, C., Liu, H., and Hu, Y. 2006. Vapor-liquid critical properties of multi-component mixtures containing carbon dioxide and N-alkanes. *Fluid Phase Equilibria* 248(2): 191–196. doi:10.1016/j.fluid.2006.08.008.
79. Shariati, A., Peters, C.J., and Moshfeghian, M. 1998. Bubble point pressures of some selected carbon dioxide + synthetic C6+ mixtures. *Journal of Chemical & Engineering Data* 43(97): 785–788. doi:10.1021/je970261x.
80. Hicks, C.P. and Young, C.L. 1975. The gas-liquid critical properties of binary mixtures. *Chemical Reviews* 75(2): 119–175. doi:10.1021/cr60294a001.
81. Cismondi, M., Rodríguez-Reartes, S.B., Milanesio, J.M., and Zabaloy, M.S. 2012. Phase equilibria of CO_2 + N-alkane binary systems in wide ranges of conditions: Development of predictive correlations based on cubic mixing rules. *Industrial and Engineering Chemistry Research* 51(17): 6232–6250. doi:10.1021/ie2018806.
82. Lee, B.I. and Kesler, M.G. 1975. A generalized thermodynamic correlation based on three parameter corresponding states. *AIChE Journal* 21(3): 510–527. doi:10.1002/aic.690210313.
83. Reamer, H.H., Sage, B.H., and Lacey, W.N. 1951. Phase equilibria in hydrocarbon systems—Volumetric and phase behavior of the methane-hydrogen sulfide system. *Industrial & Engineering Chemistry* 43(4): 976–981. doi:10.1021/ie50496a052.
84. Kohn, J.P. and Kurata, F. 1958. Heterogeneous phase equilibria of the methane-hydrogen sulfide system. *AIChE Journal* 4(2): 211–217. doi:10.1002/aic.690040217.
85. Kay, W.B. and Brice, B. 1953. Liquid-vapor equilibrium relations in ethane-hydrogen sulfide system. *Journal of Industrial and Engineering Chemistry* 45(3): 615–618. doi:10.1021/ie50519a042.
86. Kay, W.B. and Rambosek, G.M. 1953. Liquid-vapor equilibrium relations in binary systems—Propane—Hydrogen sulfide system. *Industrial & Engineering Chemistry* 45(1): 221–226. doi:10.1021/ie50517a065.

87. Leu, A.D. and Robinson, D.B. 1989. Equilibrium phase properties of the N-butane-hydrogen sulfide and isobutane-hydrogen sulfide binary systems. *Journal of Chemical and Engineering Data* 34(3): 315–319. doi:10.1021/je00057a017.
88. Reamer, H.H., Sage, B.H., and Lacey, W.N. 1953. Phase equilibria in hydrocarbon systems—Volumetric and phase behavior of N-pentane–hydrogen sulfide system. *Industrial & Engineering Chemistry* 45(8): 1805–1809. doi:10.1021/ie50524a054.
89. Reamer, H.H., Selleck, F.T., Sage, B.H., and Lacey, W.N. 1953. Phase equilibria in hydrocarbon systems—Volumetric and phase behavior of decane–hydrogen sulfide system. *Industrial & Engineering Chemistry* 45(8): 1810–1812. doi:10.1021/ie50524a055.
90. Nichita, D.V., Gomez, S., and Luna, E. 2002. Multiphase equilibria calculation by direct minimization of Gibbs free energy with a global optimization method. *Computers & Chemical Engineering* 26(12): 1703–1724. doi:10.1016/S0098-1354(02)00144-8.
91. Seto, C.J. and Beliveau, D.A. 2000. Reservoir souring in the caroline field. In *SPE 59778, SPE/CERI Gas Technology Symposium*, 1–9. Society of Petroleum Engineers. doi:10.2118/59778-ms.
92. Jacoby, R.H. and Rzasa, M.J. 1952. Equilibrium vaporization ratios for nitrogen, methane, carbon dioxide, ethane and hydrogen sulfide in absorber oil-natural gas and crude oil-natural gas systems. *Journal of Petroleum Technology* 4(4): 99–110. doi:10.2118/952099-G.
93. Wichert, E. and Aziz, K. 1972. Calculation of Z's for sour gases. *Hydrocarbon Processing* 51: 119–122.
94. Benedict, M., Webb, G.B., and Rubin, L.C. 1940. An empirical equation for thermodynamic properties of light hydrocarbons and their mixtures I. Methane, ethane, propane and *N*-butane. *The Journal of Chemical Physics* 8(4): 334–345. doi:10.1063/1.1750658.
95. Jia, W., Li, Z., Liao, K., and Li, C. 2016. Using Lee-Kesler equation of state to compute the compressibility factor of CO_2-content natural gas. *Journal of Natural Gas Science and Engineering* 34(August): 650–656. doi:10.1016/j.jngse.2016.07.032.
96. Bian, X., Du, Z., Tang, Y., and Du, J. 2011. Compressibility factor of high CO_2-content natural gases: Measurement and correlation. *Carbon Dioxide Sequestration and Related Technologies* 82–83(February): 65–86. doi:10.1002/9781118175552.ch6.
97. Hou, D., Deng, H., Zhang, H., Li, K., Sun, L., and Pan, Y. 2015. Phase behavior and physical parameters of natural gas mixture with CO_2. *Journal of Chemistry* 2015. doi:10.1155/2015/873718.
98. Tabasinejad, F., Moore, R.G., Mehta, S.A., Fraassen, K.C.V., Barzin, Y., Rushing, J.A., and Newsham, K.E. 2010. Density of high pressure and temperature gas reservoirs: Effect of non-hydrocarbon contaminants on density of natural gas mixtures. In *SPE Western Regional Meeting*. Society of Petroleum Engineers. doi:10.2118/133595-MS.
99. Sigmund, P.M., Dranchuk, P.M., Morrow, N.R., and Purvis, R.A. 1973. Retrograde condensation in porous media. *Society of Petroleum Engineering Journal* 13(2): 93–104. doi:10.2118/3476-PA.
100. Danesh, A. 1998. *PVT and Phase Behaviour of Petroleum Reservoir Fluids*. Edited by A. Danesh. Elsevier.

101. Tissot, B.P. and Welte, D.H. 1984. *Petroleum Formation and Occurrence. Second Revised and Enlarged Edition* (2nd Edition). Vol. 66. Berlin: Springer-Verlag. doi:10.1029/EO066i037p00643.
102. Nagarajan, N.R., Honarpour, M.M., Arasteh, F., and Corporation, H. 2013. The critical role of rock and fluid in unconventional shale reservoir performance. *Unconventional Resources Technology Conference*. doi: 10.2118/168864-MS.
103. Swami, V., Clarkson, C.R., and Settari, A. 2012. Non-darcy flow in shale nanopores: Do we have a final answer? In *SPE Canadian Unconventional Resources Conference*. Society of Petroleum Engineers. doi:10.2118/162665-MS.
104. Pitakbunkate, T., Balbuena, P.B., Moridis, G.J., and Blasingame, T.A. 2016. Effect of confinement on pressure/volume/temperature properties of hydrocarbons in shale. *Society of Petroleum Engineers (SPE) Journal* 21(2): 621–634. doi:10.2118/170685-PA.
105. Singh, S.K., Sinha, A., Deo, G., and Singh, J.K. 2009. Vapor-liquid phase coexistence, critical properties, and surface tension of confined alkanes. *Journal of Physical Chemistry C* 113(17): 7170–7180. doi:10.1021/jp8073915.
106. Bloomer, O.T., Gami, D.C., Parent, J.D. 1953. Physical-Chemical Properties of Methane-Ethane Mixtures, TT-. Institute of Gas Technology Research Bulletin 22; Institute of Gas Technology. 22. Research Bulletin ; TA -. Chicago: Institute of Gas Technology.
107. Dong, X., Liu, H., Hou, J., and Chen, Z. 2016. Phase behavior of hydrocarbon mixtures in the organic nanopores of unconventional gas condensate reservoirs. In *Proceedings of the 4th Unconventional Resources Technology Conference*. Tulsa, OK: American Association of Petroleum Geologists. doi:10.15530/urtec-2016=2460485.
108. Civan, F., Michel, G., and Sigal, R. 2013. Impact of fluid behavior modification under elevated pressure and temperature conditions on shale-gas/condensate reservoir engineering and production analysis. In *Society of Petroleum Engineers—SPE Canadian Unconventional Resources Conference 2013—Unconventional Becoming Conventional: Lessons Learned and New Innovations*. Vol. 2. Society of Petroleum Engineers. doi:10.2118/167186-MS.
109. Xiong, Y., Winterfeld, P., Wang, C., Huang, Z., and Yu-Shu, W. 2015. Effect of large capillary pressure on fluid flow and transport in stress-sensitive tight oil reservoirs. In *SPE Annual Technical Conference and Exhibition*. Society of Petroleum Engineers. doi:10.2118/175074-MS.
110. Teklu, T.W., Alharthy, N., Kazemi, H., Yin, X., Graves, R.M., and AlSumaiti, A.M. 2014. Phase behavior and minimum miscibility pressure in nanopores. *SPE Reservoir Evaluation & Engineering* 17(3): 396–403. doi:10.2118/168865-PA.
111. Honarpour, M.M., Nagarajan, N.R., Orangi, A., Arasteh, F., and Yao, Z. 2012. Characterization of critical fluid PVT, rock, and rock-fluid properties—Impact on reservoir performance of liquid rich shales. In *SPE Annual Technical Conference and Exhibition*. Society of Petroleum Engineers. doi:10.2118/158042-MS.
112. Du, L. and Chu, L. 2012. Understanding anomalous phase behavior in unconventional oil reservoirs. In *SPE Canadian Unconventional Resources Conference*. Society of Petroleum Engineers. doi:10.2118/161830-MS.
113. Luo, S., Lutkenhaus, J.L., and Nasrabadi, H. 2016. Use of differential scanning calorimetry to study phase behavior of hydrocarbon mixtures in nano-scale porous media. *Journal of Petroleum Science and Engineering*. doi:10.1016/j.petrol.2016.12.019.

114. Alfi, M., Nasrabadi, H., and Banerjee, D. 2016. Experimental investigation of confinement effect on phase behavior of hydrocarbons using lab-on-a-chip technology. *Fluid Phase Equilibria* 423(September): 25–33. doi:10.1016/j.fluid.2016.04.017.
115. Robinson, D.B. and Peng, D-Y. 1978. The characterization of the heptanes and heavier fractions for the GPA Peng-Robinson programs. *Research report/Gas Processors Association, RR-28; Research report (Gas Processors Association), RR-28.* Tulsa, OK: Gas Processors Association.

8

Porous Media Effects on Phase Behavior of (Unconventional) Petroleum Reservoir Fluids

8.1 Practical Significance and Implications of Porous Media Effects on Phase Behavior

The effect of very low permeability porous media or confinement on fluid phase behavior is discussed in Chapter 7 with the help of some examples from the actively pursued current area of research interest given the prominence of unconventional petroleum reservoir fluids. Note that here we consider only the shale-based fluids, i.e., liquid-rich shales (LRS) or other types of reservoir fluids such as gas condensates or wet gases or dry gases present in particularly low permeability or nano Darcy range porous media.

It is well recognized that the fluid phase behavior and PVT data that are typically measured on the bulk of the fluid (obviously in the absence of the porous media) do not represent its actual behavior and flow in the nanoscale pore spaces. This difference in the behavior in the porous media and bulk is mainly due to dominant capillary forces. This can be explained as follows. In a PVT cell starting from single-phase conditions when saturation pressure is reached in a typical PV expansion or a constant composition expansion (CCE) test, there is equilibrium between the newly formed phase and the original phase. Typically, these phases being gas and liquid, if we assign a pressure to each of them then in a PVT cell at equilibrium it can be stated that $P_{Gas} = P_{Liquid} = P_{Sat.}$ because the interface is flat[1] or the curvature is zero.[2] However, generally speaking, in the case of any porous media for that matter, a nonzero curvature of the gas–liquid interface will result in a pressure difference between the two phases (capillary pressure, P_c), such that $P_{Gas} - P_{Liquid} \neq 0$, but is $= P_c$. This can be generalized as $P_G - P_L = P_{cGL}$ or $P_V - P_L = P_{cVL}$ or $P_G - P_O = P_{cGO}$, where G or V means a gas or vapor or less dense phase, and L or O means a hydrocarbon liquid phase, oil phase, or a more dense phase, while P_c with the respective designations is the capillary pressure, which, in turn, shifts or alters the saturation pressure and other thermodynamic properties.[1,2] Note that these notations are interchangeably used in this chapter. Shapiro et al.[2] gave another plausible explanation of the so-called chromatographic effect of porous media on the fluid mixture resulting in different components produced with different rates as well as time, in a fashion somewhat similar to their elution in the order of boiling points from a gas chromatograph column. Note that this also has been stated, albeit somewhat differently by Pitakbunkate et al.[3] (see Chapter 7). For a conventional porous media (large pore sizes or radii), however, a simplification or assumption that the gas and liquid phase pressures can be treated as equal or relatively small capillary pressures can be neglected for practical purposes is a reasonable engineering approach. However, the same is not true when it comes to pore sizes that are orders of magnitude smaller than the conventional pores, which in turn result in equally high capillary pressures and thus its influence on the fluid phase behavior and thermodynamic properties can no longer be ignored or neglected in the overall management of unconventional reservoirs.

Therefore, for reservoir engineering applications, such as simulations, the data obtained in bulk need to be scaled according to the level of confinement, because as of yet capabilities to obtain all the fluid behavior information directly or in situ in a laboratory setting do not exist. Numerically, this can be accomplished by the two generally accepted methodologies: (1) directly shifting the critical constants of the fluid constituents, which, for instance, is the key input to an equations of state (EOS) simulator or (2) by including capillary pressure in EOS calculations. Note that examples of the shifted critical constants, critical temperature (T_c), and pressure (P_c) are discussed in Chapter 7. The second approach of directly including capillary pressures in the calculations appears to be more common.

As mentioned in Chapter 7, overall, all phase envelopes under confinement shrink and varying degree of "suppression" in bubble-point pressure occur in all cases, and gas separation or liberation has to overcome the relatively higher capillary pressures. This particular phenomenon causes the gas to remain in a sort of "pseudo" solution (keeping the oil somewhat light), rather than breaking through, for a much longer time despite the continued decline in pressure.[4] Alternatively, the oil remains undersaturated for a relatively larger pressure range.[1] The delay in gas breakthrough and extended undersaturated conditions has important (positive) consequences on estimated ultimate recovery of oil, in which the oil production improves,[1] which has been adequately demonstrated by inclusion of capillary pressures in numerical reservoir simulations.[1,5]

As far as retrograde fluids or gas condensates are concerned, confinement may result in either a suppression or an elevation in the dew-point pressure. The suppression of bubble point and elevation or suppression of dew point due to confinement seems to be logical when one considers typical phase envelopes. Bubble points, or the bubble-point curve, monotonically increase with increasing temperature; however, as one moves on the right on the phase envelope, i.e., with increasing temperatures, the dew-point curve that originates from the critical point is actually "dome" shaped; the top of the dome being the cricondenbar. Basically, what this means is a unique bubble-point pressure at a given temperature but same upper dew-point pressure at two different (low and high) temperatures; bracketed between the cricondentherm and critical temperature. The bubble- and dew-point curves rotating due to confinement around the critical point (see discussion in Chapter 7) as an anchor or fixed point means that capillary pressure is zero because interfacial tension (IFT) is zero since the identity of the individual phases is diminished at the critical point.

The two cases discussed in Chapter 7 show an increase in the dew-point pressures. Shapiro et al.[2] also showed an increase in the dew-point pressure under confinement by ~70 psi for gas condensates from Austria. Given the absence of reliable models or strong experimental evidence confirming the elevation or suppression of dew-point pressures, Orangi and Nagarajan[6] in their simulation work arbitrarily varied the bulk dew-point pressure of a fluid with fixed composition by ±200 psi. Generally speaking, in a gas condensate reservoir the appearance (at the dew-point pressure starting with single-phase vapor) and precipitation of retrograde condensate with pressure decline regardless of high or low permeability porous media has a negative effect on the productivity of the gas due to what is called as condensate ring or blockage near the production well. This condensate will have an adverse effect on the relative permeability of the gas that is being produced. Therefore, strictly from a productivity standpoint the premature (elevation) or delayed (suppression) dew-point pressure is simply a matter of time as to when the gas rate will be affected by the condensate blockage. Nevertheless, in the lean and rich gas condensate LRS reservoirs, despite being economically attractive, the loss of productivity due to altered dew point, and condensate banking are factors of particularly major concern that pose unique production challenges.[6] In the study conducted by Orangi and Nagarajan,[6] they concluded that regardless of elevation or suppression in the dew-point pressure, the retrograde condensate blockage seem to have a severe negative impact on the well productivity, i.e., it is a matter of time as mentioned before, elevated dew point resulting in earlier decline and accelerated negative impact and vice-versa. Also, generally speaking, although the retrograde condensate may have some mobility after reaching a critical condensate saturation, the recovery of the condensate in shale reservoirs may be more severely impacted than in conventional gas condensate reservoirs.[6]

8.2 Capillary Pressure

Let us first consider the fundamental general functional form of the capillary pressure (P_c) equation that is of significance for inclusion in the flash calculations, which is given below. This equation is a function of pore radius, r, IFT, σ, and wetting angle θ.

$$P_c = \frac{2\sigma \operatorname{Cos}\theta}{r} \tag{8.1}$$

If σ is expressed in N/m and r is in m, then the P_c value will be in N/m^2, which can be converted to the customary oilfield units of psi as follows: $101.325 \times 1{,}000\,\text{N/m}^2 = 14.7$ psi.

If Eq. 8.1 is expressed in terms of a particular gas–liquid (or oil) or vapor–liquid (oil) pair capillary pressure, then it can be written as follows:

$$P_{cGO} = P_G - P_O = \frac{2\sigma_{GO} \operatorname{Cos}\theta_{GO}}{r} \tag{8.2}$$

Simply speaking, for calculating the gas–oil capillary pressure from Eq. 8.2 will require the IFT value(s), the contact angle, and the pore radius (note that this in itself is a variable as not all pores in a real porous media has the same radii).

Let us now consider the two different practical application aspects of the capillary pressure (equation). First, from a petrophysical or initialization standpoint capillary pressure as a function of saturation is of significance to determine irreducible wetting phase saturation, capillary entry or threshold pressure, thickness of the gas–oil transition zone, depth to gas–oil contact, and in immiscible gas injection processes. Capillary pressure vs. the saturation data is typically expressed in terms of depth or height (h) vs. saturation data by using $h = \frac{P_c}{\Delta\rho g}$, where $\Delta\rho$ is the density difference between gas and oil and g is acceleration due to gravity, if P_c is in N/m^2; $\Delta\rho$ is in kg/m^3 and g (9.81) in m/s^2, then the calculated h is in m. For this application, the capillary pressures vs. saturation data are typically measured in a laboratory directly using the representative rock and fluid systems (usually in the presence of irreducible water saturation) or measured using air and mercury and then converted to the actual system. With oil being the wetting phase, the P_c vs. saturation curve is termed as the drainage curve. The variables that appear on the right side of Eq. 8.2 are in-built or somewhat implicit in the measured P_c values, and with the presumed constant values of contact angle and IFT (for a given system), the P_c values are basically a reflection of the variability in the pore radii and saturation, which is typically used in obtaining the pore size distribution. For the same rock type, it may be possible to obtain a composite P_{cGO}/σ_{GO} vs. saturation curve in cases of different gas–oil pairs (and thus IFT) values.

The second application of the capillary pressure equation is the inclusion in phase behavior calculations for unconventional systems to take into account the influence of nano-sized pores. Just to put the capillary pressure values in perspective, for a conventional system, the typical P_{cGO} values at the irreducible liquid saturation (total of irreducible water and residual oil) may be of the order of 50–60 psi (e.g., see ref. 7), whereas the same may be order(s) of magnitude higher in the case of confined systems. As far as the contact angle, θ_{GO}, is concerned, it may be reasonable to assume that in a gas–oil system oil is fully and uniformly wetting, which means a contact angle of zero, such that $P_{cGO} = P_G - P_O = \frac{2\sigma_{GO}}{r}$, because cosine of zero is one. Alternatively, another value, but nevertheless a constant, needs to be assumed because, unlike σ_{GO} (see discussion later), there is a lack of any relationship between θ_{GO} and phase compositions, which itself is an unknown in phase equilibria computations. A customary σ_{GO} of 5 mN/m (same as dyne/cm) and a pore radii of 10 nm will thus result in a P_{cGO} of $2 \times 5 \times 10^{-3}/10 \times 10^{-9} = 10^6 \text{N/m}^2$ or 145 psi.

8.2.1 Gas–Oil IFT

It is clear that in the two applications of P_c mentioned above, a constant IFT (for a given system) and variable IFT (for a given system, changing with or dependent on equilibrium phase compositions) values, respectively, and a variable pore size distribution (in both) are considered. Therefore, the practical issue or difficulty that makes the inclusion of capillary pressure in phase equilibria calculations somewhat nontrivial or iterative is primarily because σ_{GO} is function of equilibrium vapor and liquid compositions or mole fractions and the respective molar densities, which are the main unknowns. It is quite common to measure the IFT values, fairly routinely as part of standard (e.g., constant composition expansion, CCE) as well as exotic (e.g., multiple contact) PVT measurements,[8,9] by using well-established techniques such as the pendant drop, meniscus height, laser light scattering, etc. However, from a computational standpoint such as including P_c values in phase behavior calculations, it is practical and perhaps logical to use phase composition and density-dependent correlations such as the parachor methods (functional forms and references given below), which in general are very widely accepted and used in the petroleum industry. The experimental IFT values can of course be compared with the predictions to ascertain the accuracy and reliability of methods such as the parachor. Dandekar[9] presented a large database on measured IFT values for a variety of systems and also compared the performance of the parachor model, and basically concluded that reasonable predictions can be achieved for values close to 1 mN/m; however, the method largely underpredicts and overpredicts in the (very) low and high IFT regions, respectively. Based on these observations, Dandekar[9] later modified the parachor method, expressing the fixed exponent in the original method, a variable by making it dependent on the molar density difference between the liquid and the vapor phase to improve the predictive capability. The functional forms of the original parachor method[10] and its modification are provided in the following equations:

$$\sigma^{\frac{1}{E}} = \sum_{i=1}^{n} P_{\sigma i}\left(X_i\rho_M^L - Y_i\rho_M^V\right) \tag{8.3}$$

The exponent, E, which has a fixed value of 4 in the original parachor method,[10] is made a function of molar density difference in the modified method[9] as follows:

$$E = 3.583 + 0.16\left(\rho_M^L - \rho_M^V\right) \tag{8.4}$$

where σ is the vapor–liquid or gas–liquid IFT (σ_{GO} as denoted previously) in mN/m or dyne/cm; $P_{\sigma i}$ is the parachor of component i in the mixture; X_i is the mole fraction of component i in the liquid phase; ρ_M^L is the molar density of equilibrium liquid phase in g-mol/cm^3; Y_i is the mole fraction of component i in the vapor phase; and ρ_M^V is the molar density of equilibrium vapor phase in g-mol/cm^3.

The modified form of the parachor method has been recently used by Luo et al.[11] in phase behavior computations of hydrocarbon mixtures in nanoscale porous media. The parachor values, which are constants (similar to T_c, P_c) for a given component, i, are listed by Danesh[8] for the three non-hydrocarbon components and several well-defined normal alkanes. For real reservoir fluids, parachor values of crude or condensate fractions can be determined from the empirical equation proposed by Firoozabadi and Katz[12] which relates parachor and the molecular weight (MW_f being the molecular weight of a given fraction) as follows:

$$P_\sigma = -11.4 + 3.23 MW_f - 0.0022 MW_f^2 \tag{8.5}$$

Therefore, in a normal scheme of computations, iterative or otherwise, the calculation of σ_{GO} values, from the described variables, is relatively straightforward.

8.2.2 Pore Size and Pore Size Distribution

Next, as far as the calculation of capillary pressure is concerned, the issue of pore radii, however, needs to be addressed because this is an intrinsic rock characteristic, which is variable, i.e., every rock or the pore space has a pore size distribution. Additionally, the pore space geometry may not necessarily be spherical, but it may be reasonable to assume as such. According to Orangi and Nagarajan,[6] use of single pore of a given radius, thus essentially rendering this as a constant in Eq. 8.2, is well established in, for example, dew-point calculations. However, given the (wide) pore size distribution, assumptions have to be made for an "average pore radius," thus making the overall results somewhat approximate.[6] In the below paragraph, we review how the pore size has been handled in the various publications.

Wang et al.,[13] on the other hand, in their calculations, considered a hypothetical (but plausibly reflecting the tight condensate reservoirs) pore size distribution ranging between 7.5 and 50 nm with majority of the pores in 7.5–10 nm range (~60% of the volume). However, their calculations appear to progress sequentially or in stages, i.e., from large to small pores in case of light oil and from small to large pores in case of the retrograde gas. For example, the first-stage phase equilibria calculations (including capillary pressure) are done for a fixed pore size, which then switches to the next smaller or higher according to the fluid. Although Teklu et al.[14] considered different pore sizes, their calculations are individual in that they apply for a single pore size and they recommend that further work needs to be conducted in order to incorporate the heterogeneity or pore size distribution to model the reality of a rock. Nojabaei[5] compared the effect of including pore size distribution and a fixed pore size on capillary pressure and thus on oil recovery calculations for Bakken. The pore size distribution, however, is supposedly represented by 20–24 nm, whereas the fixed modal pore size is 13 nm. Note that the modal range of pore size for Bakken ranges from 10 to 50 nm.[5] As expected the bubble-point suppression due to the fixed (smaller) pore size is more than the distributed pore sizes. Ultimately, the difference in oil recovery for both the cases was found to be only 2.5%. Dong et al.[15] in their simulations also considered the various pore sizes, but separately as fixed, for a synthetic quaternary system and a Marcellus shale gas. Similarly, Fathi et al.[16] also considered a fixed average pore size of 2 nm for their phase equilibria calculations for the Marcellus shale. Pitakbunkate et al.[3] also performed phase equilibria calculations using fixed pore sizes. Finally, pertaining to the pore size (and its occupancy) Honarpour et al.[4] make a practical and valid point. In reality, the pore space will contain a hydrocarbon phase and the aqueous phase (formation water), i.e., connate water and in the case of water–wet rocks a thin film of water may coat the pore wall thereby reducing the available pore space to hydrocarbons. For example, if the pore diameter is 10 nm and a thin 1 nm water film coats the pore wall, then the effective pore diameter becomes 8 nm or a pore radius of 4 nm. This is a reasonable engineering approach to calculate the capillary pressure, based on the "effective pore radius," for inclusion in the phase equilibria computations.

8.3 Inclusion of Confinement in Phase Behavior

8.3.1 Capillary Pressure in Flash or Vapor–Liquid Equilibria (VLE) Calculations

For incorporating confinement or capillary pressure in phase equilibria or what is commonly known as VLE calculations, we first turn to the fundamental relation known as the equilibrium ratio or the partitioning coefficient, K_i, which is simply defined as $K_i = Y_i/X_i$, where Y_i and X_i are the mole fractions of a given component i (from $i = 1$ to N components) in the vapor and liquid phases, respectively. If confinement or capillary modification is introduced, then equilibrium ratio can be denoted as K_i^C such that[2]

$$K_i^C = \left(1 - \frac{P_{cGO}}{P_G}\right)K_i \tag{8.6}$$

Equation 8.6 is consistent in that if capillarity is not considered or ignored then $K_i^C = K_i$. Alternatively,

$$K_i^C = \left(1 - \frac{P_G - P_O}{P_G}\right)K_i = \left(\frac{P_G - P_G + P_O}{P_G}\right)K_i = \left(\frac{P_O}{P_O + P_{cGO}}\right)K_i \tag{8.7}$$

In the scheme of EOS computations, K_i is iteratively determined using the ratio of fugacity coefficients of component i in the liquid $\left(\Phi_i^L\right)$ and vapor $\left(\Phi_i^V\right)$ phases, respectively (due to equality of fugacities at equilibrium conditions), i.e., $K_i = \frac{\Phi_i^L}{\Phi_i^V}$, which can be used to rewrite Eq. 8.7 as follows. Note that when capillary pressure is included in VLE calculations, the fugacities need to be calculated using the respective phase pressures (which are not the same) such that $f_i^V = Y_i P_G \Phi_i^V$ and $f_i^L = X_i P_L \Phi_i^L$. At equilibrium $f_i^V = f_i^L$, which naturally results in the following equations because $Y_i P_G \Phi_i^V = X_i P_L \Phi_i^L$ and thus $K_i^C = \left(\frac{Y_i}{X_i}\right) = \left(\frac{P_L}{P_G}\right)\left(\frac{\Phi_i^L}{\Phi_i^V}\right)$ or $K_i^C = \left(\frac{P_L}{P_G}\right)K_i$, which naturally reduces to $K_i^C = K_i$ when capillary pressure is ignored or considered as zero ($P_L = P_G$) and results in Eq. 8.7 when capillary pressure is considered, $K_i^C = \left(\frac{P_L}{P_L + P_{cGL}}\right)K_i$. Note that V and G and O and L mean vapor and gas and oil and liquid phase, respectively, and have been used interchangeably. Alternatively, it can be written as

$$K_i^C = \left(\frac{P_O}{P_O + P_{cGO}}\right)\frac{\Phi_i^L}{\Phi_i^V} = \left(\frac{P_G - P_{cGO}}{P_G}\right)\frac{\Phi_i^L}{\Phi_i^V} \tag{8.8}$$

Using the above equilibrium ratio so defined by including confinement, the Rachford–Rice[17] flash function can be written as

$$\sum_{i=1}^{n} \frac{Z_i(K_i^C - 1)}{1 + n_V(K_i^C - 1)} = 0 \tag{8.9}$$

where Z_i is the feed composition (mole fraction) and n_V are the overall moles of the equilibrium vapor phase, such that n_L (overall moles of the equilibrium liquid phase) = $1 - n_V$ if the basis for the total feed moles is 1. Equation 8.9 needs to be solved iteratively, with a value of n_V between 0 and 1. After convergence is achieved, i.e., the summation expressed by Eq. 8.9 is 0, the compositions or the mole fractions of the equilibrium vapor and liquid phases can be simply determined as follows:

$$\sum_{i=1}^{n} Y_i = \sum_{i=1}^{n} \frac{Z_i K_i^C}{1 + n_V(K_i^C - 1)} = 1 \tag{8.10}$$

$$\sum_{i=1}^{n} X_i = \sum_{i=1}^{n} \frac{Z_i}{1 + n_V(K_i^C - 1)} = 1 \tag{8.11}$$

Applications or use of above equations for various VLE calculations (unconfined or bulk systems) are shown elsewhere.[8,18]

The iterative process in VLE calculations can begin with an initial guess of equilibrium ratios from the Wilson[19] equation shown below

$$K_i(or\ K_i^C) = \frac{P_{ci}}{P} \mathrm{EXP}\left[5.37(1+\omega_i)\left(1 - \frac{T_{ci}}{T}\right)\right] \tag{8.12}$$

where P_{ci} is the critical pressure of component *i*, *P* is the system pressure, ω_i is the acentric factor of component *i*, T_{ci} is the critical temperature of component *i*, and *T* is the system temperature. In Eq. 8.12, pressures and temperatures in any absolute units can be used. For bubble and dew-point calculations under confinement, the following relationships apply, which in principle are similar to the ones without confinement, the difference being due to the equilibrium ratios used:

$$\sum_{i=1}^{n} Z_i K_i^C = 1 \tag{8.13}$$

$$\sum_{i=1}^{n} \frac{Z_i}{K_i^C} = 1 \tag{8.14}$$

By definition, bubble point is the point at which the first bubble of gas is formed and the quantity of gas is infinitesimal. This means $n_V \cong 0$ and $n_L \cong 1$ (assuming the basis to be 1 mol of feed), which means what is already known is $Z_i = X_i$, and thus the equilibrium is established for the newly formed gas or vapor phase, which leads to Eq. 8.13 by using the definition of K_i or $\left(K_i^C\right) = Y_i/X_i$. The exact opposite of this applies to the case of dew-point pressure calculations.

In order to illustrate the PT flash or VLE calculation procedure that includes capillary pressure a six-component mixture of CO_2 and CH_4 through n-C_5H_{12} is used (see Table 8.1).[18] This particular mixture is flashed at 25°F and 1,000 psia from single-phase conditions, thus forming an equilibrium vapor and liquid phase. The VLE calculation procedure without including the capillary pressure is first validated by using the Peng–Robinson EOS model (PR EOS),[20] as described and illustrated by Dandekar.[18] The calculation results

TABLE 8.1

Mixture Composition and Defined Properties of Individual Components

Components	Z_i, Mole Fraction	MW, g/g-mol	T_{ci}, °F	P_{ci}, psia	ω_i	$P_{\sigma i}$
CO_2	0.0728	44.01	87.76	1070.00	0.2239	78.0
CH_4	0.5812	16.04	−116.65	667.06	0.0114	77.0
C_2H_6	0.1821	30.07	89.92	706.59	0.0993	108.0
C_3H_8	0.0728	44.10	206.02	616.07	0.1524	150.3
n-C_4H_{10}	0.0546	58.12	305.56	550.56	0.2010	189.9
n-C_5H_{12}	0.0364	72.15	385.80	488.78	0.2510	231.5

(shown in the form of equilibrium ratios) are compared with values predicted by PR EOS from a well-known commercial VLE package, which agree very closely as shown in Figure 8.1.

Next, the same six-component mixture is now subjected to VLE calculations by including capillary pressure. A constant pore size (radius) of 20 nm is considered for calculating the gas–oil capillary pressure, P_{cGO} from Eq. 8.2 in which the contact angle is assumed to be zero (fully oil or hydrocarbon liquid phase wetting). The P_{cGO} value in Eq. 8.2 is determined by Eq. 8.3 with a modified exponent (Eq. 8.4). The compositions and molar densities required in Eq. 8.3 are determined from the VLE calculations as described below. The various steps in VLE calculations are as follows:

Step 1: As an initial estimate, K_i values at 25°F and 1,000 psia are calculated from the Wilson equation and the Rachford–Rice flash function is solved (Eq. 8.9), which gives the n_V value and subsequently the equilibrium vapor and liquid compositions from Eqs. 8.10 and 8.11, respectively. Next, using these compositions the EOS model is set-up for the vapor and liquid phases

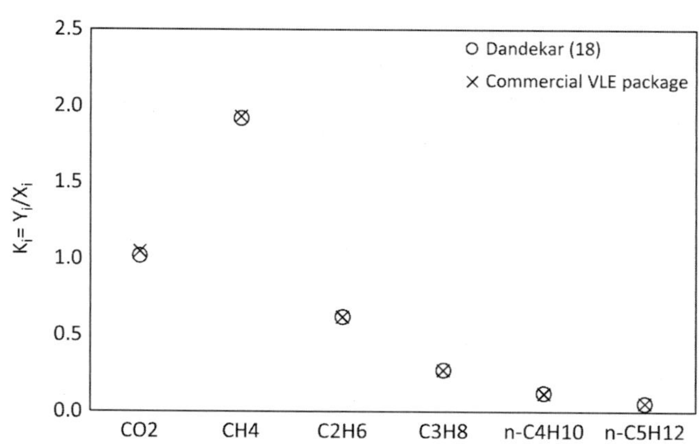

FIGURE 8.1
Validation of PT flash or VLE calculations without capillary pressure for a six-component synthetic mixture.

to calculate the fugacities of individual components and the convergence criteria of $\left[1-\frac{f_i^L}{f_i^V}\right]^2 \leq 10^{-10}$ is checked.

Step 2: Since the convergence criteria is seldom satisfied with the initial K_i values from Wilson equation, an iterative sequence is set-up that also includes the capillary pressure. However, first the value of σ_{GO} is calculated, for P_{cGO}, from the compositions and Z factors based vapor and liquid densities, respectively (from Step 1). These values are shown in Table 8.2.

Step 3: From the fugacity coefficients determined in Step 1, new K_i values are calculated from the ratio of Φ_i^L and Φ_i^V, which are then modified based on Eq. 8.6 to include the capillary pressure. The fugacity coefficients, uncorrected K_i values and the modified K_i values are shown in Table 8.3.

Step 4: The modified K_i values from Step 3 are now used to solve the Rachford–Rice flash function that gives the new equilibrium vapor and liquid phase compositions which are used to again set-up the EOS model for the vapor and liquid phases, respectively. However, while doing this the pressure of the vapor phase is adjusted by the previously calculated capillary

TABLE 8.2

Step 1 and 2 Flash Calculation Values at 25°F

Components	K_i (Eq. 8.12)	X_i, Mole Fraction	Y_i, Mole Fraction
CO_2	0.4568	0.1034	0.0472
CH_4	3.2625	0.2602	0.8490
C_2H_6	0.3205	0.2892	0.0927
C_3H_8	0.0611	0.1491	0.0091
n-C_4H_{10}	0.0132	0.1182	0.0016
n-C_5H_{12}	0.0033	0.0797	0.0003

$Z_L = 0.2192$; $Z_V = 0.7443$; $\rho_M^L = 0.01405$ g-mol/cm^3; $\rho_M^V = 0.0041$ g-mol/cm^3; $\sigma_{GO} = 3.25$ mN/m; $P_{cGO} = 47.11$ psi and $\left[1-\frac{f_i^L}{f_i^V}\right]^2 = 126.$

TABLE 8.3

Step 3 Flash Calculation at 25°F

Components	Φ_i^L	Φ_i^V	$K_i = \Phi_i^L/\Phi_i^V$	K_i^C(Eq. 8.6)
CO_2	0.6700	0.6572	1.0195	0.9715
CH_4	1.7442	0.8208	2.1251	2.0250
C_2H_6	0.2845	0.5043	0.5642	0.5376
C_3H_8	0.0735	0.3408	0.2157	0.2055
n-C_4H_{10}	0.0194	0.2312	0.0837	0.0798
n-C_5H_{12}	0.0054	0.1577	0.0343	0.0327

K_i^C values in column 5 are obtained by multiplying K_i values shown in column 4 by (1–47.11/1,000).

pressure such that $P_G = P_O + P_{cGO}$, i.e., $P_G = 1{,}000 + 47.11$ (value shown in Table 8.2), which is the value used in the determination of all EOS parameters for the vapor phase. Note that the oil phase pressure is kept the same as 1,000 psia. This step thus results in the new iterated values of fugacity coefficients, fugacities, and the equilibrium ratios $\left(K_i^C\right)$ that use the updated P_{cGO}, a new value of which is calculated in this step (see Table 8.4). Again, the convergence criteria are checked, and if it is not satisfied then the iteration proceeds to the next iteration, essentially repeating Step 4. Table 8.5, which is similar to Table 8.4 in which the convergence criteria are satisfied after 12 iterations, shows the final results for this particular fluid mixture.

Finally, the flashed equilibrium vapor and liquid phase compositions for the studied mixture with (under confinement) and without (bulk) capillary pressure are compared in Table 8.6. Although some differences are seen, the confinement effect on equilibrated fluid compositions appears to be small,

TABLE 8.4

Flash Calculation Results at 25°F after the 1st Iteration

Components	X_i, Mole Fraction	Y_i, Mole Fraction	Φ_i^L	Φ_i^V	K_i^C
CO_2	0.0739	0.0718	0.6656	0.6274	1.0314
CH_4	0.3759	0.7613	1.6150	0.8265	1.9000
C_2H_6	0.2416	0.1299	0.2841	0.4569	0.6047
C_3H_8	0.1262	0.0259	0.0768	0.2836	0.2633
n-C_4H_{10}	0.1071	0.0085	0.0211	0.1766	0.1164
n-C_5H_{12}	0.0751	0.0025	0.0061	0.1107	0.0539

$Z_L = 0.2234$; $Z_V = 0.6683$; $\rho_M^L = 0.01378\ \text{g-mol/cm}^3$; $\rho_M^V = 0.0048\ \text{g-mol/cm}^3$; $\sigma_{GO} = 1.99\ \text{mN/m}$; $P_{cGO} = 28.94$ psi and $\left[1 - \frac{f_i^L}{f_i^V}\right]^2 = 0.66$. K_i^C values in column 6 are obtained by multiplying the ratio of Φ_i^L and Φ_i^V by (1–28.94/1047.11).

TABLE 8.5

Converged Flash Calculation Results at 25°F after the 12th Iteration

Components	X_i, Mole Fraction	Y_i, Mole Fraction	Φ_i^L	Φ_i^V	K_i^C
CO_2	0.0717	0.0737	0.6631	0.6294	1.0272
CH_4	0.3961	0.7328	1.5912	0.8385	1.8500
C_2H_6	0.2309	0.1421	0.2845	0.4508	0.6153
C_3H_8	0.1208	0.0335	0.0776	0.2733	0.2769
n-C_4H_{10}	0.1050	0.0133	0.0216	0.1664	0.1264
n-C_5H_{12}	0.0753	0.0045	0.0063	0.1019	0.0604

$Z_L = 0.2247$; $Z_V = 0.6413$; $\rho_M^L = 0.01371\ \text{g-mol/cm}^3$; $\rho_M^V = 0.0049\ \text{g-mol/cm}^3$; $\sigma_{GO} = 1.77\ \text{mN/m}$; $P_{cGO} = 25.75$ psi and $\left[1 - \frac{f_i^L}{f_i^V}\right]^2 = 10^{-10}$. K_i^C values in column 6 are obtained by multiplying the ratio of Φ_i^L and Φ_i^V by (1–25.75/1025.75).

TABLE 8.6

Comparison of Flashed Equilibrium Vapor and Liquid Phase Compositions for the Studied Mixture with (Under Confinement) and without (Bulk) Capillary Pressure

	Y_i, Mole Fraction		X_i, Mole Fraction	
Component	**Unconfined (bulk)**	**Confined (20 nm)**	**Unconfined (bulk)**	**Confined (20 nm)**
CO_2	0.0743	0.0737	0.0708	0.0717
CH_4	0.7292	0.7328	0.3774	0.3961
C_2H_6	0.1444	0.1421	0.2339	0.2309
C_3H_8	0.0341	0.0335	0.1261	0.1208
n-C_4H_{10}	0.0134	0.0133	0.1113	0.1050
n-C_5H_{12}	0.0045	0.0045	0.0803	0.0753

which is mainly due to the low IFT values between the vapor and liquid phases, and relatively larger pore radius of 20 nm in consequently reducing the capillary pressure. Observations similar to these were also reported by Wang et al.[13] for the light oil and retrograde systems they studied, which are somewhat similar to the six-component mixture. Obviously, if one were to consider typical gas–(black) oil type systems then an order of magnitude higher IFT could be expected, obviously increasing the capillary pressure and thus enhancing the effect on the flashed equilibrium vapor and liquid phase compositions compared to bulk. Nevertheless, a detailed computational procedure has been provided here for phase equilibria calculations that include capillary pressure.

8.3.2 Flash or Vapor–Liquid Equilibria (VLE) Calculations using Modified Critical Temperature and Pressure

As mentioned earlier in the beginning of this chapter, another approach to incorporate the confinement effect in VLE calculations is by shifting the critical constants of the mixture constituents, i.e., T_c and P_c, which was shown by Pitakbunkate et al.[3] The methane (30.02 mol%) and ethane (69.98 mol%) binary system studied by Pitakbunkate et al.[3] is used to illustrate this. First, in order to validate the VLE calculations (in bulk) for this binary system at a given temperature (–10°F) the bubble point is calculated by the PR EOS as per the standard procedure shown by Dandekar.[18] At this chosen temperature, the calculated bubble points are 622,[18], 625, and 618 by two well-known commercial VLE packages and 629 psia reported by Pitakbunkate et al.,[3] which agree with each other quite well. Next, using the shifted critical temperature and pressure of methane and ethane the bubble-point calculations are repeated at the same chosen temperature. The shifted[3] T_c and P_c of methane and ethane for a 5-nm separation or confinement are –144.7°F, 355 psia, and 66.9°F, 355 psia, respectively. The suppressed bubble-point calculations using a well-known commercial VLE package is 400 psia which matches very well with the 402 psia value calculated by the procedure shown by Dandekar[18]; however, this is about 150 psia lower than what is reported by Pitakbunkate et al.[3] (see the phase envelope on their publications' Figure 15). Bubble-point calculations at –100°F were also repeated using the shifted T_c and P_c, again in this case the value calculated by Dandekar[18] and the well-known commercial VLE package agreed within 1 psi (179 and 178 psia, respectively), whereas the bubble point on the suppressed phase envelope of Pitakbunkate et al.[3] is 233 psia. These noted discrepancies are rather significant.

Sandoval et al.[21] also studied the phase behavior of hydrocarbon mixtures under confinement and presented an efficient and robust algorithm for tracing the entire phase envelope in the presence of capillary pressure. Their studies also confirm the "rotation" or shift of the original or bulk phase envelope around the fixed or anchor point of critical conditions, when capillary pressure is included in phase equilibria computations. One of the systems they studied is a simple binary system of methane (70 mol%) and *n*-butane (30 mol%) for which they compared the bubble-point branch of the bulk phase envelope with bubble point under 5-, 10-, 20-, and 50-nm confinement (referred to as capillary radii). Clearly, the largest confinement or smallest capillary radii of 5 nm shows the maximum shift (suppression) in the bubble-point curve compared to bulk. All their computations are based on the Soave–Redlich–Kwong (SRK) EOS model.[22] Given the discrepancy noted earlier for the methane-ethane binary system, PR EOS model-based bubble-point calculations for the methane–*n*-butane system are also carried out in bulk and by using the shifted critical temperatures and pressures. The shifted values reported in Figures 1 and 2 of Teklu et al.[14] for the 5-nm confinement were used for methane and *n*-butane, which are –164.3°F, 617.4 psia and 208.5°F and 499.8 psia, respectively. Note that the methane values differ significantly from Pitakbunkate et al.[3] Before comparing the bubble-point calculations using the shifted critical properties, the bulk bubble-point calculations are first validated at a selected temperature of –28°F. The PR EOS-based bubble point is 1,306.32 psia which compares relatively well with the 1,382 psia predicted[21] by the SRK EOS, although values using two different EOS models are compared the entire phase envelope predicted by the two EOS models are quite close and thus the comparison of 1,306.32 and 1,382 can be considered as acceptable. As is the case with Sandoval et al.'s [21] calculations, the binary interaction parameters are set to zero for the PR EOS calculations as well. Next, the PR EOS bubble-point calculations at –28°F are repeated using the shifted T_c, P_c values. Surprisingly, these calculations on the contrary result in an increase in the bubble point by more than 400 psi.

8.3.3 Bubble-Point Calculations for the Methane–*n*-Butane Binary System Including Capillary Pressure

Noting the difference observed above in bubble points calculated using the shifted critical properties, the bubble point for the methane–*n*-butane binary system at four different temperatures (–28°F, 8°F, 44°F, and 80°F) are performed by actually including the capillary pressure and later compared with the results of Sandoval et al.[21] In principle, the procedure for bubble-point calculations including capillary pressure is somewhat similar to the flash calculations described earlier for the six-component system. However, the stepwise methodology (for –28°F) is shown below. Defined properties of methane and *n*-butane can be found in Table 8.1. The capillary radius used in these calculations is 5 nm, which shows the maximum suppression in bubble point.

Step 1: The first calculation step involves the determination of normal or bulk bubble point. This value is 1,306.32 psia as mentioned before. The converged results for this step are shown in Table 8.7. The converged bubble-point calculation is an essential first step because it allows the determination of σ_{GO} and P_{cGO} to initiate the iteration sequence that includes the capillary pressure.

Step 2: Before proceeding to this step let us first go back to the fundamental definitions used earlier; we know that $f_i^V = Y_i P_G \Phi_i^V$ or by rearranging $Y_i = \frac{1}{P_G}\left[\frac{f_i^V}{\Phi_i^V}\right]$; however, at equilibrium $f_i^V = f_i^L$, thus, $Y_i = \frac{1}{P_G}\left[\frac{f_i^L}{\Phi_i^V}\right]$. At bubble point, the newly formed phase is the vapor phase for which the mole fractions should add up to one. Therefore, $\sum_{i=1}^{n} Y_i = 1 = \frac{1}{P_G}\sum_{i=1}^{n}\left[\frac{f_i^L}{\Phi_i^V}\right]$ or simply $P_G = \sum_{i=1}^{n}\left[\frac{f_i^L}{\Phi_i^V}\right]$, which basically allows us to calculate the updated pressure in an iterative bubble-point calculation that includes capillary pressure. Now, using the values of σ_{GO} and P_{cGO} from Step 1 (see Table 8.7), the liquid or oil phase pressure (P_O or P_L) is calculated as $P_O = P_G - P_{cGO}$ or P_O = 1,306.32 – 76.14 = 1,230.18 psi. Note that at this stage the P_G value is the bulk bubble point or last updated pressure which is converged, i.e., 1,306.32 psia.

Step 3: The K_i values shown in Table 8.7 are then modified to include the capillary pressure and then the vapor phase composition is updated. The liquid phase composition remains unchanged since this is the feed. Using P_G of 1,306.32 psia and P_O of 1,230.18 psia and the updated vapor phase

TABLE 8.7

Converged Bulk Bubble-Point Calculation of 1306.32 psia at –28°F

Components	X_i, Mole Fraction	Y_i, Mole Fraction	Φ_i^L	Φ_i^V	$K_i = \Phi_i^L/\Phi_i^V$
CH_4	0.7000	0.9670	0.9440	0.6833	1.3815
n-C_4H_{10}	0.3000	0.0330	0.0056	0.0507	0.1099

$Z_L = 0.2884$; $Z_V = 0.5551$; $\rho_M^L = 0.0156\ \text{g-mol/cm}^3$; $\rho_M^V = 0.0081\ \text{g-mol/cm}^3$; $\sigma_{GO} = 1.31\ \text{mN/m}$;

$P_{cGO} = 76.14$ psi and $\left[1 - \frac{f_i^L}{f_i^V}\right]^2 = 10^{-30}$.

composition the PR EOS is again set-up for the individual phases and the convergence criteria is checked. This basically constitutes the first iteration (see Table 8.8 for detailed values). Obviously, the convergence criteria is not satisfied and also the updated vapor phase compositions do not sum up to exactly one, since this will occur only at the correct converged bubble point.

Step 4: The iterative sequence described in Step 3 repeats until convergence criteria is satisfied. When that is achieved the vapor phase mole fractions will automatically add up to one. It should be noted that in the intermediate iteration steps, since the vapor phase mole fractions will not exactly add up to one the calculated σ_{GO} and P_{cGO} are considered as somewhat pseudo values, which will eventually become realistic when convergence is achieved. Table 8.9 shows the final results after multiple iterations.

Following the stepwise procedure described above, calculations of bubble-point pressure at three other temperatures for the 5-nm confinement as well as in bulk are also carried out and the results shown in Figure 8.2. Based on the convention of Sandoval et al.,[21] the 5-nm confinement bubble-point pressures shown in Figure 8.2 correspond to the liquid phase or oil phase pressures, P_O (also see Firincioglu[24]). Clearly, as seen in Figure 8.2 as the temperature increases and approaches the critical point (measured value reported by Sage et al.[23] is also shown) the suppression in the bubble point systematically decreases. The converged values of σ_{GO} and P_{cGO} at the four

TABLE 8.8

Step 3 or 1st Iteration Confined Bubble-Point Calculations at –28°F

Components	K_i (from Table 8.7)	K_i^C	Y_i, Mole Fraction
CH_4	1.3815	1.3010	0.9107
n-C_4H_{10}	0.1099	0.1035	0.0310

$\sigma_{GO} = 1.80\,\text{mN/m}$; $P_{cGO} = 104.69$ psi (both are treated as pseudo values since convergence is not yet achieved, $\left[1-\frac{f_i^L}{f_i^V}\right]^2 = 10^{-2}$, and updated Y_i values do not exactly sum up to one).

$K_i^C = K_i\,(\text{from Table 8.7}) \times \left(1-\frac{76.14}{1306.32}\right)$ and $Y_i = K_i^C \times X_i(\text{feed})$.

TABLE 8.9

Converged Confined Bubble-Point Calculation Results at –28°F

Components	X_i, Mole Fraction	Y_i, Mole Fraction	Φ_i^L	Φ_i^V	K_i^C
CH_4	0.7000	0.9754	1.0585	0.6993	1.3935
n-C_4H_{10}	0.3000	0.0246	0.0063	0.0704	0.0819

$Z_L = 0.2497$; $Z_V = 0.5973$; $\rho_M^L = 0.0154\ \text{g-mol/cm}^3$; $\rho_M^V = 0.0070\ \text{g-mol/cm}^3$; $\sigma_{GO} = 1.66\,\text{mN/m}$; $P_{cGO} = 96.31$ psi; $P_G = 1211.77$ psia; $P_O = 1{,}115.46$ psia and $\left[1-\frac{f_i^L}{f_i^V}\right]^2 = 10^{-31}$.

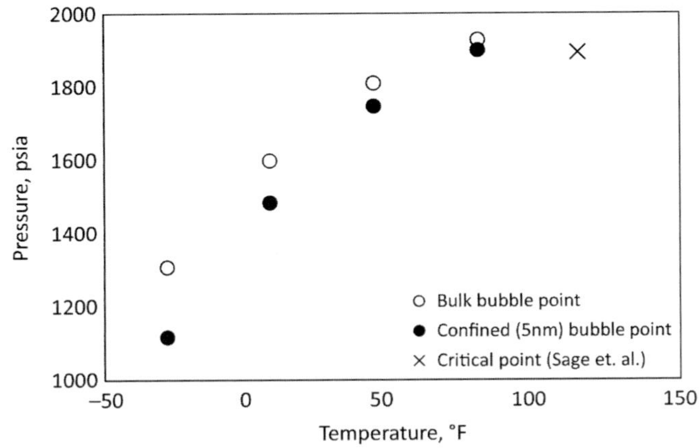

FIGURE 8.2
Comparison of PR EOS[20] predicted bulk and 5-nm confinement bubble-point pressures at various temperatures for the methane (70 mol%)–*n*-butane (30 mol%) binary system. In order to put the calculations in perspective, also shown in this plot is the measured critical point for this binary system reported by Sage et al.[23]

tested temperatures as shown in Figure 8.3 also supports this because IFT and consequently the capillary pressure reduces as one approaches the critical point, completely vanishing as phases lose their identity at the critical conditions. The calculation results for this binary system are also depicted in Figure 8.4 in the format of Sandoval et al.[21] (see their Figure 1), showing qualitative similarity. Finally, Figure 8.5 compares ΔP_b or $P_O - P_b$ (bulk)[21] vs. temperature for this binary system with the results of Sandoval et al.[21]

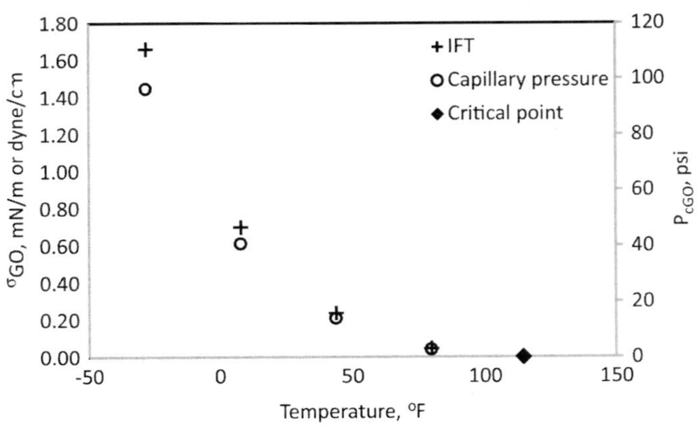

FIGURE 8.3
Gas–oil (or liquid phase) IFT and capillary pressure values as a function of temperature for the methane (70 mol%)–*n*-butane (30 mol%) binary system. Note that the corresponding pressures are implicit, i.e., they represent the 5-nm confined bubble points shown in Figure 8.2.

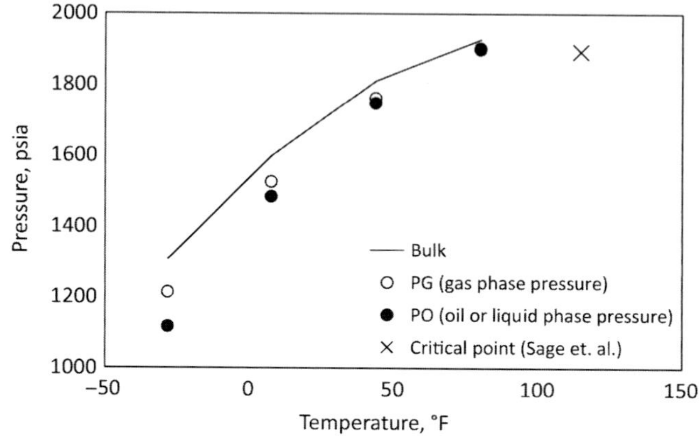

FIGURE 8.4
Bulk and confined bubble-point branch of the phase envelope of the methane (70 mol%)–*n*-butane (30 mol%) binary system that compares the respective phase pressures (see Figure 1 of Sandoval et al.[21] as a reference).

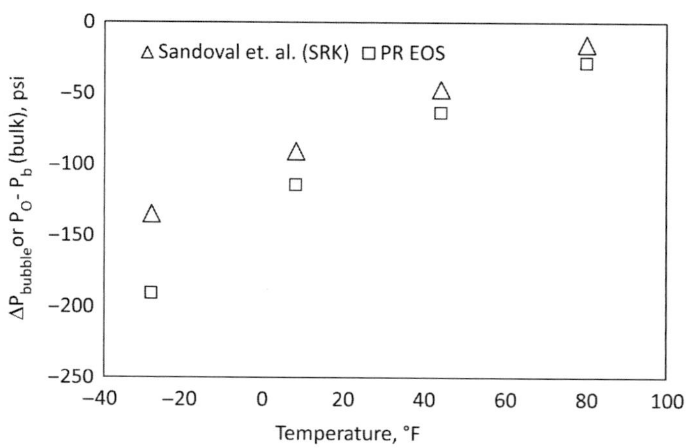

FIGURE 8.5
Comparison of ΔP_b or P_O–P_b (bulk) vs. temperature for the methane (70 mol%)–*n*-butane (30 mol%) binary system.

which shows reasonable trend wise agreement especially given the fact that two different EOS models and parachor model scaling exponents have been used.

8.3.4 Correcting Bulk PVT Data for Confinement

Currently, there is no documented systematic or otherwise methodology that exists for correcting or adjusting the bulk PVT data for pore confinement effects. The term adjustment or correction is referred to in the context of something that is typically done in the case of a black oil PVT data, i.e., a combination of CCE—separator test data and differential liberation—separator test data. In short the bulk PVT data cannot be simply adjusted by a "pore confinement multiplier". The closest and perhaps the only approach to what may be construed as adjustment is the analysis proposed by Sandoval et al.[21] that allows the determination of the saturation pressure shift (+ve or −ve) and the magnitude based only on the saturation pressure without confinement (or capillary pressure), which can be considered as the (measured or predicted) bulk saturation point. Their proposed relation basically provides a shift in the saturation pressure and relates to the capillary pressure and volumes of the original phase (liquid or vapor) and the incipient phase that is formed. Qualitatively, from an adjustment standpoint, something that is known is the fact that as the system becomes heavier and moves farther away from the critical temperature, higher is the degree of suppression in the bubble point (due to confinement), which is mainly because of the fact that the liquid phase becomes heavier thus increasing the gas–oil or gas–liquid IFT raising the capillary pressure.

Although correcting the bulk saturation pressure for pore confinement by simple relationships such as those proposed by Sandoval et al.[21] is perhaps possible, it is practically impossible to adjust or correct typical reservoir engineering parameters such as the solution GOR (R_s), oil formation volume factor (B_o) and the fluid properties such as density and viscosity below saturation pressure. Note that in the single-phase conditions the overall fluid composition is unchanged, i.e., no interphase exists and thus R_s, B_o, μ_o etc. will simply overlap each other for bulk and under confinement with the trend eventually shifting at the respective saturation pressures under different levels of confinement and in bulk (see Figure 1 in Zhang et al.[25]). Clearly, as seen in their figure none of the values are parallel to each other, that basically start branching out in the vicinity of the saturation pressures with the level of bifurcation pronouncing at lower pressures. The fundamental reason for this is the fact that due to different levels of confinement the compositional characteristics of the equilibrium vapor and liquid phases at each of the pressure steps are not the same, which in turn influence the volumetrics, i.e., the respective R_s and B_o values and, for example, the compositions and densities producing different viscosity values from compositional models such as the Lohrenz–Bray–Clark.[26] Additionally, respecting the fact that every reservoir fluid is unique in nature a simple adjustment in the form of a correction factor for pore confinement is fundamentally not possible.

In lieu of correcting the bulk PVT data for confinement, something that can be potentially done is as follows. If a representative fluid sample that is collected under conducive sampling conditions[4] is available and traditional PVT measurements are conducted (also refer to the last paragraph in this section) then these data can be used to tune or calibrate an EOS model by varying the pseudo and/or plus fraction properties. Such a tuned EOS model can then be employed in predicting the saturation pressure under a given level of pore confinement by a methodology described earlier. Subsequently, VLE or flash calculations can be carried out honoring the capillary pressure to simulate the CCE and differential liberation type sequences to produce the needed reservoir engineering parameters for reservoir simulation studies. Further improvement in the suggested approach can be made if at least the directly measured saturation pressure value under realistic pore confinement conditions is available. Based on the very recent patent by Wang et al.,[27] it appears that such measurements under reservoir type conditions and fluids may be a possibility in the very near future. Wang et al.[27] used nuclear magnetic resonance (NMR) to experimentally measure the dew-point pressures for a 96.24 mol% methane and 3.76 mol% *n*-butane binary system under confinement. If such measurements are directly made on representative shale gas or oil fluid samples under confined conditions, then these data can be used to tune an EOS model that also includes capillary pressure and then subsequently employed in predicting the rest of the needed reservoir engineering parameters, thereby increasing the overall accuracy of fluid phase behavior and property modeling. As pointed out by Whitson and Sunjerga,[28] estimation of saturation pressure (under confined conditions or even otherwise) is a critical reference parameter in optimizing the operational pressure range because large drawdown will lead to two-phase flow issues.[29]

The dew-point measurements by Wang et al.[27] were mostly at an isotherm of −18°C or −0.4°F for which they reported both the bulk and confined values obtained by NMR. As mentioned before, the confined dew-point pressure measured is perhaps the first given the fluid and the type of conditions in their experiments. They did not report the exact values of these dew-point pressures but stated that the nanopores in the shale sample used shifted up the dew point by ~115 psi above the bulk dew point. The dew-point values digitized from Figure 5 of their patent are ~1,242 psia and ~1,357 psia for bulk and under confinement, respectively, i.e., ~115 psi shift. Average pore size of the shale sample were not reported but based on the porous media characterization they stated that the porosity has significant contribution from 15 nm diameter (or 7.5 nm radius) pores. Outlined step wise below are the dew-point pressure calculations at −0.4°F for this binary system, which in principle are conceptually similar to the bulk and confined bubble-point calculations shown earlier.

Step 1: The first calculation involves the determination of bulk dew point, which is 1,246.60 psia. The converged results for this step are shown in Table 8.10. These values are used in the determination of σ_{GO} and P_{cGO} to initiate the iteration sequence that includes the capillary pressure. Note that in illustrating these calculations a pore radius of 7.5 nm is used.

Step 2: Using the fundamental definitions used earlier; we know that $f_i^L = X_i P_O \Phi_i^L$, rearranged as $X_i = \frac{1}{P_O}\left[\frac{f_i^V}{\Phi_i^L}\right]$ by invoking $f_i^L = f_i^V$ at equilibrium. At dew point the newly formed phase is the liquid phase for which the mole

TABLE 8.10

Converged Bulk Dew-Point Calculation of 1,246.60 psia at −0.4°F

Components	Y_i, Mole Fraction	X_i, Mole Fraction	Φ_i^L	Φ_i^V	$K_i = \Phi_i^L/\Phi_i^V$
CH_4	0.9624	0.5831	1.2296	0.7449	1.6506
$n\text{-}C_4H_{10}$	0.0376	0.4169	0.0098	0.1083	0.0902

$Z_L = 0.2802$; $Z_V = 0.6535$; $\rho_M^L = 0.0145$ g-mol/cm^3; $\rho_M^V = 0.0062$ g-mol/cm^3; $\sigma_{GO} = 2.49$ mN/m;

$P_{cGO} = 96.48$ psi and $\left[1-\dfrac{f_i^L}{f_i^V}\right]^2 = 10^{-29}$.

fractions should add up to one. Therefore, $\sum_{i=1}^{n} X_i = 1 = \dfrac{1}{P_O}\sum_{i=1}^{n}\left[\dfrac{f_i^V}{\Phi_i^L}\right]$ or simply $P_O = \sum_{i=1}^{n}\left[\dfrac{f_i^V}{\Phi_i^L}\right]$, allowing us to calculate the updated pressure in an iterative dew-point calculation that includes capillary pressure. With the values of σ_{GO} and P_{cGO} from Step 1 (Table 8.10), the vapor or gas phase pressure (P_G) is calculated as $P_G = P_O + P_{cGO}$ or $P_G = 1{,}246.60 + 96.48 = 1{,}343.08$ psi. Note that at this stage the P_O value is the bulk dew point or last updated pressure which is converged, i.e., 1,246.60 psia.

Step 3: The K_i values shown in Table 8.10 are modified next by including the capillary pressure and then the liquid phase composition is updated. Obviously, the vapor phase being the feed remains unchanged. Based on P_G of 1,343.08 psia and P_O of 1,246.60 psia and the updated liquid phase composition the PR EOS is again set-up for the individual phases and convergence criteria is checked, which basically marks the first iteration (see Table 8.11 for detailed values). As seen in the table, the convergence criteria are not satisfied and the updated liquid phase compositions do not sum up to exactly one as this occurs only at the correct converged dew point.

Step 4: The iterative sequence described in Step 3 repeats until convergence criteria is satisfied at which point the liquid phase mole fractions

TABLE 8.11

Step 3 or 1st Iteration Confined Dew-Point Calculations at −0.4°F

Components	K_i (from Table 8.10)	K_i^C	X_i, Mole Fraction
CH_4	1.6506	1.5320	0.6282
$n\text{-}C_4H_{10}$	0.0902	0.0837	0.4492

$\sigma_{GO} = 2.76$ mN/m; $P_{cGO} = 106.74$ psi (both are treated as pseudo values since convergence is not yet achieved, $\left[1-\dfrac{f_i^L}{f_i^V}\right]^2 = 10^{-2}$, and updated X_i values do not exactly sum up to one).

$K_i^C = K_i(\text{from Table 8.10}) \times \left(1-\dfrac{96.48}{1343.08}\right)$ and $X_i = Y_i(\text{feed})/K_i^C$

TABLE 8.12

Converged Confined Dew-Point Calculation Results at −0.4°F

Components	X_i, Mole Fraction	Y_i, Mole Fraction	Φ_i^L	Φ_i^V	K_i^C
CH_4	0.6306	0.9624	1.1886	0.7362	1.5262
$n\text{-}C_4H_{10}$	0.3694	0.0376	0.0105	0.0980	0.1018

$Z_L = 0.2760$; $Z_V = 0.6434$; $\rho_M^L = 0.0145\ \text{g-mol/cm}^3$; $\rho_M^V = 0.0066\ \text{g-mol/cm}^3$; $\sigma_{GO} = 1.84\,\text{mN/m}$;

$P_{cGO} = 71.14$ psi; $P_G = 1{,}301.24$ psia; $P_O = 1{,}230.10$ psia and $\left[1 - \dfrac{f_i^L}{f_i^V}\right]^2 = 10^{-30}$.

automatically add up to one. Table 8.12 shows the converged results after multiple iterations.

Considering the results obtained in the foregoing, although the shift in the dew-point pressure due to confinement is only ~55 psi (1,301.24–1,246.60), the absolute value of the calculated dew point under confinement compares remarkably well (a difference of ~4%) with the 1,357 psia digitized from Figure 5 of Wang et al.'s[27] patent. Again, the convention of Sandoval et al.[21] has been followed, i.e., the calculated gas phase pressure of 1,301.24 psia being the elevated dew-point pressure due to confinement. Figure 8.6 depicts the overall results for this binary system. Note that the calculations using shifted T_c and P_c of methane and n-butane were also attempted, which simply failed to converge.

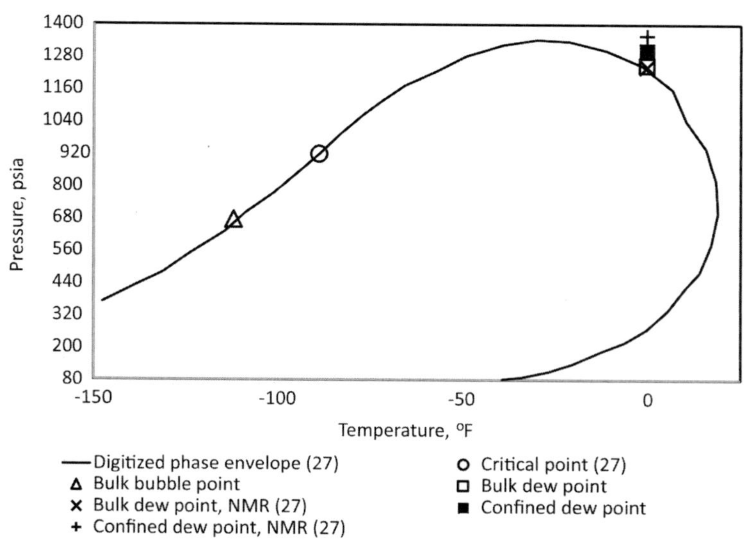

FIGURE 8.6

Comparison of various calculation results and experimental measurements[27] for the methane (96.24 mol%) and n-butane (3.76 mol%) binary system. In order to validate the calculations the bubble point of this system was also calculated at −112°F using the PR EOS. Note that the numbers in the parentheses represents the reference of Wang et al.[27]

Finally, Yang et al.[29] proposed a methodology for estimation of shale gas and oil PVT properties based on field data that may be much more readily available. Their work was inspired by (1) challenges in obtaining representative fluid samples; (2) contamination of samples by fracking and production operations and (3) scarcity of publicly reported PVT measurements due to confidentiality consequently resulting in the lack of PVT correlation studies for shale gas and oil. Note that although the methodology proposed by Yang et al.[29] implies the determination of typical or traditional PVT data, i.e., in bulk, it certainly is valuable from the standpoint of filling the data gaps and tuning or calibration of an EOS model that can then be used to include the capillary pressure to obtain the needed reservoir engineering parameters under confinement. The correlations developed and proposed by Yang et al.[29] are based on the PVT database (traditional or conventional is implied here) from the Eagle Ford area from more than 50 wells that were sampled in early production stages that span almost the entire fluid type spectrum from dry gases to black oils. They use the field GOR or the Condensate to Gas Ratio, CGR, which are likely to be readily available as the basic predictor since it conforms with typical fluid properties, i.e., (1) °API $\propto$ GOR; (2) saturation pressure $\propto$ GOR for oil systems; (3) saturation pressure $\propto$ 1/GOR for gas systems; (4) methane mole fraction in the reservoir fluid $\propto$ GOR; (5) C_{7+} mole fraction in the reservoir fluid $\propto$ 1/GOR and (6) a proprietary gas index of surface gas $\propto$ 1/GOR. Unfortunately, no numerical values on plots relating each of these properties to GOR and specifics of the gas index were provided in their publication; however, the concept and the given examples demonstrate the methodology that can be adapted to estimate the in situ reservoir fluid composition (much needed in compositional modeling including the confinement effect) and corresponding PVT data from readily available field information.

8.4 Handling of Porous Media Effects on Phase Behavior in a Compositional Reservoir Simulator

Currently, to the best of our knowledge, commercial compositional reservoir simulators that are typically used in the petroleum industry do not have the option of incorporating porous media effects on phase behavior. Stimpson and Barrufet[30] have also recognized this and have stated that confinement effects may be a significant factor. Most of the efforts in incorporating confinement effects have focused on the development of customized in-house reservoir simulators that often assume somewhat simplified conditions such as homogeneity and single uniform pore sizes. Therefore, this appears to be the current state-of-the-art as far as compositional reservoir simulation for confined or unconventional systems is concerned. A common theme that emerges from these in-house simulation studies[25,30,31] is the fact that inclusion of capillary pressure has a positive effect on the cumulative oil recovery, which is mainly attributed to the suppressed bubble-point pressure leaving "excess" gas in solution resulting in decreased oil density and viscosity improving the oil mobility.

References

1. Firincioglu, T., Ozgen, C., and Ozkan, E. 2013. An excess-bubble-point-suppression correlation for black oil simulation of nano-porous unconventional oil reservoirs. *Society of Petroleum Engineers*. doi:10.2118/166459-MS.
2. Shapiro, A.A., Potsch, K., Kristensen, J.G., and Stenby, E.H. 2000. Effect of low permeable porous media on behavior of gas condensates. *Society of Petroleum Engineers*. doi:10.2118/65182-MS.
3. Pitakbunkate, T., Balbuena, P.B., Moridis, G.J., and Blasingame, T.A. 2016. Effect of confinement on pressure/volume/temperature properties of hydrocarbons in shale. *Society of Petroleum Engineers (SPE) Journal* 21(2): 621–634. doi:10.2118/170685-PA.4.
4. Honarpour, M.M., Nagarajan, N.R., Orangi, A., Arasteh, F., and Yao, Z. 2012. Characterization of critical fluid PVT, rock, and rock-fluid properties—Impact on reservoir performance of liquid rich shales. In *SPE Annual Technical Conference and Exhibition*. Society of Petroleum Engineers. doi:10.2118/158042-MS.
5. Nojabaei, B. 2015. Phase behavior and flow analysis of shale reservoirs using a compositionally-extended black-oil approach. *PhD Thesis*, The Pennsylvania State University.
6. Orangi, A., and Nagarajan, N.R. 2015. Unconventional shale gas-condensate reservoir performance: Impact of rock, fluid, and rock-fluid properties and their variations. *Unconventional Resources Technology Conference*. doi:10.15530/URTEC-2015-2170061.
7. www.npd.no/engelsk/cwi/pbl/wellbore_documents/426_34_10_3_CORE_ANALYSIS_REPORT.PDF. Date accessed June 11, 2017.
8. Danesh, A. 1998. *PVT and Phase Behaviour of Petroleum Reservoir Fluids*. Edited by A. Danesh. Elsevier.
9. Dandekar, A.Y. 1994. *Interfacial Tension and Viscosity of Petroleum Reservoir Fluids*. Edinburgh: Heriot-Watt University.
10. Weinaug, C.F. and Katz, D.L. 1943. Surface tensions of methane-propane mixtures. *Industrial & Engineering Chemistry* 35(2): 239–246. doi:10.1021/ie50398a028.
11. Luo, S., Lutkenhaus, J.L., and Nasrabadi, H. 2016. Use of differential scanning calorimetry to study phase behavior of hydrocarbon mixtures in nano-scale porous media. *Journal of Petroleum Science and Engineering*. doi:10.1016/j.petrol.2016.12.019.
12. Firoozabadi, A. and Katz, D.L. 1988. Surface tension of reservoir crude oil/gas systems recognizing the asphalt in the heavy fraction. *SPE Reservoir Engineering* 3: 265–272. doi:10.2118/13826-PA.

13. Wang, L., Yin, X., Neeves, K.B., and Ozkan, E. 2016. Effect of pore-size distribution on phase transition of hydrocarbon mixtures in nanoporous media. *SPE Journal* 21(6). doi:10.2118/170894-PA.
14. Teklu, T.W., Alharthy, N., Kazemi, H., Yin, X., Graves, R.M., and AlSumaiti, A.M. 2014. Phase behavior and minimum miscibility pressure in nanopores. *SPE Reservoir Evaluation & Engineering* 17(3): 396–403. doi:10.2118/168865-PA.
15. Dong, X., Liu, H., Hou, J., and Chen, Z. 2016. Phase behavior of hydrocarbon mixtures in the organic nanopores of unconventional gas condensate reservoirs. In *Proceedings of the 4th Unconventional Resources Technology Conference.* Tulsa, OK: American Association of Petroleum Geologists. doi:10.15530/urtec-2016-2460485.
16. Fathi, E., Elamin, A., and Ameri, S. 2013. Simulation of multicomponent gas flow and condensation in Marcellus shale reservoir. Society of Petroleum Engineers. doi:10.2118/164538-MS.
17. Rachford, H.H. Jr. and Rice, J.D. 1952. Procedure for use of electronic digital computers in calculating flash vaporization hydrocarbon equilibrium. *Journal of Petroleum Technology* 4(10). doi:10.2118/952327-G.
18. Dandekar, A.Y. 2013. *Petroleum Reservoir Rock and Fluid Properties.* CRC Press.
19. Wilson, G. 1968. A modified Redlich-Kwong EOS, application to general physical data calculations. In *Annual AIChE National Meeting.* Cleveland, OH: American Institute of Chemical Engineers.
20. Robinson, D.B. and Peng, D.-Y. 1978. The characterization of the heptanes and heavier fractions for the GPA Peng-Robinson programs. Research report/Gas Processors Association, RR-28; Research report (Gas Processors Association), RR-28. Tulsa, OK: Gas Processors Association.
21. Sandoval, D.R., Yan, W., Michelsen, M.L., and Stenby, E.H. 2016. The phase envelope of multicomponent mixtures in the presence of a capillary pressure difference. *Industrial & Engineering Chemistry Research* 55(22): 6530–38. doi: 10.1021/acs.iecr.6b00972.
22. Soave, G. 1972. Equilibrium constants from a modified Redlich-Kwong equation of state. *Chemical Engineering Science* 27(6): 1197–1203. doi:10.1016/0009-2509(72)80096-4.
23. Sage, B.H., Budenholzer, R.A., and Lacey, W.N. 1940. Phase equilibria in hydrocarbon systems methane–n butane system in the gaseous and liquid regions. *Industrial & Engineering Chemistry* 32(9): 1262–1277. doi: 1021/ie50369a047.
24. Firincioglu, T. 2013. Bubble point suppression in unconventional liquids rich reservoirs and its impact on oil production. *PhD Thesis,* Colorado School of Mines.
25. Zhang, Y., Yu, W., Sepehrnoori, K., and Di, Y. 2017. Investigation of nanopore confinement on fluid flow in tight reservoirs. *Journal of Petroleum Science and Engineering* 150(Supplement C): 265–271. doi:10.1016/j.petrol.2016.11.005.
26. Lohrenz, J., Bray, B.G., and Clark, C.R. 1964. Calculating viscosities of reservoir fluids from their compositions. *Journal of Petroleum Technology* 16(10). 1171–1176. doi:10.2118/915-PA.
27. Wang, H., Sun, B., Seltzer, S.J., Zintsmaster, J.S., and Alboudwarej, H. 2017. Pore Fluid Phase Behavior Measurements. US Patent 2017/0285215 A1, issued 2017.
28. Whitson, C.H. and Sunjerga, S. 2012. PVT in liquid-rich shale reservoirs. In *SPE Annual Technical Conference and Exhibition.* Society of Petroleum Engineers. doi: 10.2118/155499-MS.

29. Yang, T., Basquet, R., Callejon, A., Van Roosmalen, J.J., and Bartusiak, B. 2014. Shale PVT estimation based on readily available field data. *Unconventional Resources Technology Conference*. doi:10.15530/URTEC-2014-1884129.
30. Stimpson, B.C. and Barrufet, M.A. 2016. Effects of confined space on production from tight reservoirs. *Society of Petroleum Engineers*. doi:10.2118/181686-MS.
31. Yan, B., Wang, Y. and Killough, J.E. 2017. A fully compositional model considering the effect of nanopores in tight oil reservoirs. *Journal of Petroleum Science and Engineering* 152(Supplement C): 675–682. doi:10.1016/j.petrol.2017.01.005.

9

Compositional and Phase Behavior Effects in Conventional and Exotic Heavy Oil EOR Processes

9.1 CO_2 Induced Hydrocarbon Liquid–Liquid Phase Split and Phase Behavior Type

In the discussion presented in Section 7.5.1 of Chapter 7, it was stated that binary systems of CO_2 and certain alkanes exhibit a rather intricate and nontrivial phase behavior as it pertains to the liquid–liquid (LL) split (see Table 7.2). Note that these are model or synthetic binary mixtures or expressed as pseudo binaries that are purposefully prepared to study the complex phase behavior. However, as far as CO_2–crude oil systems are concerned, the nontrivial phase behavior and in particular the LL phase split can be considered as occurrence or outcome of injecting CO_2 in enhanced oil recovery (EOR) applications. As mentioned in Chapter 7, in such applications, the oil intrinsically does not contain high CO_2 but the overall mol% of CO_2 may reach as high as 60%–70% in the total mixture later during the injection process, potentially promoting the complex phase behavior. For this reason, we refer to this as "CO_2 induced" LL phase split/immiscibility, and/or the complex phase behavior, something that has been observed for certain types of heavier oils[1–7] and their custom made fractions,[8,9] at relatively lower temperatures. The pseudo binary concept used in Chapter 7 can also be applied to CO_2–crude oil systems in order to better understand the intricate phase behavior. As a matter of fact, the CO_2–crude oil fractions' phase behavior reported by Lucas et al.[8] and Kokal and Sayegh[9] can be considered as direct pseudo binaries (directly referred to as such in Lucas et al.[8]), since in both publications the *n*-alkane molecular weight equivalency with the actual fraction molecular weights has been used for the fractions for comparison. As a matter of fact, Kokal and Sayegh[9] used the binaries of CO_2–$nC_{12}H_{26}$ (*n*-dodecane) and CO_2–$nC_{33}H_{68}$ (*n*-tritriacontane) to compare the phase behavior with CO_2–Fraction 1 and CO_2–Fraction 2, respectively, given the similar molecular weights of the two *n*-alkanes and the fractions. Similarly, the four different oil fractions, namely F1–F4, that were studied by Lucas et al.[8] were considered as analogs of *n*-octane, *n*-undecane, *n*-hexadecane, and *n*-octacosane, respectively, given the molecular weight similarities.

For crude oils, the pseudo binary concept is as follows. CO_2 is treated as a discrete component, whereas rest of the components are recombined as one pseudo component for which the molecular weights are calculated to place or compare them according to the well-defined CO_2–normal alkane binary system phase behavior types. Note that the molecular weights for the crude oil systems[1–7] are CO_2 free, i.e., the given values are calculated after removing the inherently present CO_2 in the crude oil and the phase behavior type identified for crude oil systems is based on the analogy with *n*-alkanes which also is confirmed (if available) from the experimental data reported in the relevant cases/references. As far as the fractions are concerned, the molecular weights are values that are actually or directly reported.[8,9] Table 9.1 which is nearly identical to Table 7.2 categorizes the phase behavior type and other observed characteristics for different crude oils and oil fractions. As can be seen in Table 9.1, for the most part the heavier oils seem to fall in the Type III category of phase behavior type.

TABLE 9.1

Categorization of Various CO_2–Crude Oil and Oil Fraction Systems According to Phase Behavior Types and Other Observed Characteristics

Crude Oils or Fractions	Molecular Weight, lbm/lbm–mol	Phase Behavior Type
Crude oil (Wasson)[1] Reservoir or test temperature = 105°F	122.6–127.7	Type II Exhibits a three-phase lower liquid, upper liquid, and vapor at high CO_2 concentrations
Crude oil (Maljamar)[2] Reservoir or test temperature = 90°F	183.7	Borderline Type IV Same experimental observations as Wasson
Crude oil (West Texas)[3] Reservoir or test temperature = 94°F	161.6	Type II Experiments indicated two liquid phases, an oil-rich liquid and a CO_2-rich liquid at high CO_2 concentrations
Crude oil (Sulimar Queen)[4] Reservoir or test temperature = 95°F and 138°F	194.8	Construed as Type IV Upper and lower liquid formations not explicitly reported
Crude oil (Spraberry)[4] Reservoir or test temperature = 95°F and 138°F	180.9	Construed as Type IV Same as Sulimar Queen
Crude oil (West Sak)[5] Reservoir or test temperature = 80°F	229.7	Type III Experimental evidence of immiscible upper and lower liquid phase

(*Continued*)

TABLE 9.1 (*Continued*)

Categorization of Various CO_2–Crude Oil and Oil Fraction Systems According to Phase Behavior Types and Other Observed Characteristics

Crude Oils or Fractions	Molecular Weight, lbm/ lbm–mol	Phase Behavior Type
Crude oil (A)[6] Reservoir or test temperature = 105.9°F	150.1	Type II Experiments indicated two liquid phases, an oil-rich liquid and a CO_2-rich liquid at high CO_2 concentrations
Crude oil (A, B1, B2, C1, C2 and F)[7] Reservoir or test temperature = 94°F–106°F	129–151.8	Type II Experiments indicated two liquid phases, an oil-rich liquid and a CO_2-rich liquid at high CO_2 concentrations. Equilibrium vapor phase for oil B2 and C2 at lower pressures.
Crude oil (D)[7] Reservoir or test temperature = 105°F	182.4	Type II to Type IV Experiments indicated two liquid phases, an oil-rich liquid and a CO_2-rich liquid at high CO_2 concentrations. Equilibrium vapor phase at lower pressures.
Crude oil fractions (F1 and F2)[8] Reservoir or test temperature = 68°F–176°F	117.7 and 160.9	Type I Stated as such by authors; however, according to Table 7.2 these should be Type II
Crude oil fraction (F3)[8] Reservoir or test temperature = 68°F–176°F	231.1	Type III Stated as such by authors and also consistent with Table 7.2; experimental evidence of LLE and VLLE at high CO_2 concentrations
Lone Rock Heavy oil lighter Fraction 1[9] Reservoir or test temperature = 69.5°F–283.7°F	170	Type II According to Table 7.2
Lone Rock Heavy oil heavier Fraction 2[9] Reservoir or test temperature = 69.5°F–283.7°F	465	Type III According to Table 7.2; experimental evidence of LLE and VLE at high CO_2 concentrations
Crude oil (Lone Rock heavy oil)[9] Reservoir or test temperature = 69.5°F–283.7°F	386	Type III According to Table 7.2; experimental evidence of LLE, VLLE (narrow) and VLE at high CO_2 concentrations
Crude oil (deasphalted "De-oil" from Lone Rock heavy oil)[9] Reservoir or test temperature = 69.5°F–283.7°F	347	Type III According to Table 7.2; experimental evidence of VLLE (narrow) at high CO_2 concentrations

A pseudo binary concept is used for the crude oils for expressing the *n*-alkane equivalent molecular weight (see description in the text), whereas, actual reported values are listed for the oil fractions.[8,9] The superscripts represent the literature references.

9.2 Experimental Observations and EOS Modeling of Hydrocarbon Vapor–Liquid–Liquid or Liquid–Liquid Phase Split

9.2.1 Experimental Observations

The experimental observations as they pertain to the phase envelopes or PT boundaries in conjunction with the CO_2mol% in the selected oils are portrayed in a series of figures for some of the oils.

Figures 9.1–9.3 present the CO_2 and phase behavior data of West Texas,[3] Wasson,[1] and Maljamar[2] separator oils. The descriptive individual figure

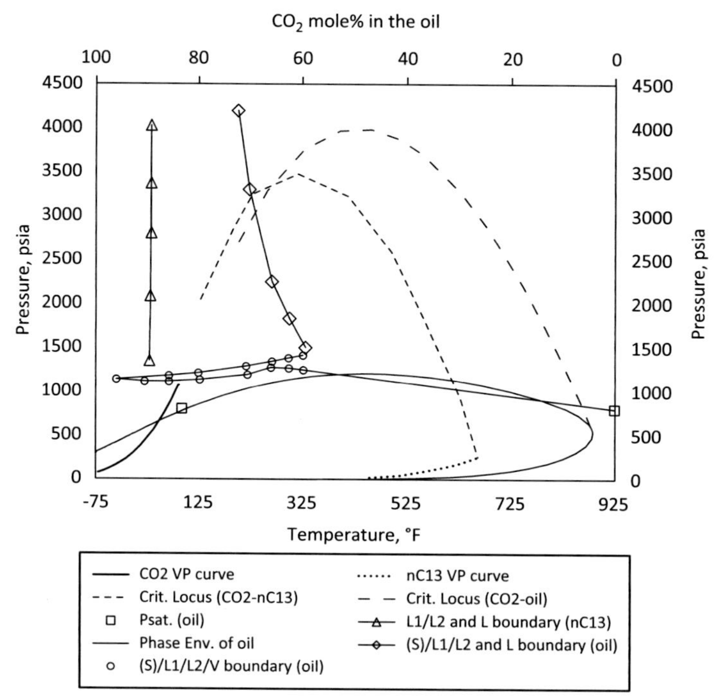

FIGURE 9.1

CO_2–West Texas oil phase behavior (treated as a pseudo binary, and Type II based on molecular weight cutoff) compared with CO_2–$nC_{13}H_{28}$ binary system (representing a Type II to Type III transition). The measured oil data used to construct the plot are from Shelton and Yarborough[3] and data for the $nC_{13}H_{28}$ system are digitized from the plot presented by Vitu et al.[10] The oil saturation or bubble point pressure is at 94°F (shown as a PT point as well as at 0 mol% CO_2). The touchdown point of the phase envelope (predicted by PR EOS[11]) of the original oil (no CO_2) with the critical locus of CO_2–oil marks the oil critical point. The L1 (CO_2 rich) and L2 ($nC_{13}H_{28}$ rich) and L boundary for the CO_2–$nC_{13}H_{28}$ system are shown as PT points, whereas these boundaries for the CO_2–oil system are shown as CO_2mol% vs. pressure since these measurements are at a constant reservoir temperature of 94°F. Also, shown for completeness is the observed solid (S) asphaltene phase, thus indicating the existence of a four-phase equilibrium.

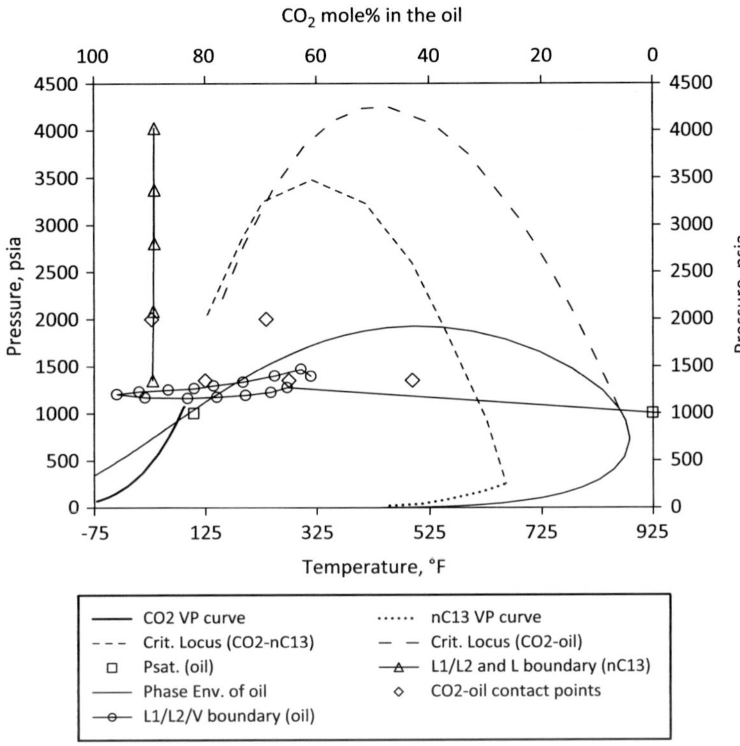

FIGURE 9.2
CO_2–Wasson oil phase behavior (treated as a pseudo binary, and Type II based on molecular weight cutoff) compared with CO_2–$nC_{13}H_{28}$ binary system (representing a Type II to Type III transition). The measured oil data used to construct the plot is from Gardner et al.[1] and data for the $nC_{13}H_{28}$ system is digitized from the plot presented by Vitu et al.[10] The oil saturation or bubble point pressure is at 105°F (shown as a PT point as well as at 0 mol% CO_2). The touch-down point of the phase envelope (predicted by PR EOS[11]) of the original oil (no CO_2) with the critical locus of CO_2–oil marks the oil critical point. The L1 (CO_2 rich) and L2 ($nC_{13}H_{28}$ rich) and L boundary for the CO_2–$nC_{13}H_{28}$ system is shown as PT points, whereas these boundaries for the CO_2–oil system are shown as CO_2mol% vs. pressure since reservoir temperature of 105°F is constant. Also shown are four CO_2mol% vs. pressures at which Gardner et al.[1] conducted single contact experiments in which two liquid and/or a vapor phase were observed (note that the solid precipitate is neglected).

captions explain how the data have been presented and compared with a CO_2–normal alkane system; however, the salient and somewhat common features of the two liquid phases (one is CO_2 rich while the other is oil rich) that are in equilibrium when the oil is brought in contact with CO_2 are discussed. All three oil systems are at low temperatures in the range of 90°F–105°F, respectively, and result in a narrow region or envelope that marks the equilibria between the two liquid phases and the vapor phase. The presence of a solid asphaltene phase also has been identified; however, for the most part it has been qualitatively described. These plots also

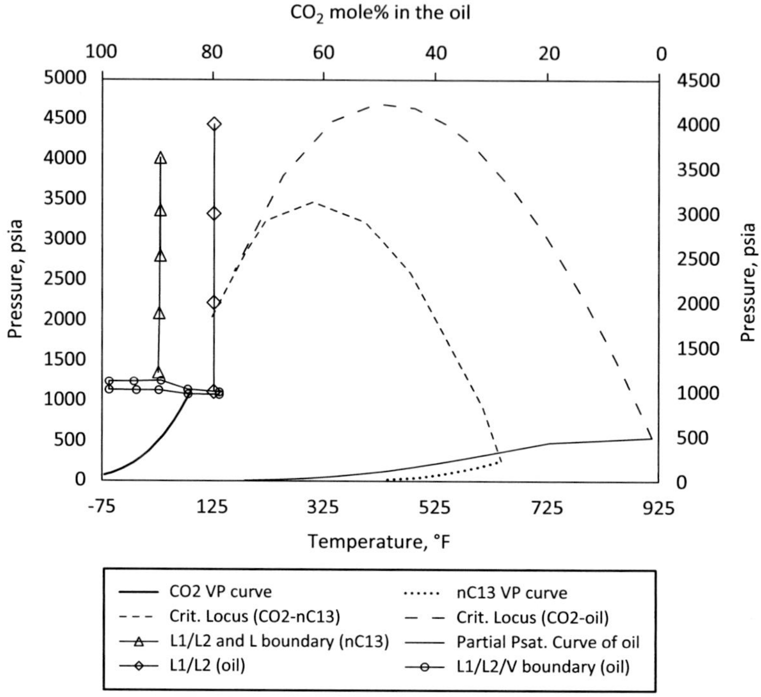

FIGURE 9.3
CO_2–Maljamar separator oil phase behavior (treated as a pseudo binary, and presenting a Type IV based on molecular weight cutoff) compared with CO_2–$nC_{13}H_{28}$ binary system (presenting a Type IV behavior, see Table 7.2 and the description). The measured oil data used to construct the plot are from Orr et al.[2] and data for the $nC_{13}H_{28}$ system are digitized from the plot presented by Vitu et al.[10] Note that Orr et al.[2] also caption the phase behavior as "Phase behavior of binary mixtures of CO_2 and Maljamar separator oil." The touchdown point of the partial bubble point curve (predicted by PR EOS[11]) of the separator oil (no CO_2) with the critical locus of CO_2–oil marks the oil critical point. The L1 (CO_2 rich) and L2 ($nC_{13}H_{28}$ rich) and L boundary for the CO_2–$nC_{13}H_{28}$ system are shown as PT points. The L1/L2/V and L1/L2 data are shown as CO_2 mol% vs. pressure since temperature of 90°F is constant. Orr et al.'s[2] CO_2 and oil contact experiments revealed two liquid phases (no vapor) between 1,000 and 4,000 psia, whereas the vapor phase appeared at 970 psia, which is approximately marked by the point at which the L1–L2 and L1/L2/V meet.

indicate that the three phases (or four counting asphaltene) coexist between 1,000 and 1,500 psia and at high CO_2 concentrations (~60%+; 80% in the case of Maljamar) in the mixture. At relatively higher pressures (up to 9,000 psia[3]), the vapor phase obviously disappears when only two liquid phases in equilibrium remain, which is something that also is the case with CO_2 and normal alkanes n-$C_{13}H_{28}$ and above (see Vitu et al.[10]). The liquid – liquid (LL) phase coexistence is typically a vertical line projecting upward from the end of the three-phase equilibria (see Figures 9.1–9.3). Orr et al.[2] have stated that liquid–liquid–vapor and LL equilibria in the CO_2–oil systems is a low-temperature phenomenon occurring at temperatures not too far above the CO_2 critical temperature of ~88°F and less than about 120°F. Note that all three oils as well as those studied by Turek et al.[6,7] (see Table 9.1) fall in this particular temperature range. A somewhat similar behavior also is seen while reviewing the CO_2 and higher normal alkane phase envelopes shown by Vitu et al.[10] In particular, the experimental data of Fall and Luks[12] for the CO_2–$nC_{13}H_{28}$ binary system showing the presence of a CO_2-rich and $nC_{13}H_{28}$-rich phase fall right into the ~100°F temperature and 1,200 psia pressure range.

Figure 9.4 compares the critical locus of the CO_2–nC_8H_{18} binary system with the measured pressure–composition–temperature data reported by Lucas et al.[8] for the lightest oil fraction, F1, which from a molecular weight standpoint is analogous to nC_8H_{18} and exhibits a Type II phase behavior according to the classification shown in Table 7.2. Although this system does not result in any LL or VLLE split and shows only a VL transition or bubble point at various temperatures, it is compared for completeness and to show the subsequent transition into the more complicated phase behavior that occurs with increasing molecular weights. Figure 9.4 shows the experimental pressure–composition–temperature data reported by Yu et al.[13] for

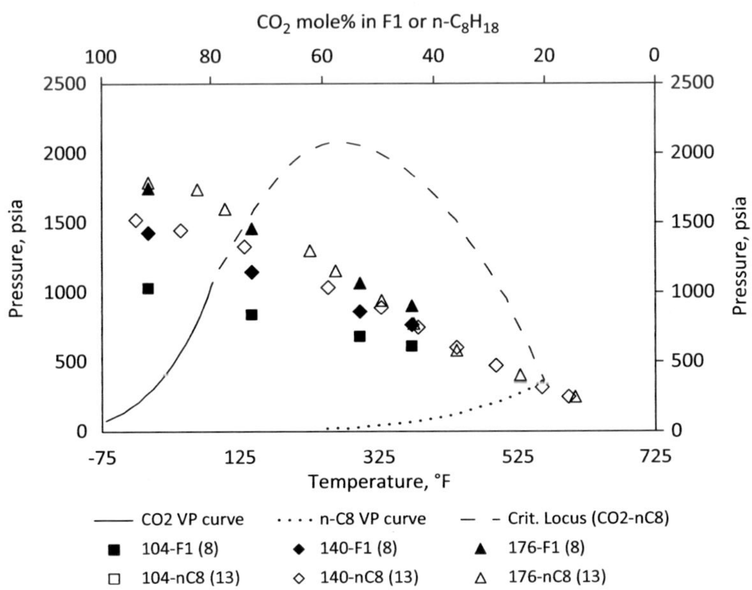

FIGURE 9.4
CO_2–F1 oil fraction phase behavior (treated as a pseudo binary, and presenting a Type II based on molecular weight cutoff) compared with CO_2–nC_8H_{18} binary system (presenting a Type II behavior, see Table 7.2 and the description). The measured pressure–composition–temperature (numbers in the legends) data used to construct the plot are from Lucas et al.[8] (F1 fraction) and Yu et al.[13] (nC_8H_{18}), respectively. The critical locus data are digitized from a curve presented by Yu et al.[13] that contains several experimental PT points all along the locus. Numbers in parentheses in the legends indicate the references.

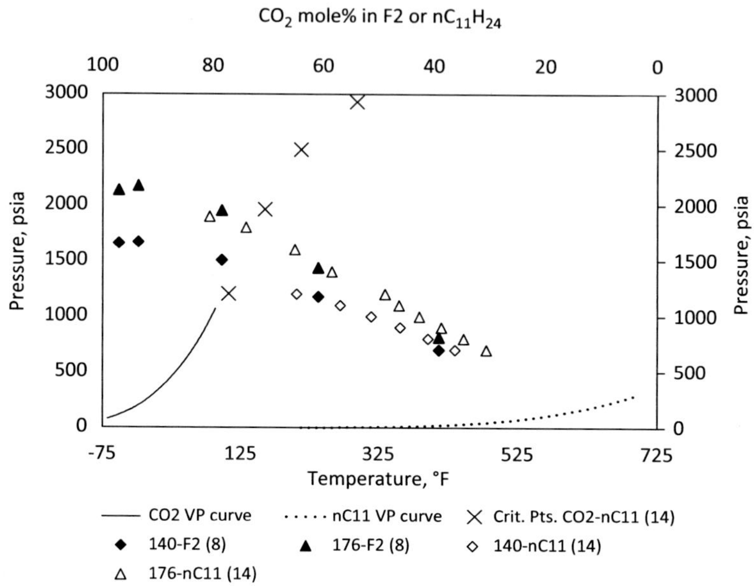

FIGURE 9.5
CO_2–F2 oil fraction phase behavior (treated as a pseudo binary, and presenting a Type II based on molecular weight cutoff) compared with CO_2–$nC_{11}H_{24}$ binary system (presenting a Type II behavior, see Table 7.2 and the description). The measured pressure–composition–temperature (numbers in the legends) data used to construct the plot are from Lucas et al.[8] (F2 fraction) and Camacho-Camacho et al.[14] ($nC_{11}H_{24}$), respectively. The four critical points shown are extrapolated from experimental data of Camacho-Camacho et al.[14] by applying the condition of equilibrium ratios of 1 for CO_2 and $nC_{11}H_{24}$. Numbers in parentheses in the legends indicate the references.

CO_2–nC_8H_{18} binary system aligning quite well with the similar type of data measured by Lucas et al.,[8] but for a CO_2–F1 system which confirms the validity of the pseudo binary concept based on the molecular weight cutoff. Figure 9.5 shows a similar analogy between the CO_2–F2 and CO_2–$nC_{11}H_{24}$ binary systems, which again confirms the molecular weight cutoff-based pseudo binary concept.

In order to further demonstrate the pseudo binary concept, Figures 9.6a, b and 9.7 show the comparison between CO_2–F3 oil fraction system and the CO_2–$nC_{16}H_{34}$ binary that exhibits a Type III phase behavior. First, the pressure–composition phase behavior at 104°F and 140°F is compared in Figure 9.6a, b showing the mole fraction of CO_2 in the F3 oil fraction and $nC_{16}H_{34}$, respectively, which includes the data for the latter from two different publications.[15,16] As seen in the plots, the VLE data align reasonably well for both temperatures. Note that in the case of the low temperature of 104°F the VLE data of CO_2–F3 begin to bifurcate from a CO_2mol% of 73.9 into VLLE (low pressure) and LLE (high pressure), respectively. As expected, the LL immiscibility (with the vapor phase at lower pressures) is a typical

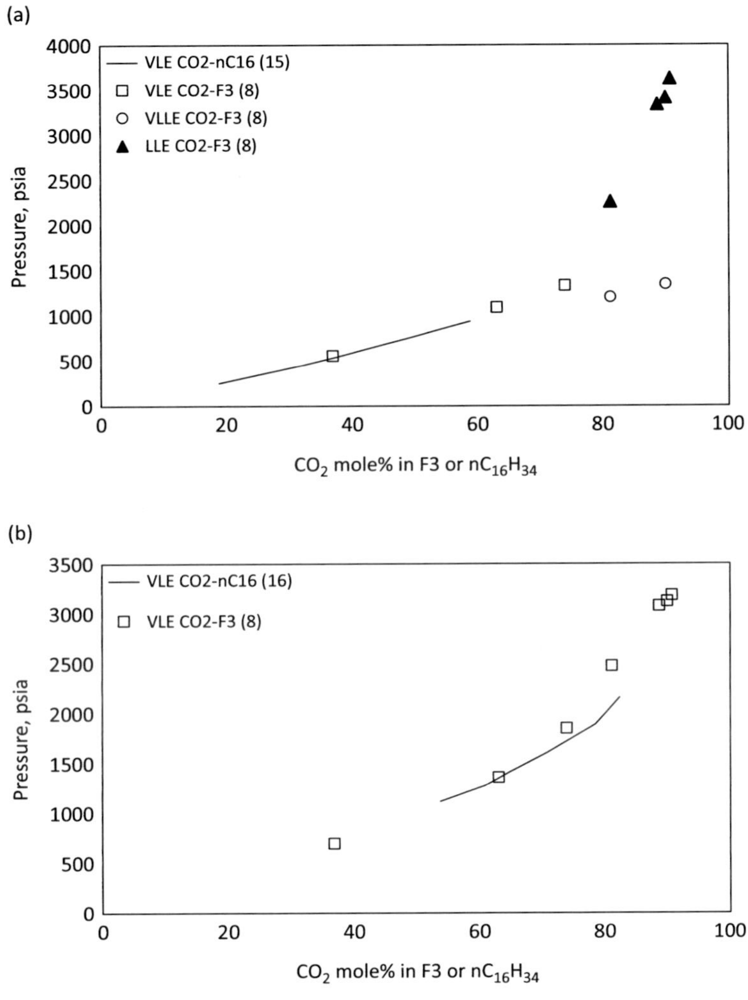

FIGURE 9.6
(a, b) Comparison of CO_2–F3[8] and CO_2–$nC_{16}H_{34}$[15,16] pressure–composition phase behavior. Top and bottom plots (a and b) correspond to temperatures of 104°F and the 140°F, respectively. Numbers in parentheses in the legends indicate the references.

characteristic of systems containing relatively high CO_2 concentration at lower temperatures. Although not stated as such by Lucas et al.,[8] but one of the liquid phases is richer in CO_2, whereas the other is richer in F3. The analogous behavior of the CO_2–F3 and CO_2–$nC_{16}H_{34}$ systems also is evident from Figure 9.7, especially when one compares the LLE region of CO_2–F3 and CO_2–$nC_{16}H_{34}$, which for the most part coincides with each other reasonably well (again note that this occurs at relatively lower temperatures and higher CO_2 concentrations). The highest pressure recorded by Lucas et al.[8]

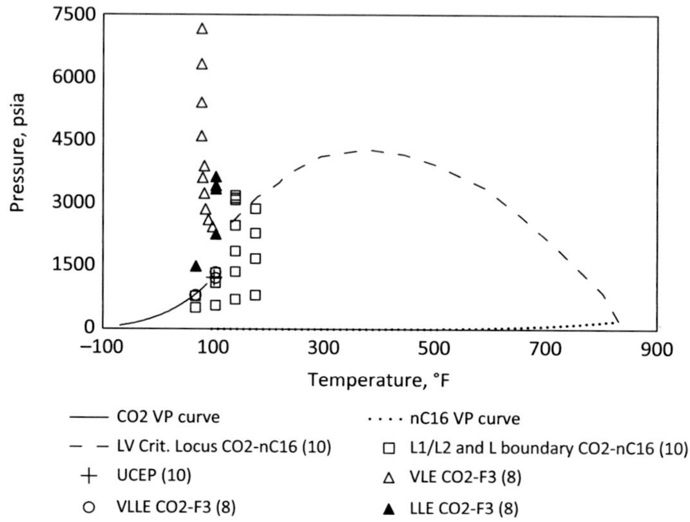

FIGURE 9.7
CO_2–F3 oil fraction phase behavior (treated as a pseudo binary, and presenting a Type III based on molecular weight cutoff) compared with CO_2–$nC_{16}H_{34}$ binary system (presenting a Type III behavior, see Table 7.2 and the description). The measured pressure–temperature data representing VLE, VLLE, and LLE for the CO_2–F3 oil fraction in the temperature range of 68°F– 176°F and 36.9–90.7 mol% used to construct the plot is from Lucas et al.,[8] whereas the CO_2–$nC_{16}H_{34}$ LV critical locus and the L1/L2 and L boundary experimental data are digitized from the plot presented in Vitu et al.[10] Note that the triangle markers should be considered as continuation of the solid dashed or broken curve, i.e., a transition from LV to L1/L2 or liquid–liquid immiscibility, which is typical at lower temperatures, high pressures, and higher CO_2 concentrations. Numbers in parentheses in the legends indicate the references.

at 104°F in the LLE region is only ~3,600 psia; however, if they were to continue the experiment at higher pressures then this LL immiscibility would continue, basically following the L1/L2 and L boundary as exhibited by the CO_2–$nC_{16}H_{34}$ system. Right beneath the LLE region of CO_2–F3 system, at relatively lower pressures, is where the VLLE phase behavior is observed, and as the pressure goes down further the system exhibits a typical VLE behavior. In other words, if one were to trace this backwards, then the transition would be VLE to VLLE to LLE because the vapor phase would simply begin to disappear due to the elevated pressures. Finally, as expected, most of the higher temperature and lower CO_2 mol% (shift to the right of the CO_2 VP curve) VLE data points for the CO_2–F3 system appear in the typical two-phase region underneath the LV critical locus of the CO_2–$nC_{16}H_{34}$ binary system.

Figure 9.8 compares the phase behavior of CO_2 and the lighter fraction (labeled as LRHO F1) of the Lone Rock heavy oil with CO_2–$nC_{12}H_{26}$ since both have molecular weights of 170. As seen by the pressure–composition plot, the CO_2–$nC_{12}H_{26}$[17–19] VLLE and VLE data are interspersed quite well with the CO_2–LRHO F1 VLE data,[9] consistently following the behavior according to the temperatures, which again confirms the validity of the pseudo binary concept. It is also interesting to note that the CO_2–$nC_{12}H_{26}$ system exhibits a VLLE behavior in a fairly narrow low temperature (0.8°F–18.8°F) and pressure (~300–400 psia) range as has been observed by Hottovy et al.[17] The two liquid phases that are immiscible and in equilibrium with each other as well as the vapor phase can be distinguished by the CO_2mol% in $nC_{12}H_{26}$- and CO_2-rich liquid phases, respectively (see Figure 9.8). Again, it should be noted that this VLLE behavior is something that is confined to relatively lower temperatures and higher overall CO_2 concentrations in the mixture. Finally, considering the alignment of the data portrayed in Figure 9.8, it can be stated that if Kokal and Sayegh[9] were to actually conduct the CO_2–LRHO F1 experiments in the 0.8°F–18.8°F temperature range then it is quite plausible that this system would exhibit a VLLE behavior similar to the one observed by Hottovy et al.[17] for the CO_2–$nC_{12}H_{26}$ binary.

The three other oil samples that were tested by Kokal and Sayegh[9] are heavier oils that have molecular weights in the range of 347–465 (see Table 9.1) and according to the molecular weights cutoff exhibit a Type III phase behavior. In all three of these samples, VLE, VLLE, and LLE were observed

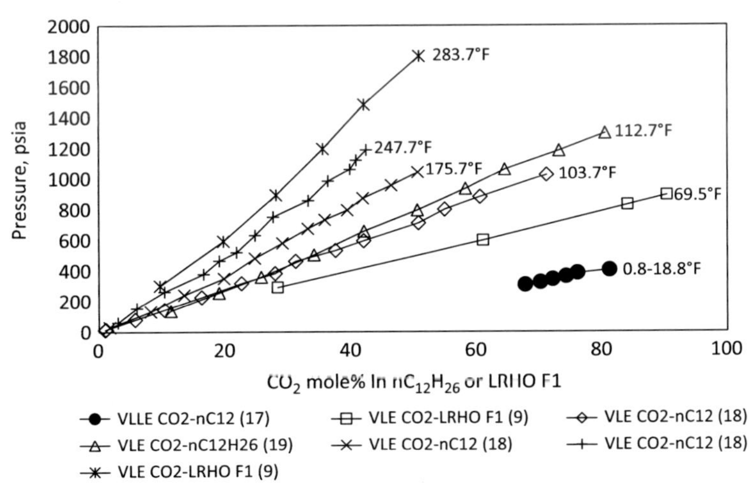

FIGURE 9.8

CO_2–LRHO F1 (Lone Rock heavy oil lighter fraction 1, see Kokal and Sayegh[9]) pressure–composition phase behavior (treated as a pseudo binary, and presenting a Type II based on molecular weight cutoff) compared with CO_2–$nC_{12}H_{26}$ binary system (presenting a Type II behavior, see Table 7.2 and the description). The measured pressure–composition data representing VLLE (0.8°F–18.8°F) and VLE (69.5°F–283.7°F), and as identified by the respective legends and data labels, used to construct the plot is from the reference in the legend parentheses. The composition used for the VLLE data is CO_2mol% in the $nC_{12}H_{26}$ rich liquid phase, which is in equilibrium with an immiscible CO_2-rich liquid phase (not shown) that contains ~93–97 mol% CO_2, in a narrow pressure and temperature range (see Hottovy et al.[17]). Also note that each of the solid circle represents one data point, i.e., a specific P–T–X_{CO_2} in $nC_{12}H_{26}$ rich liquid phase condition.[17]

at the low temperature of 69.8°F. Similar to Lucas et al.[8] and Kokal and Sayegh,[9] Al Ghafri et al.[20] presented phase behavior data for CO_2 and synthetic dead oils, low GOR live oil, and a high GOR live oil resembling Qatari crude oil, which also exhibited VLE, VLLE, and VLE behavior at 76.7°F. The oils they tested were 100% synthetic and were reconstituted by blending 17 different pure components in various proportions that included *n*-alkanes, branched, and cycloalkanes and aromatics. Since these are purely synthetic oils, they are excluded in Table 9.1; however, their phase behavior type is described as follows. Based on the molar blend compositions and the individual molecular weights, the overall mixture molecular weights for the dead oil, low GOR live, and high GOR live oil are 220.4, 140.1, and 89.9, respectively. According to the molecular weight cutoffs shown in Table 7.2, the phase behavior type of the dead oil and low GOR oil is Type III and II, respectively; however, the high GOR oil too exhibiting VLLE as well as LLE behavior despite having a low molecular weight of 89.9 appears to be somewhat unusual. However, given that this value falls in between Types I and II, it is quite plausible that this particular system actually represents a transition inclined toward Type II, in which a potential exists for the occurrence of VLLE and LLE behavior (see Figure 9.8 and the discussion in Section 7.5.1 of Chapter 7). Nevertheless, it should be noted that these experimental studies demonstrate that the LLE behavior in particular is a typical feature at relatively low temperatures and high CO_2 concentrations, which is also the case with other oils discussed previously.

Liphard and Schneider[21] noted that the placement of LLE region is dependent on the molecular weights of alkanes or oils based on the pseudo binary concept; in other words, the temperature at which the LLE region exists increases with the carbon number of the alkane or the oil (via the molecular weight analogy). Similarly, being somewhat more specific, Orr et al.[2] concluded that oils with high molecular weight C_{14+} fraction would exhibit the LLE behavior at relatively higher temperatures as opposed to oils with low molecular weight C_{14+} fractions; the high and low values ranging between 422 and 226 belonging to squalene and *n*-hexadecane, respectively.[2] Accordingly, if the temperature and oil molecular weight data based on Table 9.1 for oils that exhibit the LLE region are plotted (see Figure 9.9), then the data basically show a lot of scatter without any correlation and on the contrary shows a somewhat opposite trend. Note that the sets of horizontals in Figure 9.9 are the three different oils having three different molecular weights but the same temperature at which LLE behavior is observed, which is 69.8°F[9] and the 76.7°F[20], respectively. The vertical line on the other hand in Figure 9.9 corresponds to the same molecular weight oil fraction F3 but exhibiting LLE behavior at two different temperatures of 68°F and 104°F, respectively.[9]

Therefore, perhaps a different way, or in conjunction with temperature, to portray the effect of oil molecular weight is to compare the range of CO_2mol% with the oil molecular weight at which LLE behavior occurs

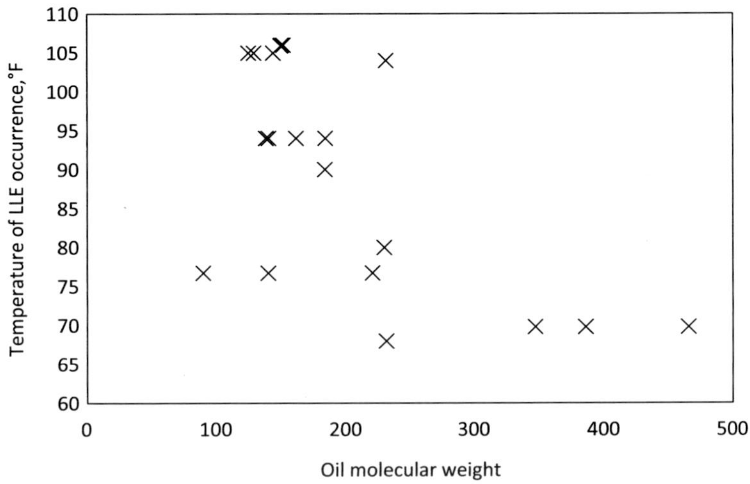

FIGURE 9.9
Temperature of LLE occurrence vs. molecular weight for oils listed in Table 9.1 and the synthetic dead oil, low and high GOR oils resembling Qatari crude oil.[20]

at a given temperature. Alternatively, the range of pressures with the oil molecular weight at which LLE behavior occurs at a given CO_2 concentration can also be compared to understand the effect of oil molecular weight or its heaviness. Figures 9.10 and 9.11 compare the VLE, VLLE, and LLE regions as a function of CO_2mol% for the LRHO and its variants[9] and the surrogates of Qatari crude oil.[20] Note that in both plots molecular weight is used to differentiate the phase behavior. In general, both figures are qualitatively similar to Figure 9.6 a in that the bifurcation of the VLE region into VLLE (at low pressure) and LLE (at high pressure) occurs at a relatively high CO_2mol%. As far as the LRHO and its variants are concerned, the VLE and VLLE behavior is quantitatively similar regardless of the molecular weight because all oils are essentially dead and do not contain any lighter components that would influence the phase behavior. However, that is not the case with the various Qatari crude oil samples, i.e., the VLE region appears at relatively higher pressures for the high GOR sample which is to be expected, whereas for the synthetic dead oil the VLE occurrence is at much lower pressures and is somewhat similar to the LRHO samples. The VLLE region though is qualitatively similar in the case of all Qatari samples as well. A closer inspection of the LLE regions in Figures 9.10 and 9.11 reveals that relatively higher CO_2 concentration are necessary to produce the LLE in higher molecular weight or heavy oils. Another important observation that can be made from Figure 9.11 is the fact that increased lighter component mole fractions, i.e., in high GOR fluid will obviously tend to elevate the LLE pressures when data at constant CO_2 mole fractions are compared.

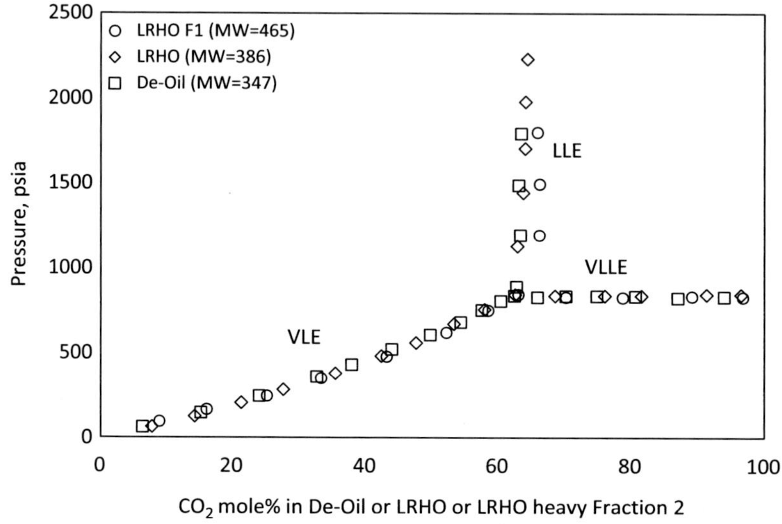

FIGURE 9.10
VLE, VLLE, and LLE pressures vs. CO_2mol% in the deasphalted oil (De-oil), Lone Rock Heavy Oil (LRHO), and LRHO heavier fraction 2 at 69.5°F. The data used to construct the plot to compare the phase behavior based on the different molecular weights (see legend for values) are digitized from Kokal and Sayegh.[9]

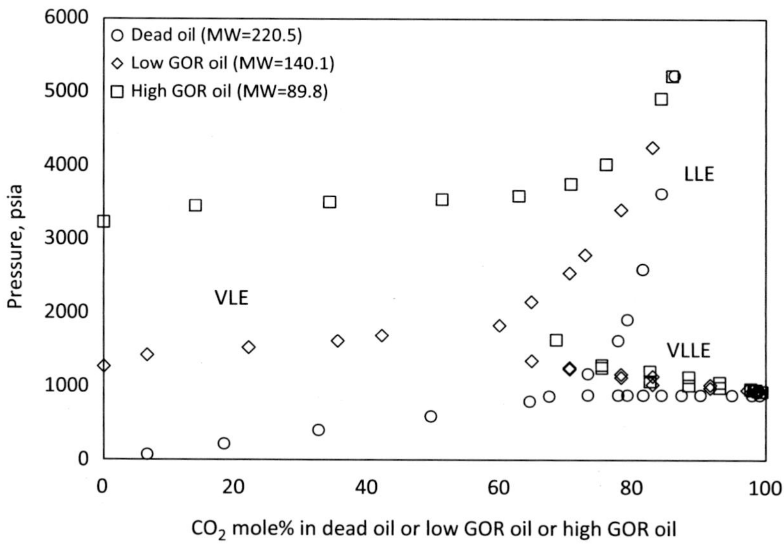

FIGURE 9.11
VLE, VLLE, and LLE pressures vs. CO_2mol% in the synthetic dead oil, low, and high GOR oil resembling Qatari crude oil, at 76.7°F. The data used to construct the plot to compare the phase behavior based on the different molecular weights (see legend for values) are tabulated by Al Ghafri et al.[20]

9.2.2 Stability Analysis and EOS Modeling

As pointed out by Whitson and Brule,[22] one of the most difficult aspects of VLE calculations using an EOS model is the splitting of a given mixture into two or more phases at a given temperature and pressure, which could be vapor–liquid (VLE); liquid–liquid (LLE), and vapor–liquid 1–liquid 2 (VLLE). The occurrence of LLE and VLLE, as shown earlier, is common in CO_2–heavy oil mixtures at high CO_2 concentrations and relatively low temperatures. Generally, based on the compositional characteristics of a given fluid it is not difficult to identify a bubble point (liquid like) or a dew point (vapor like) system[23]; however, the former resulting in typically the VLE and LLE behavior. Therefore, a bubble point calculation normally precedes the VLLE and LLE calculations.

For simple binary systems, Baker et al.[24] showed how the thermodynamic stability can be graphically established, i.e., if a given composition has a lower Gibbs energy as a single phase (stable) or it will decrease by splitting into two or more phases (unstable)[22] at a specific temperature and pressure. Therefore, stability analysis deals with the questions of a mixtures' ability to attain lower energy by splitting into two or more phases[22] or by remaining at a lower Gibbs energy as a single phase, thus answering the important question as far as the number of phases is concerned.

The stability analysis of a simple CH_4 and CO_2 binary system VLE that has been studied by Pedersen et al.[25] is shown in the following equation using the normalized Gibbs energy function g^* as defined by Whitson and Brule.[22] In this example, the overall composition considered is 40 mol% CH_4 and 60 mol% CO_2:

$$g^* = \frac{G}{RT} = \sum_{i=1}^{N} Z_i LN f_i(Z) \tag{9.1}$$

where R is the universal gas constant, T is the temperature, Z_i is the mole fraction of component i, and f_i is the fugacity of component i, which is a function of the compressibility factor (Z). Since there can be three real roots from a cubic EOS model, the root that gives the lowest overall Gibbs energy from the following expression[23] is chosen for calculating the fugacity. Note that this expression is specific to PR EOS model[11]:

$$\frac{G}{RT} = Z - 1 - LN(Z - B) + \frac{A}{2\sqrt{2}B} LN\left[\frac{Z + (1 - \sqrt{2})B}{Z + (1 + \sqrt{2})B}\right] \tag{9.2}$$

where A and B are PR EOS parameters and Z is the compressibility factor. Fugacity of component i is calculated as follows:

$$f_i(Z) = Z_i P \,\mathrm{EXP}\left[\frac{b_i}{b_m}(Z - 1) - LN(Z - B) - \frac{A}{B(\delta_2 - \delta_1)}\left(\frac{2\psi_i}{\psi} - \frac{b_i}{b_m}\right) LN\left(\frac{Z + \delta_2 B}{Z + \delta_1 B}\right)\right] \tag{9.3}$$

where P is the pressure and other variables in the exponent have been defined elsewhere.[23]

For this binary system (or any other for that matter) at a specific temperature and pressure (–43.6°F and 290 psia in this case), the calculations proceed as follows. Mole fractions of CH_4 from 0 to 1 or 1 to 0 (for CO_2) are used to set up the PR EOS (see Dandekar[26]) and fugacities are calculated based on the previously defined criteria for each of the mole fractions. Next, the normalized Gibbs energy function from Equation 9.1 is calculated for each of the mole fractions. Note that the two end points on the mole fraction scale mean 100% CO_2 and 0% CO_2, respectively, i.e., in terms of fugacities corresponding to pure CO_2 and pure CH_4, respectively. Also note that these mole fractions representing various mixtures are treated as a "phase" rather than identifying them as vapor or liquid. Finally, the mole fractions and the g^* values are plotted as shown in Figure 9.12 which also is known as the Gibbs energy surface.[22] Clearly, the energy surface shows two distinct low energy coordinates or lobes, namely (x, g_L^*) and (y, g_V^*) indicating that a CH_4–CO_2 mixture of certain overall composition at –43.6°F and 290 psia is unstable as a single phase and its Gibbs energy can be reduced by splitting into two equilibrium vapor and liquid phases. Alternatively, as stated by Whitson and Brule[22] the equilibrium for this (or any other) binary system is established by drawing a tangent through (x, g_L^*) and (y, g_V^*), with x and y representing the equilibrium liquid and vapor phase compositions, respectively. A valid tangent plane cannot intersect the Gibbs energy surface anywhere else except at the two shown coordinates.[22] The vertical lines dropping down from (x, g_L^*) and (y, g_V^*) to the composition axis means the mole fractions of CH_4 and CO_2 in the equilibrium liquid and vapor phase will be ~0.05, 0.95, and 0.48, 0.52, respectively, that satisfy the equal fugacity criterion. The overall composition

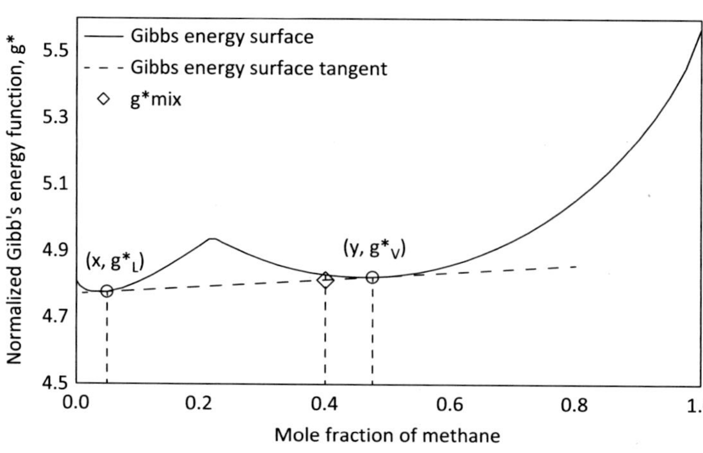

FIGURE 9.12
Gibbs energy surface plot for a CH_4–CO_2 binary system at –43.6°F and 290 psia.

of 40 mol% CH_4 and 60 mol% CO_2 lies between the equilibrium compositions which means this mixture is unstable, splitting into two equilibrium phases with mixture Gibbs energy given by[22]

$$g^*_{mix} = n_V g^*_V + n_L g^*_L = n_V g^*_V + (1 - n_V) g^*_L \tag{9.4}$$

where n_V and n_L are overall vapor and liquid mole fractions, and a sum of them is equal to 1. The point of intersection of the vertical line dropping down from g^*_Z (i.e., for the 40 mol% CH_4 and 60 mol% CO_2) and the tangent represents the lower attained mixture Gibbs energy g^*_{mix}. In this particular case, the g^*_{mix} value of 4.814 gives n_V of 0.826, which can also be calculated from the Lever rule, i.e., based on the ratio of the distance from Z to y and x to y. These values determined from Gibbs energy minimization are nearly identical to the PT flash calculations performed in a traditional manner. Finally, since this is a binary system any overall or feed composition that lies between the two lobes of x and y will result in the same equilibrium vapor and liquid phase compositions (degrees of freedom $F = C - P + 2 = 2 - 2 + 2 = 2$; temperature and pressure); however, the overall mole fractions, i.e., n_V and n_L will be different.

Similar to the CH_4–CO_2 binary system, the stability analysis also is applied to the CO_2–F3 and CO_2–heavy oil mixtures studied by Lucas et al.[8] and Kokal and Sayegh[9], respectively. Both systems exhibit VLE behavior at relatively high temperatures of 176°F and 284°F, respectively. Figures 9.13 and 9.14 show the normalized Gibbs energy function vs. CO_2 mole fractions at 176°F and 500 psia and 284°F and 725 psia for the CO_2–F3 and CO_2–heavy oil systems. In principle, both mixtures show features similar to the CH_4–CO_2 system, i.e., a valid tangent can be drawn that allows the determination of equilibrium vapor and liquid phase compositions. Clearly, the vapor phase is mostly composed of 99+ mol% CO_2, whereas the liquid phase is ~70 mol% F3 or the heavy oil as expected. These equilibrium vapor and liquid phase compositions are consistent with the bubble points reported[8,9] and agree with the predicted compositions determined by traditional flash calculations carried out in a commercial PVT package. Since a pseudo binary concept is applied for both binaries any feed with overall composition lying between x and y will result in the same equilibrium vapor and liquid phase compositions but different n_V and n_L values. Finally, a much noticeable contrast between Figure 9.12 and Figures 9.13 and 9.14 is obvious in that the change in the Gibbs energy function with increasing CO_2 mole fractions for F3 and the heavy oil is somewhat smooth, increasing gradually, with a slight hump at higher CO_2 concentrations, making it difficult to precisely draw the tangent. The reason this seems to occur is due to the vast differences between the fugacities of CO_2 and F3 and CO_2 and the heavy oil as compared to the CH_4 and CO_2 system.

The exact same methodology as described above can be theoretically extended to identify liquid–liquid and vapor–liquid 1–liquid 2 split, i.e.,

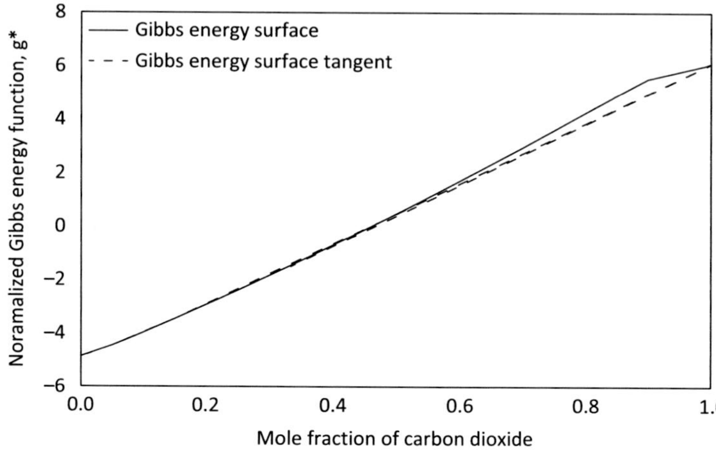

FIGURE 9.13
Gibbs energy surface plot for the CO_2–F3 pseudo binary system[8] at 176°F and 500 psia.

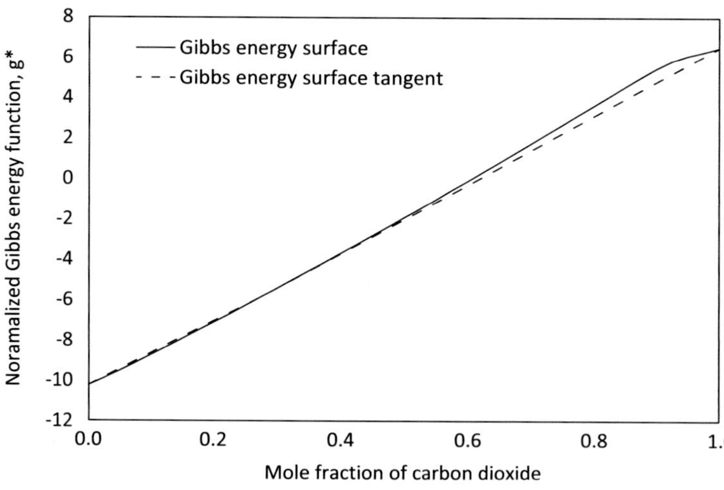

FIGURE 9.14
Gibbs energy surface plot for the CO_2–Lone Rock Heavy Oil (LRHO) pseudo binary system[9] at 284°F and 725 psia.

two lobes and three lobes on the Gibbs energy surface. However, one of the most important pitfalls of this technique is false two-phase solutions that satisfy the equal fugacity criterion but yield only a local minima in the Gibbs energy, which are difficult to detect and are commonly associated with CO_2–heavy oil mixtures.[22] Additionally, the contrasts in the CO_2 and heavy component fugacities makes it nearly impossible to clearly distinguish the lobes. The distinct Gibbs energy surfaces as shown in

Baker et al.[24] and Whitson and Brule[22] for LLE and VLLE (noted as correct solutions) are conceptual or theoretical, and in reality it is nearly impossible to identify these lobes unless the "expected" Gibbs energy surface is zoomed (e.g., see Figures 9–12 in Baker et al.,[24] which are actually calculated values). Similarly, the Gibbs energy surface for the CO_2–F3 and CO_2–heavy oil binary systems where LLE and VLLE behavior is experimentally observed shows a fairly smooth curve without any lobes (not shown here). Whitson and Brule[22] have also pointed out that the graphical interpretation of Baker et al.,[24] although useful in principle in determining phase stability, is not practical to implement as a numerical algorithm, even if a pseudo binary concept is assumed, whereas other drawbacks include the inapplicability of the graphical interpretation to systems containing more than three components.

Michelsen[27] proposed an algorithm to determine the stability of a given mixture; the salient features of which as described by Whitson and Brule[22] are shown below, followed by examples of the CH_4–CO_2 binary system and CO_2–LRHO pseudo binary system. Note that the algorithm given below is for VLE type analysis and is easily applied to multicomponent systems. For a given mixture a second phase, i.e., a "vapor-like" or "liquid-like," is located and stability or instability of the mixture (at a given temperature and pressure) having composition Z_i is determined based on the characteristics of the located phase as per the criteria shown in Whitson and Brule[22]. The stepwise procedure as described[22] is summarized in the flowchart in Figure 9.15. In the stability analysis flowchart or the algorithm that follows it, it is important to note that each test for the search of the second phase is conducted separately. Whitson and Brule[22] note that the converged k_i values from the stability test are the best initial estimates for the flash calculations if the tested mixture is found to be unstable. As an example, the stability analysis for the CH_4–CO_2 and CO_2–LRHO mixtures is shown in Tables 9.2 and 9.3, respectively. The overall compositions for both these mixtures and the test conditions along with the final completed analysis are also included in the tables.

Based on the criteria presented,[22] a given mixture at given temperature and pressure is unstable if $S_L > 1$ and a trivial solution is obtained for the vapor-like second phase and vice versa, i.e., $S_V > 1$ and a trivial solution for liquid-like second phase. This also means that the two probable lobes on the Gibbs energy surface plot. Clearly, both mixtures are determined to be unstable for the given conditions. Finally, if flash calculations are now carried out for both these mixtures using the better initial estimates of the equilibrium ratios (11th column for CH_4–CO_2 binary and the 9th column for CO_2–LRHO in Tables 9.2 and 9.3), then convergence of Rachford–Rice[29] flash function is achieved quickly.[22] The final converged flash calculation values for both these mixtures are shown in Tables 9.4 and 9.5, respectively, which are nearly identical to those determined from Gibbs energy surface plots based on Baker et al.[24] (see Figure 9.12 and 9.14).

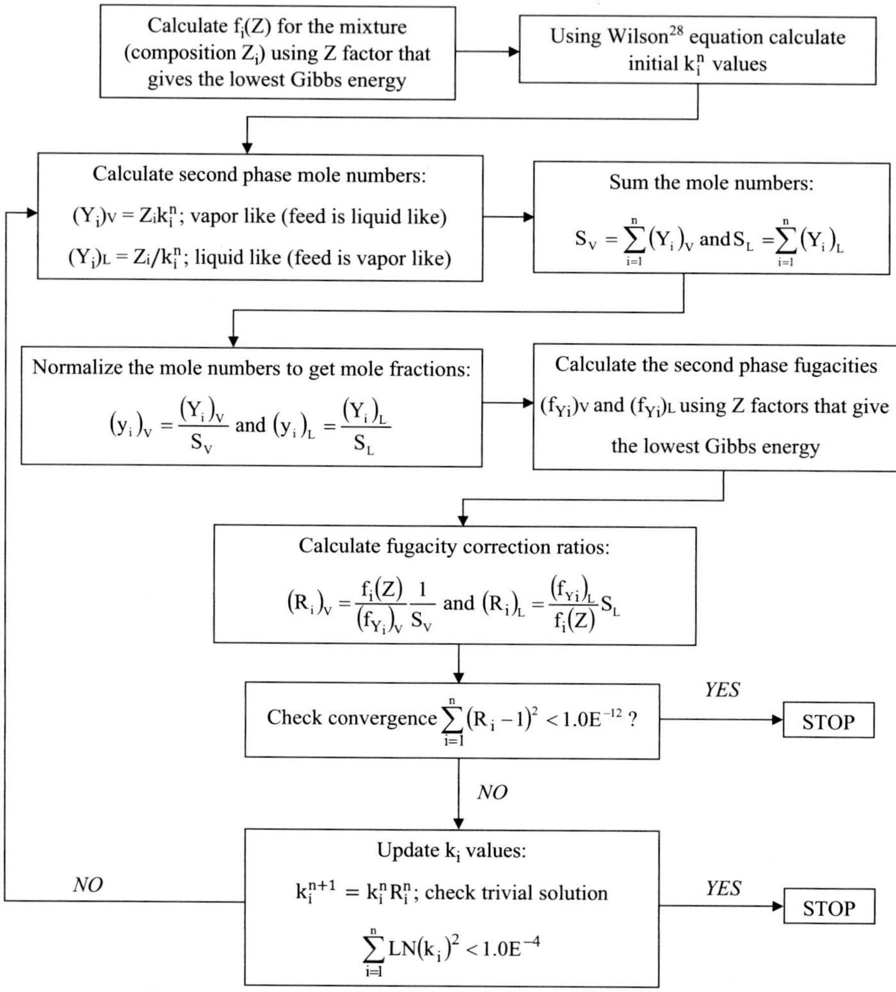

FIGURE 9.15
Flowchart illustrating the Michelsen[27] stability algorithm based on the stepwise methodology outlined by Whitson and Brule.[22] Results shown in Table 9.2 and 9.3 are obtained by following this flowchart.

9.2.3 Handling of Multiple Phases in a Compositional Reservoir Simulator

In CO_2 injection-based low-temperature floods, the formation of a third CO_2-rich liquid phase necessitates the inclusion of three-phase equilibria or flash calculations in compositional reservoir simulation. However, given the increased computational time and convergence issues associated with flash calculations involving three phases, these simulations are sometimes approximated using simple two-phase equilibria. Given the narrow region

TABLE 9.2

Stability Analysis of a 60 mol% CO_2 and 40 mol% CH_4 Mixture at –43.6°F and 290 psia Using the Michelsen Algorithm[27] as per the Flowchart shown in Figure 9.15

Comp.	$f_i(Z)$	k_i''	$(y_i)_V$	$(Y_i)_V$	$(Y_i)_L$	$(y_i)_L$	$(R_i)_V$	$(k_i)_V$	$(R_i)_L$	$(k_i)_L$
CO_2	137.5	0.463	0.600	0.600	1.116	0.971	1.000	1.000	1.000	0.538
CH_4	108.6	5.967	0.400	0.400	0.033	0.029	1.000	1.000	1.000	12.147
	$Z = 0.822$		$S_V = 1$		$S_L = 1.149$		$\varepsilon = 3.7E^{-11}$	T.S.	$\varepsilon = 3.6E^{-13}$	

Mixture is unstable (also see Figure 9.12).
T.S., Trivial Solution.

TABLE 9.3

Stability Analysis of a 80 mol% CO_2 and 20 mol% Lone Rock Heavy Oil (LRHO) Mixture[9] at 284°F and 725 psia Using the Michelsen algorithm[27] as per the Flowchart shown in Figure 9.15

Comp	$f_i(Z)$	k_i''	$(Y_i)_V$	$(y_i)_V$	$(Y_i)_L$	$(y_i)_L$	$(R_i)_V$	$(k_i)_V$	$(R_i)_L$	$(k_i)_L$
CO_2	1,577.3	8.361	2.381	0.999	0.800	0.800	1.000	2.977	1.000	1.000
LRHO	$1.05E^{-5}$	$2.97E^{-7}$	$2.13E^{-7}$	$8.96E^{-8}$	0.200	0.200	1.000	$1.07E^{-6}$	1.000	1.000
	$Z =$ 0.245		$S_V =$ 2.381		$S_L = 1$		$\varepsilon =$ $1.8E^{-18}$		$\varepsilon =$ $8.9E^{-12}$	T.S.

Mixture is unstable (also see Figure 9.14).
T.S., Trivial Solution.

TABLE 9.4

Flash Calculation Results for the 60 mol% CO_2 and 40 mol% CH_4 Mixture at –43.6°F and 290 psia Using the Better Initial Estimates of the Equilibrium Ratios

Components	X_i, Mole Fraction	Y_i, Mole Fraction	f_i^L	f_i^V
CO_2	0.9582	0.5105	117.855	117.855
CH_4	0.0418	0.4895	131.638	131.638
	$Z_L = 0.0399$ $n_L = 0.1998$	$Z_V = 0.8385$ $n_V = 0.8002$		

TABLE 9.5

Flash Calculation Results for the 80 mol% CO_2 and 20 mol% Lone Rock Heavy Oil (LRHO) Mixture[9] at 284°F and 725 psia Using the Better Initial Estimates of the Equilibrium Ratios

Components	X_i, Mole Fraction	Y_i, Mole Fraction	f_i^L	f_i^V
CO_2	0.2858	0.99999	662.361	662.361
LRHO	0.7142	$5.357E^{-7}$	$2.647E^{-5}$	$2.647E^{-5}$
	$Z_L = 0.5883$ $n_L = 0.2800$	$Z_V = 0.9118$ $n_V = 0.7200$		

in which three phases are found in equilibrium and thus possibly existing over a small part of the reservoir, the loss in accuracy by approximating a two-phase behavior may be small and acceptable. Also, the mole fraction of the CO_2-rich phase tends to be relatively small. Okuno et al.[30] have discussed the pitfalls of approximating the three-phase equilibria with two phases in a compositional reservoir simulation and demonstrated its importance. Li and Firoozabadi[31] have also recognized the effect of three different phases having different properties and mobilities on flow through the reservoir and thus oil recovery. Although a standalone three-phase flash calculation option may exist in commercial PVT modules, commercial compositional reservoir simulators capable of handling VLLE and four phase flow simulation (water being the fourth phase) are not commonplace. Therefore, customized three-dimensional compositional reservoir simulators that are capable of handling four phases have been successfully used (e.g., UTCOMP).[30] The basic work-flow of phase equilibria calculations in compositional reservoir simulators consists of stability analysis (similar to the one described earlier) to confirm the presence of three phases followed by flash calculations.[30] Based on the determined phase compositions, partial differential equations can then be set-up along with the rock and fluid properties and eventually solved by numerical methods such as finite differences to obtain pressure, saturation, flowrates, etc., as a function of position and time. The flow aspects of compositional reservoir simulation are beyond the scope of this chapter; however, the fundamental relationships and steps that govern the three-phase flash calculations are provided next.

Conceptually, three-phase split or flash calculations are similar to two phase in that the thermodynamic equilibrium is based on the equality of fugacities of components in the different phases. For a typical (e.g., CO_2 induced) three-phase system comprised of a gas phase (y), hydrocarbon-rich liquid phase (x), and a CO_2-rich liquid phase (z), we have

$$f_{yi} = f_{xi} = f_{zi}; \quad i = 1 \text{ to } N \tag{9.5}$$

where f_{yi}, f_{xi}, and f_{zi} are fugacities of components i in the said phases. Since this is a three-phase flash calculation, two different equilibrium ratios also need to be defined, i.e., $k_{yi} = y_i/x_i$ and $k_{zi} = z_i/x_i$. Using another fundamental relationship of equilibrium ratios,

$$LN\left(k_{y_i}\right) = LN\left(\Phi_{xi}\right) - LN\left(\Phi_{y_i}\right) \text{and } LN\left(k_{zi}\right) = LN\left(\Phi_{xi}\right) - LN\left(\Phi_{zi}\right); \quad i = 1 \text{ to } N \tag{9.6}$$

In the above equation, the fugacity coefficients of individual components in the three different phases are determined from an EOS model. Fugacities are subsequently calculated from these fugacity coefficients from, for example, $f_{xi} = x_i \Phi x_i P$, to determine the achievement of thermodynamic equilibrium. Similar to the two-phase flash calculations, three-phase flash calculations actually require the solution of two Rachford–Rice flash functions,[29,31] which are derived as follows. Assuming an overall feed fraction of 1 mol as the basis, overall and component based molar balance equations can be written as

$$n_x + n_y + n_z = 1 \tag{9.7}$$

$$n_i = x_i n_x + y_i n_y + z_i n_z \tag{9.8}$$

where n_i is the mole fraction of component i in the feed. Next, the substitution of previously defined equilibrium ratios allows the elimination of y_i and z_i from Equation 9.8:

$$n_i = x_i n_x + k_{yi} x_i n_y + k_{zi} x_i n_z \tag{9.9}$$

collecting all the x_i terms on the right-hand side and eliminating n_x by $1 - n_y - n_z$,

$$x_i = \frac{n_i}{k_{y_i} n_y + k_{zi} n_z + 1 - n_y - n_z} = \frac{n_i}{1 + n_y\left(k_{y_i} - 1\right) + n_z\left(k_{zi} - 1\right)} \tag{9.10}$$

Finally, using $y_i = k_{yi} x_i$ and $z_i = k_{zi} x_i$ two different Rachford–Rice flash functions can be set up

$$\begin{aligned} RR_y &= \sum_{i=1}^{N} \left(y_i - x_i\right) \\ &= \sum_{i=1}^{N} \frac{k_{y_i} n_i}{1 + n_y\left(k_{y_i} - 1\right) + n_z\left(k_{zi} - 1\right)} - \frac{n_i}{1 + n_y\left(k_{y_i} - 1\right) + n_z\left(k_{zi} - 1\right)} = 0 \end{aligned} \tag{9.11}$$

$$\begin{aligned} RR_z &= \sum_{i=1}^{N} \left(z_i - x_i\right) \\ &= \sum_{i=1}^{N} \frac{k_{zi} n_i}{1 + n_y\left(k_{y_i} - 1\right) + n_z\left(k_{zi} - 1\right)} - \frac{n_i}{1 + n_y\left(k_{y_i} - 1\right) + n_z\left(k_{zi} - 1\right)} = 0 \end{aligned} \tag{9.12}$$

$$RR_y = \sum_{i=1}^{N} \frac{n_i\left(k_{y_i} - 1\right)}{1 + n_y\left(k_{y_i} - 1\right) + n_z\left(k_{zi} - 1\right)} = 0 \tag{9.13}$$

$$RR_z = \sum_{i=1}^{N} \frac{n_i\left(k_{zi} - 1\right)}{1 + n_y\left(k_{y_i} - 1\right) + n_z\left(k_{zi} - 1\right)} = 0 \tag{9.14}$$

Although the above two Rachford–Rice flash functions appear to be "deceptively simple," their solution is not trivial.[32] Theoretically, when k_{yi} and k_{zi} are constant (say for a given iteration), both RR_y and RR_z are a function of two independent variables n_y and n_z (moles of gas phase and CO_2-rich liquid phase, respectively). The solution of RR_y and RR_z is, however, not simple but can be obtained by a method suggested by Haugen et al.[32] For example, n_y can be varied within the bounds of 0 to 1 and a corresponding value of n_z determined for each n_y value such that RR_y is zero. Similarly, n_z can be varied within the bounds of 0 to 1 and a corresponding value of n_y determined such that RR_z is zero. When these individual solutions are plotted as shown in Figure 9.16, the intersection point of the two lines is where both RR_y and RR_z flash functions are simultaneously satisfied,[32] i.e., certain corresponding n_y and n_z values are obtained. In other words, the three-phase flash solution must lie inside the solution domain bounded by the equilateral triangle[32] (see Figure 9.16). Note though that this solution constitutes just one iteration, which has to be repeated multiple times until a preset convergence criterion of equal fugacities is achieved.

Finally, a basic methodology of carrying out three-phase flash calculations using the above is shown in the flowchart in Figure 9.17. Note that

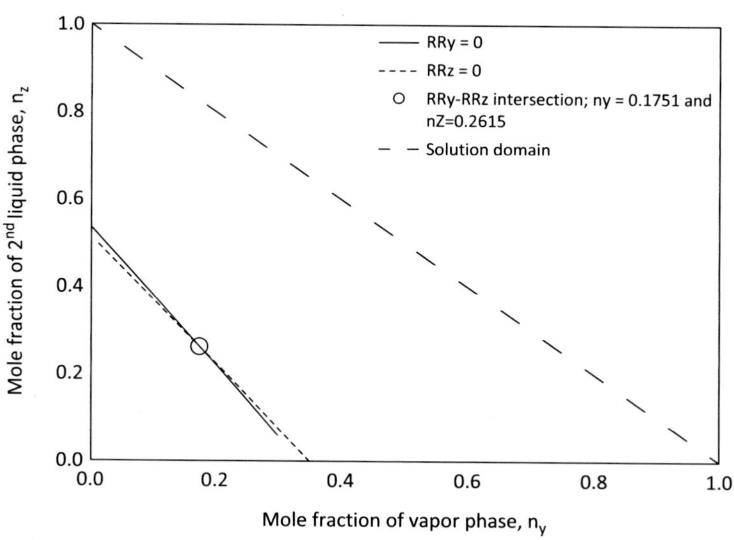

FIGURE 9.16
Simultaneously satisfied Rachford–Rice flash functions,[29,32] RR_y and RR_z, for a gas condensate mixture that splits into a vapor phase and two liquid phases at –97.6°F and 754.2 psia.[25] This is a final converged solution, which results in a vapor phase mole fraction (n_y) of 0.1751; the second liquid phase mole fraction (n_z) of 0.2615 and a hydrocarbon or oil-rich mole fraction (n_x) of 0.5633 (see the intersection point). The second liquid phase, which is low in hydrocarbons, can be construed as a CO_2-rich liquid phase. Also note that typically the CO_2-rich liquid phase mole fraction tends to be quite small and thus difficult to discern from the RR_y–RR_z intersection and thus the gas condensate example is shown to illustrate the method.

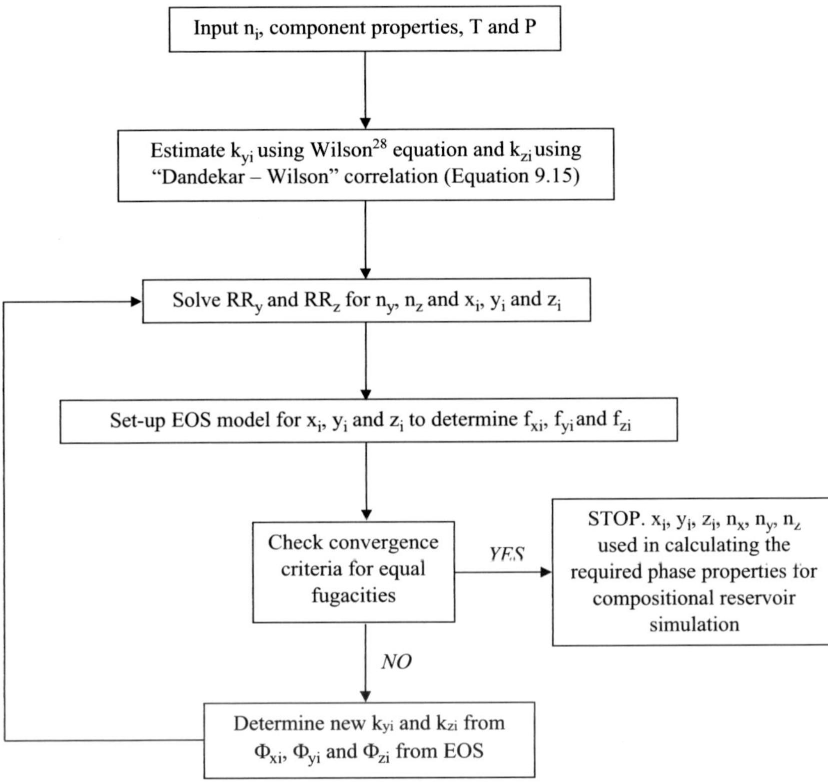

FIGURE 9.17
Flowchart depicting the basic methodology for conducting three-phase split or flash calculations. Note—variables in the flowchart have been defined in Section 9.2.3.

the convergence criteria will be rarely achieved based on the estimated values of k_{yi} and k_{zi} but are shown merely from a theoretical standpoint which means an iterative sequence based on an EOS model will almost always be necessary.

9.3 Practical Significance of CO_2 Induced Heavy Oil Phase Behavior

As far as the practical significance of the CO_2–oil phase behavior from an EOR standpoint is concerned, CO_2 plays an important role in that miscibility is developed by extraction of hydrocarbon components from the oil into a CO_2-rich phase, thus contributing or promoting recovery. Even in the liquid–liquid region, CO_2 can efficiently extract hydrocarbons (as heavy as C_{24}) from the oil.[1,2] Huang and Tracht[33] have also argued that swelling and stripping of hydrocarbons from the oil by a CO_2-rich liquid phase are the dominant mechanisms for tertiary recovery in low-temperature displacements in the liquid–liquid regions.[2] Turek et al.[6] based on their experiments also concluded that the CO_2-rich liquid phase, which is a typical feature at relatively low temperatures and high CO_2 concentrations is capable of extracting a significant portion of all but the heaviest hydrocarbon components present in the original reservoir oil, and is quite evident from the data presented in Figure 9.18. Similarly, Ngheim and Li[34] also stated the significance of the CO_2-rich liquid phase in enhancing the recovery in CO_2 floods, which is due to the efficient extraction of the C_{5+} fractions of the oil. In CO_2 floods in such types of low-temperature systems, Shelton and Yarborough[3] state that the asphaltenic material (if the oil is categorized as such) and part of the oil-rich liquid phase will be left as a residual saturation.

Given the intricate phase behavior especially for CO_2–oil systems, it is appropriate to review the compositional as well as fluid property characteristics (such as density and viscosity) of the two liquid phases that are formed when CO_2 comes in contact with oil. The first such plot (Figure 9.18) shows the equilibrium ratio of the various components for the CO_2–crude oil systems that are shown in Table 9.1. Consistent with the previous subsection, we have defined the equilibrium ratio, k_{zi} as z_i (CO_2 rich)/x_i (oil rich), where z_i and x_i are the mole fraction of a given component in the two stated phases.

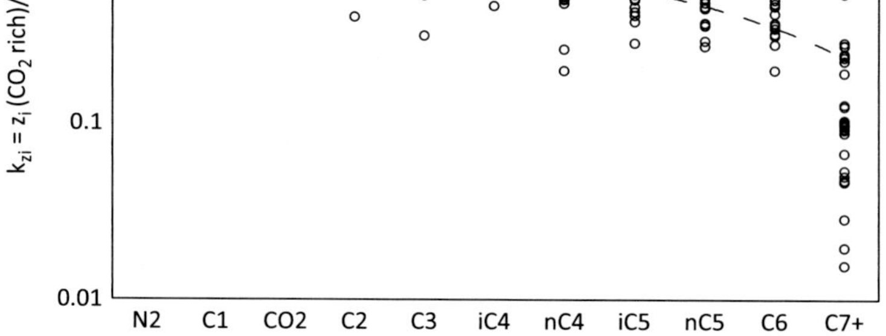

FIGURE 9.18
Partitioning or equilibrium ratio (k_{zi}) of different components into the CO_2-rich and oil-rich phases for CO_2—crude oil mixtures shown in Table 9.1. All equilibrium ratios are calculated based on the experimental compositional data and used to construct the plot. The mol% of CO_2 added to the oil ranges from 65% to 95% and covers a temperature and pressure range of 94°F–106°F and 1,084–2,205 psia, respectively.

All k_{zi} values were directly calculated based on the reported compositional data and are plotted in Figure 9.18. The average range of equilibrium ratio appears to be fairly narrow (~between 0.5 and 1.5), which in fact is an indication of somewhat similar component partitioning between the two phases; in other words, this also shows the effective extractive power of CO_2 in bringing components from the oil phase, i.e., the former eventually becoming the CO_2-rich liquid phase. Although some of the data appear to fall outside the average range, a reasonable correlation to estimate the much needed k_{zi} values in three-phase flash calculations can be developed. Currently no such correlation exists and the developed k_{zi} correlation would thus serve as a counterpart of the Wilson's correlation[28] that is used to estimate k_{yi} values. The plot in Figure 9.18 and the collected data are in fact the basis of the proposed Dandekar–Wilson's correlation to estimate k_{zi} values; the details of which are shown in the last section of this chapter.

Another elegant way of elucidating the extractive power of CO_2 is by cross-plotting the mol% of various components in the CO_2-rich and oil-rich liquid phases, respectively. This is shown in Figure 9.19 using the same data set that has been used to construct Figure 9.18, except the Maljamar[2] separator oil. Clearly, majority of the data at least up to C_6 scatter around the 45° line; however, for C_{7+} much of the data fall somewhat below the 45° line, which is to be expected since more of the C_{7+} would be in the oil-rich phase. However, the CO_2-rich phase also contains a noticeable amount of C_{7+} (extracted from the contacted oil) and thus the relative closeness of the data to the 45° line.

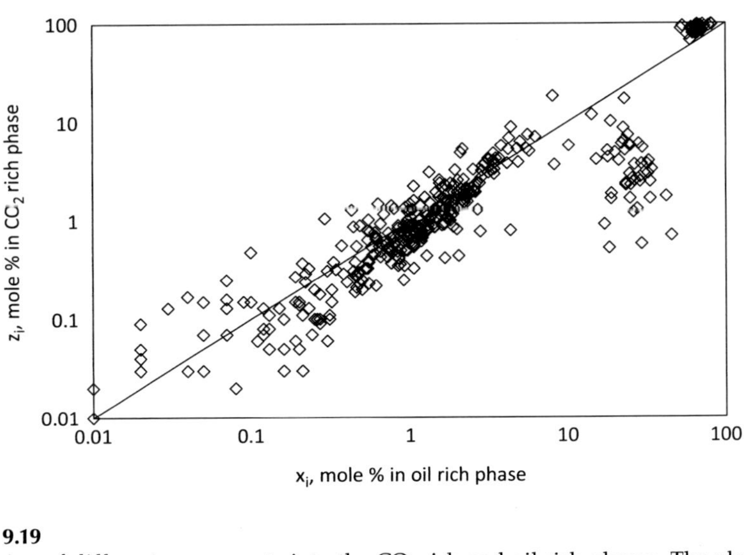

FIGURE 9.19
Partitioning of different components into the CO_2-rich and oil-rich phases. The plot is constructed on the basis of experimental compositions reported for the various CO_2–crude oil systems shown in Table 9.1. Note that the same data are used in constructing Figure 9.18 (except the Maljamar separator oil[2]).

In Figure 9.19, also notable is the nearly equal presence of CO_2 in both the phases (top right corner of the plot).

Orr et al.[2] reported comprehensive compositional data (in mass%, converted to mol% for Figure 9.19) on the CO_2 and oil-rich liquid samples after CO_2 contact with Maljamar separator oil. For a component range as wide as C_5–C_{24}, most data points place around the 45° line, thus indicating a one to one correlation or similar compositional distribution between the two phases. These data are the only exception in that it is a separator oil,[2] while the others are live oils, with major differences being in the lighter fractions. Nevertheless, this diversity of oils contacted by CO_2 at the favorable conditions again demonstrate the remarkable extractive capacity of CO_2, which is much desired in EOR applications.

Finally, Figure 9.20 compares the densities of the CO_2 and oil-rich phases, respectively, for some of the same systems as discussed earlier; note though that for the Wasson oil the densities are that of the C_{7+} fraction instead, which are still representative of the characteristics of the two phases. Again, the "similarities" in the two densities are essentially a reflection of the compositional characteristics of the two liquid phases; in particular, see the nearly close to one ratio of the densities for the CO_2–Maljamar separator oil system, and other oils as well as the CO_2–$nC_{13}H_{28}$ system[12] which is shown as a reference. Limited reliable experimental data on viscosities of CO_2 and oil-rich liquid phases are available; however, the values reported by Orr et al.[2] for the

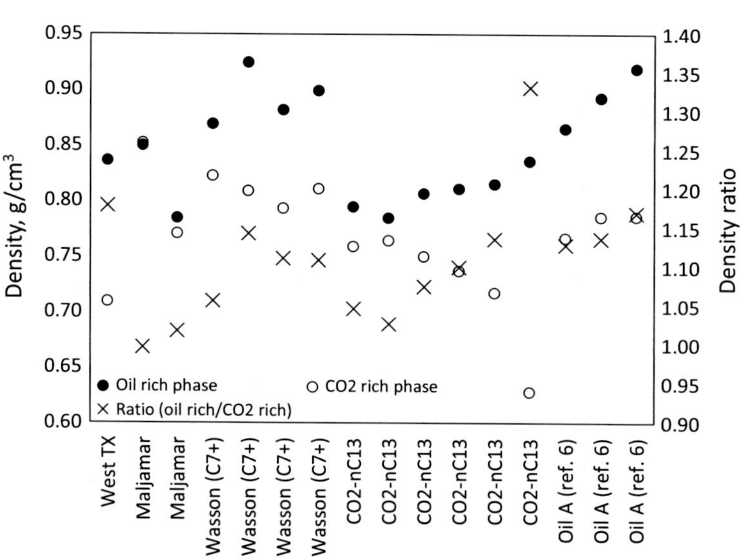

FIGURE 9.20
Comparison of densities of the CO_2-rich and oil-rich phases. Densities for the Wasson oil are not for the total phase but only represent the C_{7+} fraction. Plot is constructed using the data reported in Refs. 1, 2, 3, 6 for crude oils, and 12 for CO_2–$nC_{13}H_{28}$ binary.

CO_2–Maljamar separator oil system indicate that CO_2 has a much more profound effect on the viscosities than densities, which is primarily controlled by compositions or component exchange/extraction. For example, the Maljamar separator oil has a viscosity of 2.8 cP at 94°F; however, after bringing the oil in contact with 79.3 mol% CO_2 at 90°F and 2,164 psia, the CO_2-rich and oil-rich liquid phases have a viscosity of 0.185 and 0.76 cP[2], respectively. The viscosity of the CO_2-rich phase is ~2.5 times that of pure CO_2, whereas the viscosity of the oil-rich phase is reduced by ~3.7 times compared with the original oil at the test conditions. As rightly pointed by Orr et al.,[2] these limited data suggest that the partitioning of the components between phases, i.e., CO_2 extracting components as heavy as C_{24} from the oil thus making the CO_2-rich phase more viscous and consequently, the oil-rich phase less viscous, which in turn may have an important impact on displacement behavior.[2] Density and viscosity conclusions somewhat similar to those aforementioned were also drawn by Turek et al.[7]

9.4 Compositional Changes in a Heavy Oil in Exotic EOR Processes—Microbial and In Situ Combustion

Microbial and in situ combustion EOR processes are particularly targeted for enhancing the recovery of heavier oils. One of the mechanisms via which recovery can be enhanced is attributed to favorable compositional changes or upgrading of the oil in the pore spaces resulting in viscosity as well as density reduction. It is not the purpose of this section to provide a detailed discussion on the methods as such but to focus only on the compositional aspects of the oil, which is of significance from a phase behavior standpoint.

9.4.1 Microbial Enhanced Oil Recovery (MEOR)

The basic strategy of MEOR is identifying the microorganisms or microbes that are inherent in the reservoir and then stimulating them by injecting suitable nutrients so that the microbes proliferate and enhance the microbial action. It may also be necessary in some cases to inject the microbes also with the nutrients, which is referred to as the traditional MEOR method.[35] In both strategies, the goal is to create favorable in situ conditions that will result in bioproducts such as the biomass, which is conducive for compositional changes in the oil thereby reducing viscosity and interfacial tension and/or gases such as CO_2, CH_4, and N_2 also promoting the oil viscosity reduction as well as swelling of the oil.[36] However, MEOR as a method has not been widely applied. Only a handful of cases of successful results have been reported.[36] One of the primary reasons for lack of widespread use or success, compared to the traditional methods, is possibly due to the delicate nature of the process that is underpinned on the microorganisms. Jimoh[36] has provided a discussion on the primary constraints on MEOR, which include pore size, temperature, pressure, salinity, and pH of the formation water. These constraints according to their respective influence (e.g., temperature being one of the most significant) will impact the efficacy of the process by limiting the microbial action. Also, most of these constraints are due to natural given factors; most of which cannot be altered and/or due to the improper identification of the microbes present and the nutrients, making the entire MEOR process selectively successful with a handful of positive examples. Nevertheless, laboratory studies in which MEOR has shown a potential are discussed in the next section, where a summary of compositional changes observed are included.

Ghotekar[37] carried out chemical and microbial characterization studies of Alaska North Slope (ANS) West Sak viscous oil to gauge the MEOR potential. A two-component microbial formulation and a Bushnell Haas broth (nutrient) were used to treat the oil. Oil composition before and after the treatment were measured, which revealed a slight reduction in the C_{25+} mass%, a slight increase in the C_{12}–C_{24} but a decrease in C_6–C_{11} components that was attributed to the microbial action (see Figure 9.21). The slight decrease in the C_{25+} post-treatment had a reasonably significant impact on the API gravity (+8.8°API) as well as viscosity decrement of 17 cP. Incremental oil recovery based on the microbial action ranged from ~7% to 13% in six different corefloods. Somewhat similar to Ghotekar,[37] Jimoh's[36] results also showed a degradation through drastic reduction in the concentrations of C_{15}, C_{16}, C_{18}, C_{20}, and C_{28}, upgrading the tested North Sea oil to some extent. Jimoh states that the alkane (heavier) degradation is probably from the effect of biosurfactant production. Results obtained by Sakthipriya et al.[38] for a waxy and model oil

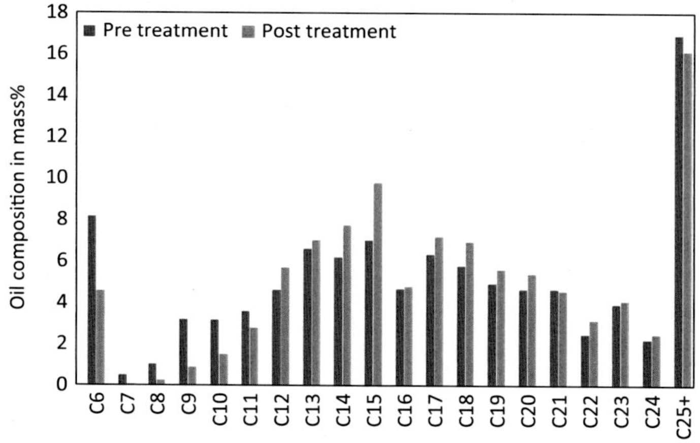

FIGURE 9.21
Pre- and post-microbial treatment composition of Alaska North Slope (ANS) West Sak viscous oil. Plot is constructed based on the gas chromatographic data reported by Ghotekar.[37]

also are in line with the previous observations. Although the authors did not provide (bio)degradation in terms of carbon numbers; based on the Fourier Transform Infrared Spectroscopy (FTIR) tests they reported the most significant reduction in the high molecular weight paraffins, and the corresponding viscosity reduction of the order of 60%. In principle, Gudina et al.[39] and Shibulal et al.[40] also reported the similar results that demonstrated the degradation of heavy alkanes resulting in the reduction of viscosity, which are conducive to heavy oil EOR due to microbial treatment.

9.4.2 In Situ Combustion (ISC)

The basic strategy of ISC involves the injection of air or oxygen that mixes with the oil in situ which is heated with downhole heaters. The supply of air is controlled such that the ignited oil burns at a controlled rate. Just the way catalysts are used in the refinery in a fluid catalytic cracking unit, they can also be employed in the ISC process to enhance the efficiency of cracking. The high temperature generated in ISC results in production of gases such as CO_2, N_2, CO, CH_4 and even higher carbon number gases, and visbreaking of the (heavy) oil, i.e., reduction in the viscosity (as well as density), which is attributed to compositional upgrading in a natural reactor (the reservoir).[41] Somewhat similar to MEOR, the application of ISC also is not very widespread—the two noteworthy successful field cases that are most often cited are in Romania and India (relevant references can be found in www.onepetro.org). A coreflood analog of laboratory experiments to determine the oil upgrading potential of an ISC process is the combustion tube experiment. Based on selected literature on combustion tube and other reactor experiments, the compositional changes in the oil before and after ISC are summarized next.

Hart et al.[41] conduced a Toe to Heel Air Injection (THAI), assisted with Ni–Mo/Al_2O_3 catalyst, experiments on a 13°API and 490 cP (at 20°C) viscosity heavy oil from Whitesands oilfield in Alberta. Other specific details of their test can be found in their paper. Hart et al.[41] presented detailed TBP (simulated) distillation of the original heavy oil and the produced upgraded oil at three different test temperatures of 350°C, 400°C, and 425°C. The reported TBP distillation data are converted into single carbon number (SCN) fractions and eventually expressed in mol% using the generalized SCN specific gravity and molecular weight.[22,42] Figure 9.22 shows a comparison of the original and upgraded oil at 400°C, spanning a SCN distribution from C_5–C_{36}, which clearly demonstrates a significant upgrade (reduction or cracking of heavy molecules); in particular, see the increase in C_9–C_{20} (peaking at C_{11}) and then a decrease from C_{21}–C_{36} in the post-ISC oil. Besides the compositional change, it is also important to note the yield of the upgraded oil which ranges from ~93% to 97% of the heavy oil with the balance being gaseous products and coke. The compositional upgrade also is conducive to an increase in the API by 2°–5°, whereas the viscosity reduction is in the

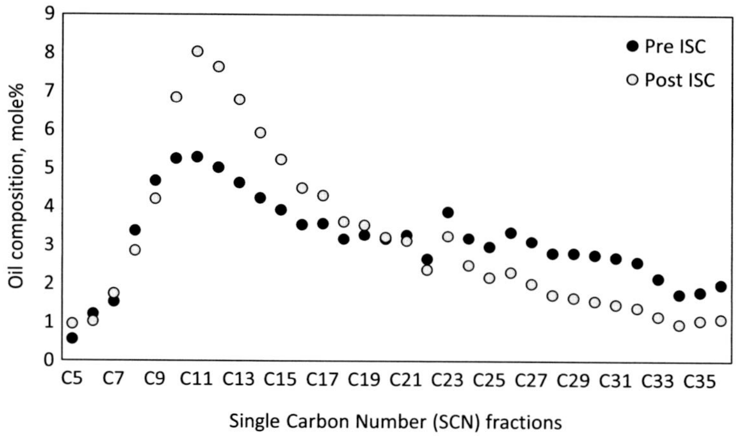

FIGURE 9.22
Comparison of pre- and post-ISC composition of Canadian Whitesands heavy oil.[41] The reported TBP distillation data of Hart et al.[41] are expressed in terms of SCN fractions according to the boiling temperature interval[42] and then converted into mol%, using the generalized specific gravity and molecular weights,[22,42] which is used in constructing this plot.

range of 20–110 cP.[41] Another notable compositional change reported[41] was in terms of asphaltene reduction in the range of 29%–43%. Clearly, the higher asphaltene content in the original oil contributes to low API gravity, thus high viscosity, which obviously reduces due to the upgrading of oil where asphaltene degradation plays an important role.

Yusuf et al.[43] carried out catalytic aquathermolysis experiments on a ~15, 000 cP (at 70°C) viscosity Omani heavy oil and obtained similar results in terms of upgrading. Their experiments varied somewhat with a reaction temperature of 277°C, a Ni–Mo catalyst, and glycerol and water as hydrogen donors, which is somewhat similar to hydrocracking where hydrogen is added to heavier hydrocarbon molecules or feedstock that is lacking in hydrogen. Figure 9.23 compares the grouped composition in terms of C_6–C_{12}, C_{12}–C_{14}, and C_{14}–C_{17+} (per the boiling point interval of Katz and Firoozabadi[42]) for the original oil and the upgraded oil, which shows an increase in the C_6–C_{12} fractions and a complete reduction of the C_{14}–C_{17+}. The role played by asphaltenes in oil upgrading is clear in this case as well with a reduction of ~17%,[43] thus making the original oil less viscous by a reduction of ~67%.[43]

Zhang et al.[44] conducted combustion tube experiments in the range of 340°C–360°C on a 0.985 specific gravity (at 20°C) and 38,670 cP heavy oil from Liahoe oilfield in China. The results they obtained are similar to the two previous studies. Although the authors did not provide numerical compositional values of the original and ISC upgraded oil, comparison of the reported gas chromatograms clearly indicate a shift in the Flame Ionization Detector (FID) trace toward lower retention times for the upgraded oil. This clearly shows a thermal cracking of the heavier molecules thereby increasing the lighter

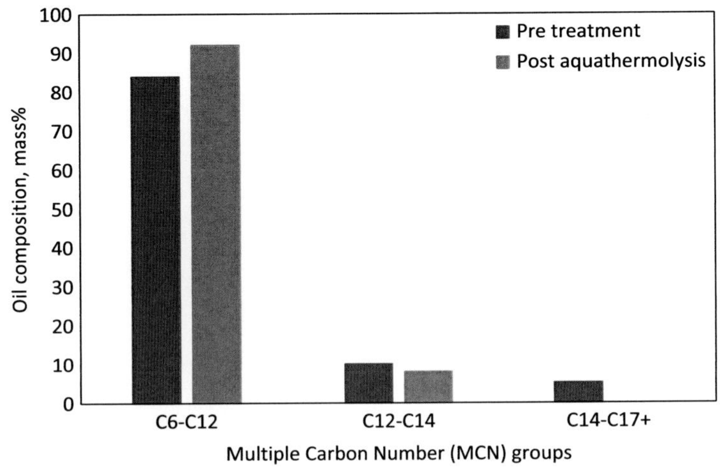

FIGURE 9.23
Comparison of original and optimized condition aquathermolysis treated Omani heavy oil.[43] The reported boiling point range and % components data of Yusuf et al.[43] is expressed in terms of MCN groups C_6–C_{12}, C_{12}–C_{14} and C_{14}–C_{17+} per the boiling point interval of Katz and Firoozabadi,[42] which is used in constructing this plot. Note that the oil compositions are directly based on the gas chromatographic data and are in terms of groups and thus plotted as mass% instead of mol%.

components; the authors report this as a change in the main peaks from the original oil C_{28}–C_{30} to upgraded oil C_{13}–C_{15}. A complimentary change in the outlet gas composition containing lighter hydrocarbon components from C_1 to C_{6+} compared to the ~22% O_2 and 78% N_2 in the inlet gas also was reported in their study. The upgraded oil resulted in a specific gravity of 0.97 and viscosity of 1,411 cP at 20°C and 50°C, respectively. Their study did not include any pre and post-ISC asphaltene comparison; however, the authors state that the cracking of resins and asphaltenes contributes to the formation of typical saturates and aromatics found in typical conventional crude oils.

9.5 Case Studies and Recent Advances

As stated by Haugen et al.,[32] three-phase flash calculations are not yet matured enough and are still emerging and thus their implementation in commercial compositional reservoir simulators is not very common. Recent publications such as those of Zhu et al.[45] continue to make advances that specifically focus on the efficiency and robustness of three-phase flash calculations; however, their inclusion in commercial compositional reservoir simulators is still limited. The reader is referred to Zhu et al.[45] for the several case studies of three-phase flash calculations they have presented together with a summary of others reported in the literature.

One of the major obstacles in carrying out three-phase flash calculations from a conventional methodology is the availability of a reasonable set of equilibrium ratios to initiate the iterative sequence. Although the Wilson[28] equation or correlation is widely used to obtain the equilibrium ratios between the vapor phase and the hydrocarbon (rich) liquid phase, no method currently exists to estimate the initial equilibrium ratios between the CO_2-rich liquid phase and the hydrocarbon-rich liquid phase. As part of the recent advances in three-phase flash calculations we have attempted to fill this gap by developing a new correlation to obtain the initial estimates of the equilibrium ratios between the CO_2-rich and the hydrocarbon-rich liquid phase, k_{zi}. The developed correlation hereafter referred to as the "Dandekar–Wilson" correlation is essentially a modified functional form of the well-known Wilson equation. Although the newly developed and proposed correlation has not yet been tested in three-phase flash calculations, it is anticipated that it will be valuable in at least obtaining the initial set of k_{zi} values to commence the iteration process such as the one outlined by the flowchart in Figure 9.17.

The Dandekar–Wilson correlation is developed on the basis of extensive experimental LLE and VLLE data reported in the literature[1,3,6,7] for the various CO_2–crude oil systems. The experimental database covers the following ranges; 94°F–106°F (typical temperatures conducive for the presence of the CO_2-rich liquid phase), 1,084–2,204 psia, 65–95 mol% CO_2 in oil, oil molecular weights of 125–182 and discrete components ranging from N_2, CO_2, and C_1–nC_5 and SCN C_6 and a plus fraction C_{7+}. Since Wilson correlation requires the critical temperature, critical pressure, and acentric factor of all components, these values for the C_{7+} were estimated from the Riazi and Daubert[46] correlations, whereas the C_6 properties and those of the well-defined components were taken from Katz and Firoozabadi[42] and Danesh[23], respectively. Using the more than 400 data points of experimental k_{zi} values (calculated from the reported CO_2 and hydrocarbon-rich compositions) and the given parameters and properties of the database oils, the following k_{zi} correlation was obtained by regressing the various correlation coefficients:

$$k_{zi} = \text{EXP}\left[\begin{array}{l} -7.558 + 0.852\left(\left(1+\omega_i\right)\left(1-\dfrac{T_{ci}}{T}\right)\right) + \\ 0.0151T + 0.000117P + 0.00158MW_{oil} - 0.01624X_{CO_2} \end{array}\right] \tag{9.15}$$

where ω_i are T_{ci} are the acentric factor and critical temperature of component i, T and P are reservoir temperature and pressure in °R and psia, respectively, MW_{oil} is the oil molecular weight (either a reported value or calculated based on compositions can be used), and X_{CO_2} is the mol% of CO_2 added to the oil. This obviously means that all these values are required (essential data) in

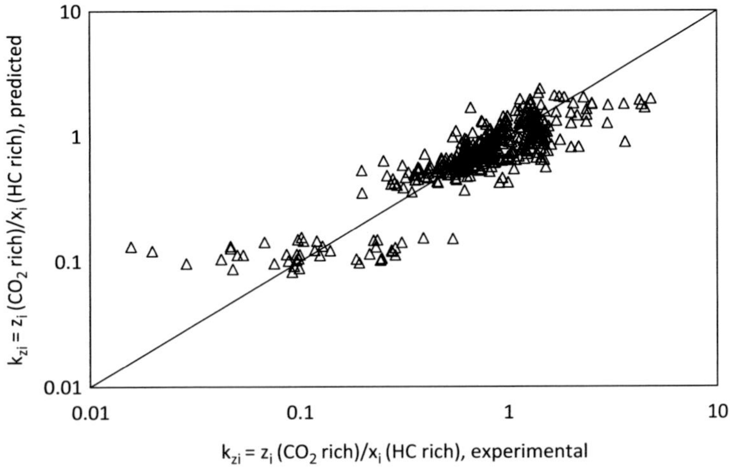

FIGURE 9.24 Comparison of experimental equilibrium ratio k_{zi} values and those predicted by the developed and proposed "Dandekar–Wilson" correlation (Equation 9.15). Note that the HC or hydrocarbon-rich phase has been alternatively stated or referred to as oil-rich phase.

obtaining the estimates of k_{zi} for a given CO_2–crude oil system expected to split into three phases. Clearly, a correlation as generalized (and easy to use) as this will certainly not be perfect, but we believe that it does capture the k_{zi} values reasonably well with an average absolute deviation of less than 30%. Finally, Figure 9.24 compares the experimental k_{zi} values with those calculated from the proposed correlation (Equation 9.15), which shows some outliers but for the most part the data lies on or around the 45° line.

References

1. Gardner, J.W., Orr, F.M., and Patel. P.D. 1981. The effect of phase behavior on CO_2-flood displacement efficiency. *Journal of Petroleum Technology* 33(11): 2067–2081. doi:10.2118/8367-PA.
2. Orr, F.M., Yu, A.D., and Lien. C.L. 1981. Phase behavior of CO_2 and crude oil in low-temperature reservoirs. *Society of Petroleum Engineers Journal* 21(4): 480–492. doi:10.2118/8813-PA.
3. Shelton, J.L. and Yarborough, L. 1977. Multiple phase behavior in porous media during CO_2 or rich-gas flooding. *Journal of Petroleum Technology* 29(9): 1171–1178. doi:10.2118/5827-PA.
4. Siagian, U.W.R. 1997. A laboratory study of the extraction of hydrocarbons from crude oil by high pressure carbon dioxide, *MS Thesis*, New Mexico Institute of Mining and Technology.
5. Sharma, G.D. 1990. Development of effective gas solvents including carbon dioxide for the improved recovery of West Sak oil, US DOE Final Report.

6. Turek, E.A., Metcalfs, R.S., Yarborough, L., and Robinson, R.L. Jr. 1984. Phase equilibria in CO_2- multicomponent hydrocarbon systems: Experimental data and an improved prediction technique. *Society of Petroleum Engineering Journal.* doi:10.2118/9231-PA.
7. Turek, E.A., Metcalfe, R.S., and Fishback. R.E. 1988. Phase behavior of several CO_2/West Texas-reservoir-oil systems. *SPE Reservoir Engineering.* doi:10.2118/13117-PA.
8. Lucas, M.A., Borges, G.R., da Rocha, I.C.C., Santos, A.F., Franceschi, E., and Dariva, C. 2016. Use of real crude oil fractions to describe the high pressure phase behavior of crude oil in carbon dioxide. *The Journal of Supercritical Fluids* 118: 140–147. doi:10.1016/j.supflu.2016.08.004.
9. Kokal, S.L. and Sayegh, S.G. 1993. Phase behavior and physical properties of CO_2-saturated heavy oil and its constitutive fractions: Experimental data and correlations. *Journal of Petroleum Science and Engineering* 9(4): 289–302. doi:10.1016/0920-4105(93)90060-R.
10. Vitu, S., Privat, R., Jaubert, J.N., and Mutelet, F. 2008. Predicting the phase equilibria of CO_2 + hydrocarbon systems with the PPR78 model (PR EOS and K_{ij} calculated through a group contribution method). *Journal of Supercritical Fluids* 45(1): 1–26. doi:10.1016/j.supflu.2007.11.015.
11. Robinson, D.B. and Peng, D-Y. 1978. The characterization of the heptanes and heavier fractions for the GPA Peng-Robinson programs. Research report/Gas Processors Association, RR-28; Research report (Gas Processors Association), RR-28. Tulsa, Okla: Gas Processors Association.
12. Fall, D.J. and Luks, K.D. 1985. Liquid-liquid-vapor phase equilibria of the binary system carbon dioxide + N-tridecane. *Journal of Chemical and Engineering Data* 30(3): 276–79. doi:10.1021/je00041a012.
13. Yu, J., Wang, S., and Tian, Y. 2006. Experimental determination and calculation of thermodynamic properties of CO_2+octane to high temperatures and high pressures. *Fluid Phase Equilibria* 246(1): 6–14. doi:10.1016/j.fluid.2006.05.014.
14. Camacho-Camacho, L.E., Galicia-Luna, L.A., Elizalde-Solis, O., and Martínez-Ramírez, Z., 2007. New isothermal vapor–liquid equilibria for the CO_2+ n-Nonane, and CO_2+ n-Undecane systems. *Fluid Phase Equilibria* 259(1): 45–50. doi:10.1016/j.fluid.2007.04.022.
15. Tanaka, H., Yamaki, Y., and Kato, M. 1993. Solubility of carbon dioxide in pentadecane, Hexadecane, and pentadecane + hexadecane. *Journal of Chemical & Engineering Data* 38(3): 386–388. doi:10.1021/je00011a013.
16. D'souza, R., Patrick, J.R., and Teja, A.S. 2018. High pressure phase equilibria in the carbon dioxide—n-hexadecane and carbon dioxide—Water systems. *The Canadian Journal of Chemical Engineering* 66(2): 319–323. doi:10.1002/cjce.5450660221.
17. Hottovy, J.D., Luks, K.D., and Kohn, J.P. 1981. Three-phase liquid-liquid-vapor equilibriums behavior of certain binary carbon dioxide-N-paraffin systems. *Journal of Chemical & Engineering Data* 26(3): 256–58. doi:10.1021/je00025a009.
18. Henni, A., Jaffer, S., and Mather A.E. 2018. Solubility of N_2O and CO_2 in n-Dodecane. *The Canadian Journal of Chemical Engineering* 74(4): 554–557. doi:10.1002/cjce.5450740418.
19. Gardeler, H., Fischer, K., and Gmehling, J. 2002. Experimental determination of vapor–liquid equilibrium data for asymmetric systems. *Industrial & Engineering Chemistry Research* 41(5): 1051–56. doi:10.1021/ie0103456.

20. Al Ghafri, S.Z., Maitland, G.C., and Trusler, J.P.M. 2014. Experimental and modeling study of the phase behavior of synthetic crude oil+CO_2. *Fluid Phase Equilibria* 365: 20–40. doi:10.1016/j.fluid.2013.12.018.
21. Liphard, K.G. and Schneider, G.M. 1975. Phase equilibria and critical phenomena in fluid mixtures of carbon dioxide + 2,6,10,15,19,23-Hexamethyltetracosane up to 423 K and 100 MPa. *The Journal of Chemical Thermodynamics* 7(9): 805–814. doi:10.1016/0021-9614(75)90088-9.
22. Whitson, C.H., Brulé, M.R., and Society of Petroleum Engineers of AIME. 2000. *Phase Behavior.* SPE monograph, Henry L. Doherty series, v. 20; Monograph (Society of Petroleum Engineers of AIME), Henry L. Doherty series; 32892 v. 20. Richardson, Tex.: Henry L. Doherty Memorial Fund of AIME, Society of Petroleum Engineers.
23. Danesh, A. 1998. *PVT and Phase Behaviour of Petroleum Reservoir Fluids.* Edited by A. Danesh. Elsevier.
24. Baker, L.E., Pierce, A.C., and Luks, K.D. 1982. Gibbs energy analysis of phase equilibria. *Society of Petroleum Engineering Journal.* doi:10.2118/9806-PA.
25. Pedersen, K.S., Christensen, P.L., and Shaikh, J.A. 2015. *Phase Behavior of Petroleum Reservoir Fluids.* 2nd edition. Boca Raton, FL: CRC Press, Taylor & Francis Group.
26. Dandekar, A.Y. 2013. *Petroleum Reservoir Rock and Fluid Properties.* CRC Press.
27. Michelsen, M.L. 1982. The isothermal flash problem. Part I. Stability. *Fluid Phase Equilibria* 9(1): 1–19. doi:10.1016/0378-3812(82)85001-2.
28. Wilson, G. 1968. A modified Redlich-Kwong EOS, Application to general physical data calculations. In *Annual AIChE National Meeting.* Cleveland, OH: American Institute of Chemical Engineers.
29. Rachford, H.H. Jr. and Rice, J.D. 1952. Procedure for use of electronic digital computers in calculating flash vaporization hydrocarbon equilibrium. *Journal of Petroleum Technology* 4(10). doi:10.2118/952327-G.
30. Okuno, R., Johns, R.T., and Sepehrnoori, K. 2010. Three-phase flash in compositional simulation using a reduced method. *SPE Journal.* doi:10.2118/125226-PA.
31. Li, Z. and Firoozabadi, A. 2012. Initialization of phase fractions in Rachford–Rice equations for robust and efficient three-phase split calculation. *Fluid Phase Equilibria* 332: 21–27. doi: 10.1016/j.fluid.2012.06.021.
32. Haugen, K.B., Firoozabadi, A., and Sun, L. 2010. Efficient and robust three-phase split computations. *AIChE Journal* 57(9): 2555–2565. doi:10.1002/aic.12452.
33. Huang, E.T.S. and Tracht, J.H. The displacement of residual oil by carbon dioxide, paper SPE 4735 presented at the *SPE Symposium on Improved Oil Recovery,* Tulsa, April 22–24, 1974.
34. Nghiem, L.X. and Li, Y.K. 1986. Effect of phase behavior on CO_2 displacement efficiency at low temperatures: Model studies with an equation of state. *SPE Reservoir Engineering.* doi:10.2118/13116-PA.
35. Lazar, I., Petrisor, I.G., and Yen, T.F. 2007. Microbial enhanced oil recovery (MEOR). *Petroleum Science and Technology* 25(11): 1353–1366. doi:10.1080/10916460701287714.
36. Adetunji, J.A. 2012. Microbial enhanced oil recovery, *PhD Thesis,* Aalborg University.
37. Ghotekar, A. 2007. Chemical and microbial characterization of North Slope viscous oils for MEOR application. *MS Thesis,* University of Alaska Fairbanks.

38. Sakthipriya, N., Doble, M., and Sangwai. J.S. 2015. Fast degradation and viscosity reduction of waxy crude oil and model waxy crude oil using bacillus subtilis. *Journal of Petroleum Science and Engineering* 134: 158–66. doi:10.1016/j.petrol.2015.08.002.
39. Gudiña, E.J., Pereira, J.F.B., Rodrigues, L.R., Coutinho, J.A.P., and Teixeira, J.A. 2012. Isolation and study of microorganisms from oil samples for application in microbial enhanced oil recovery. *International Biodeterioration & Biodegradation* 68: 56–64. doi:10.1016/j.ibiod.2012.01.001.
40. Shibulal, B., Al-Bahry, N.S., Al-Wahaibi, M.Y., Elshafie, E.A., Al-Bemani, S.A., and Joshi, J.S. 2018. Microbial-enhanced heavy oil recovery under laboratory conditions by Bacillus Firmus BG4 and Bacillus Halodurans BG5 isolated from heavy oil fields. *Colloids and Interfaces*. doi:10.3390/colloids2010001.
41. Hart, A., Wood, J., and Greaves, M. 2017. In situ catalytic upgrading of heavy oil using a pelletized Ni-Mo/Al_2O_3 catalyst in the THAI process. *Journal of Petroleum Science and Engineering* 156: 958–965. doi:10.1016/j.petrol.2017.06.067.
42. Katz, D.L. and Firoozabadi, A. 1978. Predicting phase behavior of condensate/crude-oil systems using methane interaction coefficients. *Journal of Petroleum Technology*. doi: 10.2118/6721-PA.
43. Yusuf, A., Al-Hajri, R.S., Al-Waheibi, Y.M., and Jibril, B.Y. 2016. In-situ upgrading of Omani heavy oil with catalyst and hydrogen donor. *Journal of Analytical and Applied Pyrolysis* 121: 102–112. doi:10.1016/j.jaap.2016.07.010.
44. Zhang, X., Liu, Q., and Che, H. 2013. Parameters determination during in situ combustion of liaohe heavy oil. *Energy & Fuels* 27(6): 3416–3426. doi:10.1021/ef400095b.
45. Zhu, Di., Eghbali, S., Shekhar, C., and Okuno, R. 2018. A unified algorithm for phase-stability/split calculation for multiphase isobaric-isothermal flash. *SPE Journal*. doi:10.2118/175060-PA.
46. Riazi, M.R. and Daubert, T.E. 1980. Simplify property predictions. *Hydrocarbon Processing* 59(3): 115–116.

10

Flow Assurance in EOR Design and Operation[1]

10.1 Flow Assurance

Oil and gas are located both offshore (in shallow and deep water) and onshore. Gas and oil are likely to occur at various depths from few hundred meters to few thousand meters. Reservoir rock and fluid characteristics, pressure, and temperature vary from field to field. Development and production strategies are not same in all the fields. They are customized based on their specific parameters. A big network comprising of wells, wellheads, pipelines, and processing facilities is required to produce, transport, and process the crude oil and gas. One objective of prudent development and production field development plan (FDP) is maximization of recovery from underground with minimum operational and flow-related problems. Production is taken as primary, secondary, and tertiary phases. During primary phase, natural energy of the reservoir facilitates the production. Generally, reservoir pressure declines with production if not associated with strong aquifer support. This decline in reservoir pressure adversely affects the oil production. In order to maintain the reservoir pressure, water and gas are injected in the reservoir to maintain the reservoir pressure. Water injection helps to maintain the reservoir pressure but at the same time increases the production. Thus, the water production increases with increasing the emulsion. A significant volume of oil is left inside the reservoir after secondary recover. Tertiary recovery techniques such as, chemical, miscible CO_2, and thermal processes are applied to further enhance the recovery from reservoirs. These injected fluids modify the properties of reservoir fluid. They improve the recovery but byproducts like emulsion also created which adversely affects the flow performance. They have to be controlled and mitigated for efficient production. Thus, flow assurance is critical to the life cycle of oil and gas field. Thus, flow assurance focuses on the whole engineering and production life cycle from the reservoir through refining, to ensure with high confidence that the reservoir fluids can be moved from the reservoir to the refinery smoothly and without

[1] Rohaizad B. Norpiah contributed to this chapter.

interruption. Byproducts like, wax, asphaltenes, hydrate, emulsion, scaling, and corrosion due to reservoir souring are important challenges encountered during the production and injection phases.

10.2 Wax

10.2.1 Chemistry

Wax present in crude oil and condensate consists of straight (*n*-paraffins), branched (iso-paraffins), and cyclic paraffin (napthenes). The chemical formula is C_nH_{2n+2} for normal and iso-paraffin and C_nH_{2n} for cyclo-paraffin. Figure 10.1 shows the chemical structure of the three types of petroleum waxes.

The structure of the paraffin molecules will impact the morphology and growth rate of the wax crystals. *N*-paraffins produce large crystals (macrocrystalline) without much interference due to the linear molecule structure. Branched (iso) and cyclic paraffins produce small(microcrystalline) crystals due to interference from the branches. Both types of crystals are present in crude oil and condensate.[1] Table 10.1 shows a comparison of the basic properties of petroleum waxes.

During cooling, waxes crystalizes in the descending order of their melting point temperature from the highest to lowest (Figure 10.2).

10.2.2 Phase and Physical Properties

At high temperatures, waxes are dissolved in oil, but at low temperatures, waxes crystallize starting with the component with the highest carbon number. The temperature when wax initially precipitates is called the wax appearance temperature (WAT). All types of waxes precipitate in crude oil but most crude oils contain more than 75 wt% *n*-alkanes according to Turner and Musser.[1,2]

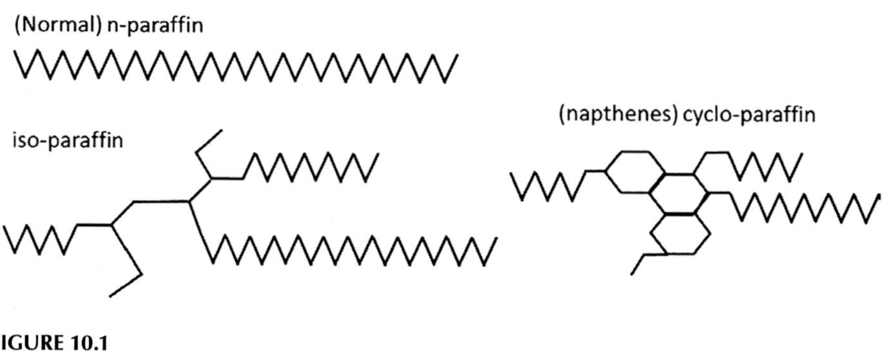

FIGURE 10.1
Schematic diagram of chemical structure of petroleum waxes.

TABLE 10.1
Basic Properties of Petroleum Waxes

Properties	*n*-Paraffins	iso-Paraffins and Cyclic Paraffin
Crystal type	Macro-crystalline	Microcrystalline
Crystal characteristics	large, well-formed crystals, rod, needle, plate shape Brittle	smaller irregular crystals, spherical shape Elastic
Carbon number range	C17–C30	>C30–C60
Molecular weight range	350–600	300–2,500
Opacity	Translucent	Opaque
Melting point range, °C	40–60	60–90

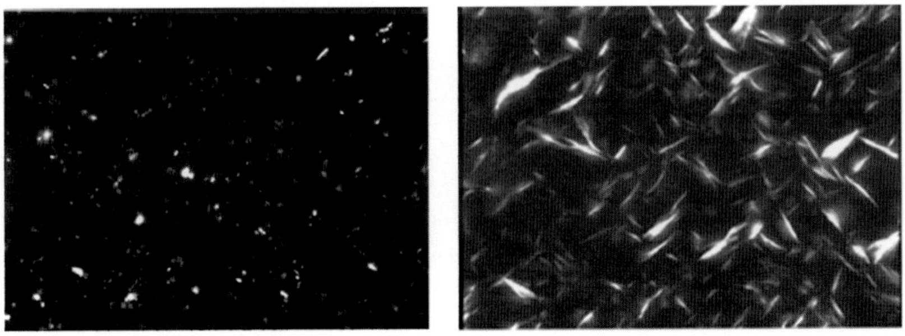

FIGURE 10.2
Microcrystalline and macocrystalline wax crystals.

Dissolved gas (light ends) reduces WAT by lowering the concentration of wax components. The effect of solution gas on dead oil and condensate is a strong function of gas oil ratio and pressure.

A typical wax phase envelope is shown in Figure 10.3.

WAT decreases as more solution gas dissolved in dead oil up to bubble-point pressure. Above bubble point, liquid density decreases with increasing pressure while wax content is unchanged causing WAT to increase with pressure. The consequences of wax precipitation are as follows:

- Increased viscosity due to the presence of suspended solids.
- Solids deposit onto the surface which reduces the flow area and increases frictional pressure drop. The surface of deposited wax is rough and enhances pressure drop.

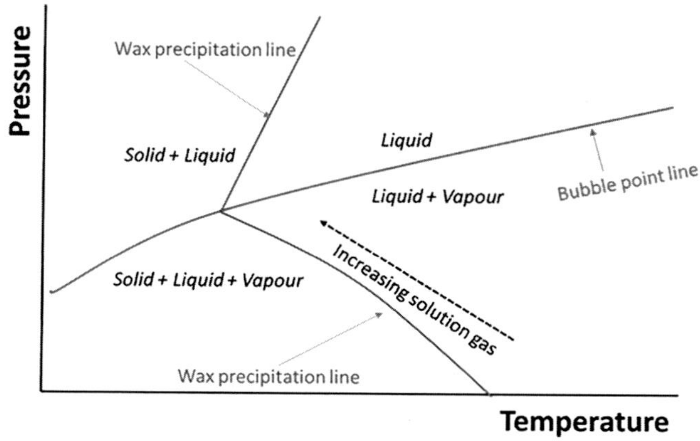

FIGURE 10.3
Wax phase envelope.

- A solid gel can form when the flow is stopped.
- In storage tanks, wax solids settle to the bottom to form a sludge that reduces the working capacity.

If a significant amount of wax is present, the viscosity–temperature behavior of a waxy crude oil or condensate will exhibit non-Newtonian behavior. The viscosity is dependent on temperature above WAT, while it is dependent on shear rate (flowrate) below WAT. Figure 10.4 illustrates the viscosity behavior of a crude oil with 3.5 wt% wax, WAT of 42°C, and pour point temperature of 18°C.

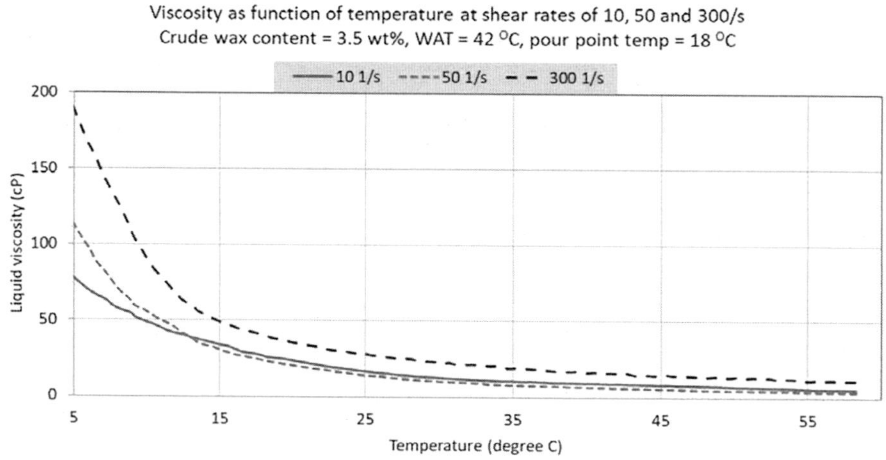

FIGURE 10.4
Waxy crude oil viscosity–temperature behavior.

The temperature loss of fluids in a well and pipeline is due to the heat transfer to the surroundings.

A typical temperature profile of a fluid flowing in a pipeline is shown in Figure 10.5.

Once the fluid temperature drops to WAT, wax starts to precipitate. Wax solids deposit on the cold surface of pipeline inner wall surface. The dominant mechanism for wax deposition is diffusion. A concentration gradient between dissolved waxes and wax in solution at the wall provides the driving force to deposit wax. The concentration of wax in solution is determined from the inverse of solid wax concentration (mass%) versus temperature points on a wax precipitation curve (WPC). The WPC curve is commonly generated using a thermodynamic model tuned to match dead oil WAT. Several models, such as Lira-Galeena, Pedersen, Countinho, etc., are described in the literature.

Figure 10.6 shows a WPC and the corresponding wax in solution content (dissolved) curve.

Shear dispersion where waxes deposit by the velocity variation of solid waxes is a secondary mechanism for wax deposition. The other deposition mechanisms, such as Brownian and gravity settling, do not play a significant role in wax deposition in wells and pipelines. The layers of wax buildup over time increases pressure drop and can impair flow.

At lower temperatures than WAT, wax crystals interact structurally to form more viscous oil and eventually a gel at pour point temperature. Similar to the effect of solution gas in reducing WAT, pour point temperature is also reduced with solution gas dissolved in crude oil.

In storage tanks and equipment such as separators where fluid velocities are low, waxes settle at the bottom. Such sludgy deposits contain entrapped crude oils, asphaltene, scale, water, corrosion solids, and sand.

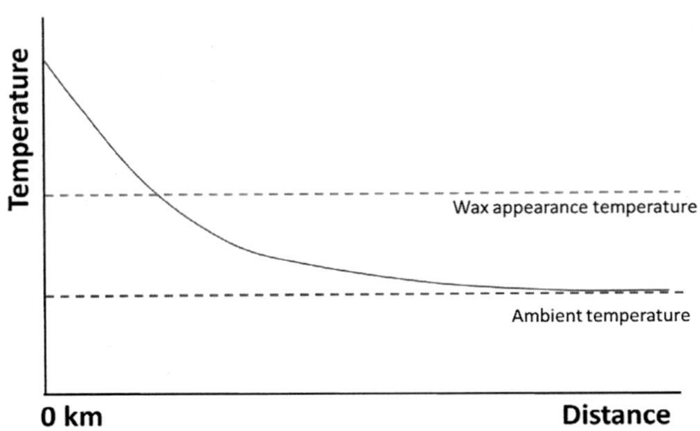

FIGURE 10.5
Typical flowing fluid temperature profile in a pipeline.

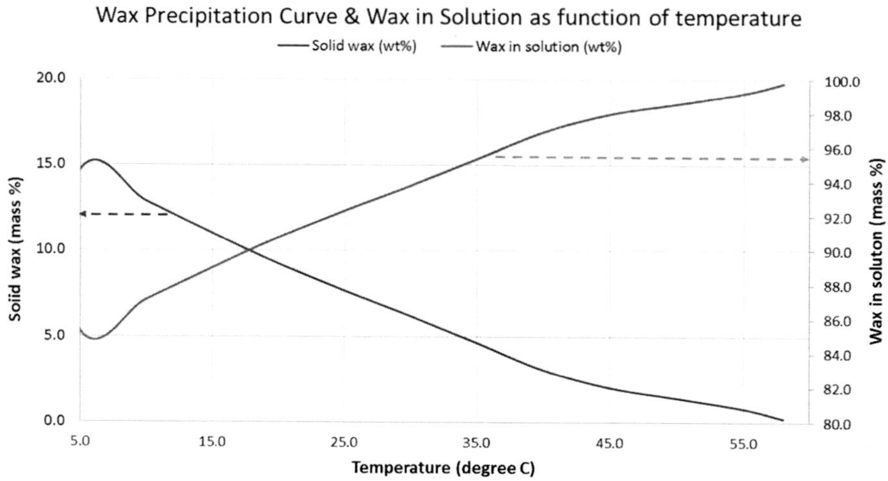

FIGURE 10.6
WPC and the corresponding wax in solution content (dissolved) curve.

A gelled crude oil exhibits complex rheological behavior. When shear is applied to a gel, it has three characteristic regions: elastic, creep, and fracture. In the elastic region, gel strength is recovered when stress is removed. The region before fracture occurs is creep where gel strength partially recovers when stress is removed. The transition point between elastic and creep is the static yield stress where a gel fracture initially occurs. In the last region, the gel structure is continuously fractured with applied stress. Sufficient pressure (restart pressure) needs to be applied to gelled crude oil to break it which requires a high enough design pressure of a pipeline or well. There are many factors that influence gel strength including cooling rate, final temperature, shear rate or absence of shear while cooling and the duration at gel condition.

10.2.3 Laboratory Testing and Modeling

The overall workflow to design and conduct an engineering study of wax issues is shown in Figure 10.7.

In this section, the laboratory tests to provide the data for engineering studies and usage of models will be covered briefly.

10.2.3.1 Deposition Modeling—Wax Composition

N-paraffin composition (mass%) from C17 to C100 is used in a wax thermodynamic model to predict WAT, wax precipitation envelope (WPE), and produce a WPC. A gas chromatograph calibrated with *n*-alkanes mixture is used to measure these data. Figure 10.8 shows an example of *n*-paraffin HTGC data.

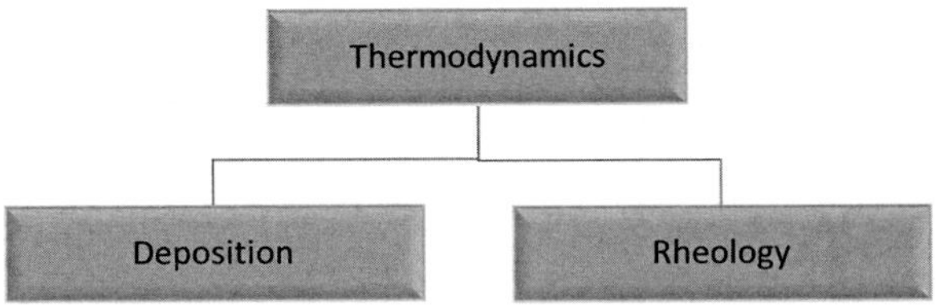

FIGURE 10.7
Workflow to conduct an engineering study on wax.

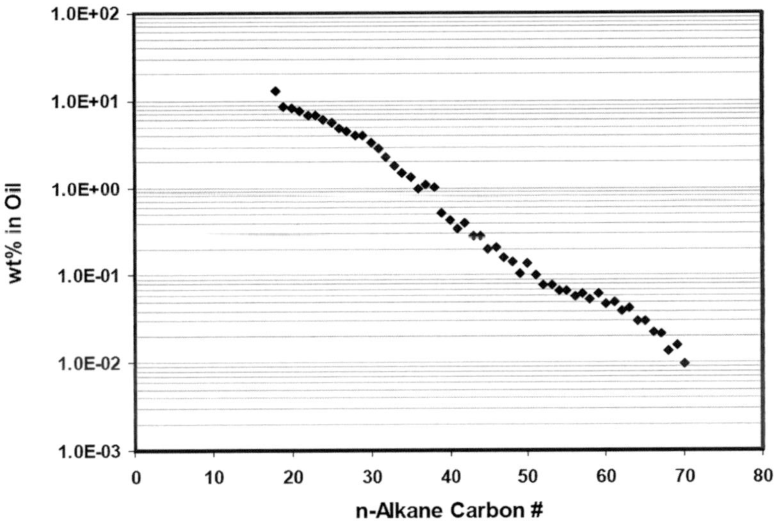

FIGURE 10.8
n-Paraffin HTGC composition data.

WPC is a key input to predict wax deposition rate, amount of deposited wax, and the effect of wax on viscosity. Various models for the wax-oil slurry and wax deposition have been developed over the years.

10.2.3.2 Deposition Modeling—WAT

WAT is the temperature of wax precipitation onset. It can be predicted from thermodynamic model, nearly all of which is based on *n*-alkanes.

A stabilized (dead) oil WAT value is normally used to calibrate a wax thermodynamic model. Both cross-polarized microscopy (CPM) and differential scanning calorimetry (DSC) method should be used to measure WAT. It is important to note that a DSC measures all detectable liquid to solid transitions. The measured WAT may comprise of microcrystalline (branched and/or cyclic paraffin or a mixture and more than one WAT could be detected depending on the relative amounts of different types of waxes.

DSC also provides data to calculate WPC from the heat release rate versus temperature values.

For DSC, a cooling rate of between 1.0 and 3.0°C/min should be used to avoid subcooling effect that can delay the formation of wax if fast cooling was used. A cooling rate of 0.5–1.0°C/min should be used for CPM. Design is usually based on CPM result as it is more sensitive and closer to thermodynamic WAT than DSC value. It should be noted that microcrystalline wax crystal formation could be undetectable by CPM due to crystal sizes being smaller than the resolution limit of 1 μm. The most recent study of various methods to measure WAT were conducted by Japper-Jaafar et al.[3] The authors concluded that precipitation rate, type of wax (micro or macrocrystalline), cooling rate, and imposed shear rate greatly affect WAT results from CPM, micro-DSC, and rheometer. Furthermore, they considered that CPM being a visual method could give a subjective WAT value, while the other two methods if applied with a low cooling rate of 1°C/min is more sensitive than using a CPM method.

10.2.3.3 Deposition—Wax Disappearance Temperature

The last residue of precipitated paraffin redissolving in liquid is the wax disappearance temperature (WDT). Heating of a sample cooled below WAT yields WDT. WDT represents a true solid–liquid equilibrium point but nearly of all of modeling for wax mitigation is based on WAT. CPM, DSC, and other experimental methods used to measure WAT can also be used to measure WDT.

The significance of WDT in design is its value that determines the dissolution of precipitated wax back to solution by heating or increasing temperature. However, due to the effect of aging in wax deposits where wax content increases over time, WDT underestimates the temperature required to redissolve waxes by heating.

10.2.3.4 Pour Point Temperature

The temperature below which liquid ceases to flow is the pour condition temperature. ASTM D-5853 (ASTM)[4] is the standard used to measure crude oil pour point temperature. In this standard, a sample cooled in an ice bath is manually removed every 3°C and held horizontally. If the liquid sample does not flow for 5 s, 3°C is added to the current ice bath temperature and is recorded as pour point temperature.

The standard provides two procedures to determine maximum (upper) pour and minimum (lower) pour point temperature. The upper value is obtained when a sample is cooled from 60°C, while the lower value is obtained when the sample is re-heated to 105°C and followed by cooling. The two procedures provide a range of pour point temperature with different thermal history, i.e., cooling (maximum) and cooling, heating and cooling (minimum).

The upper pour point is normally used for design as it represents the conditions where oil is cooled from the reservoir to surface facilities and pipeline.

The transition from a viscous liquid to a gel may be visually observed but the precise value cannot be measured due to the limitation of ASTM D5853, namely the 3°C interval between observations. The upper point and lower pour temperature procedure has a repeatability of 3°C and 6°C, respectively, at 95% confidence level.

The main shortcomings of ASTM D5853 are low precision (3°C) due to manual operation, relatively poor repeatability, and reproducibility due to operator uncertainties.

10.2.3.5 Rheology—Gelation Temperature

A gelation temperature represents the point where liquid becomes a gel/solid under controlled rheological conditions. Parameters such cooling rate, applied stress, and shear affect gelation temperature.

The network of precipitated waxes in a flowing crude is disrupted by shear stress and becomes a solid at a gelation temperature as compared to pour point temperature where gelation occurs at zero shear stress conditions. Gelation temperature is measured in a rheometer by two methods. The first method applies oscillatory protocol which allows storage and loss modulus to be measured while cooling the sample[5]. The intersection of the two moduli that coincide at gelation temperature represents the transition between liquid viscous behavior and solid elastic behavior.

The second method is simpler than the oscillatory protocol. A low constant shear stress (0.1 Pa) is applied while cooling a sample and measuring shear rate.[6] This method has better repeatability and reproducibility than the oscillatory method.

The application of the rheological method to design waxy oils requires an evaluation of influencing parameters in the system being designed under static and flowing cooling conditions.

10.2.3.6 Rheology—Viscosity

The characteristics of Newtonian and non-Newtonian viscosity and yield stress are shown in Figure 10.9.

The major rheological characteristics of a waxy oil are as follows:

- Shear-dependent viscosity below WAT. Most waxy crude oils exhibit shear thinning behavior where viscosity decreases with higher shear rates.
- Formation of a gel at pour point or gelation temperature. The gel has a yield stress value that exists at zero shear rate.
- Viscosity, yield stress, and oscillatory properties (modulus) may change with time, the so-called thixotropic behavior.

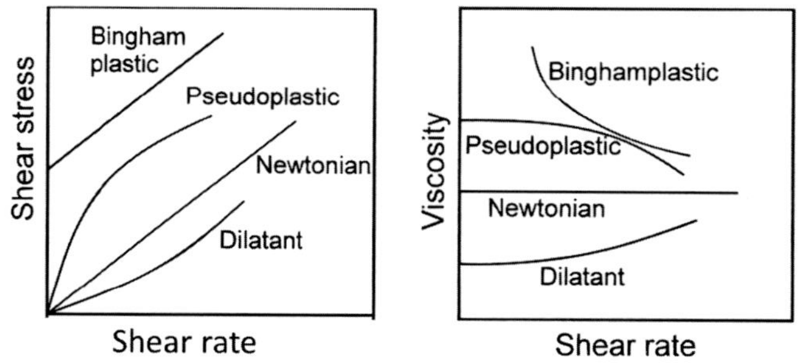

FIGURE 10.9
Newtonian and non-Newtonian Rheological characteristics.

The relationship between a Newtonian liquid viscosity and temperature is modeled by the Eyring equation. Crude oil above WAT behaves as a Newtonian fluid where viscosity remains the same at all shear rates. However, below WAT, precipitated wax solids cause viscosity to change according to shear rates. By definition, shear rate is the rate of change of velocity at which one layer of fluid passes over an adjacent layer. It is dependent on volumetric flowrate and the spacing between the layers. For a Newtonian fluid flowing within a pipe, the shear rate is given by

$$\dot{\gamma} = \frac{4Q}{\pi r^3}.$$

where

γ = shear rate, 1/s

Q = volumetric flowrate, m^3/s

r = pipe radius, m

There are several models that predict the non-Newtonian behavior of waxy crude oils. An example of one model used incommercial software is given as follows:

$$\eta = \eta_{liq}\left[\exp(D\Phi_{wax}) + \frac{E\Phi_{wax}}{\sqrt{\frac{\partial v_x}{\partial_y}}} + \frac{F\Phi_{wax}^4}{\frac{\partial v_x}{\partial_y}}\right]$$

where η_{liq} is the viscosity of the liquid phase with no precipitated wax, Φ_{wax} is the volume fraction of precipitated solid, and D, E, F are the fitting parameters.

Experiments are conducted in a rheometer by cooling a sample pretreated to remove thermal history by heating above WDT and mixing. The temperature range should cover the maximum temperature from reservoir temperature to minimum operating temperature which is normally the minimum ambient temperature. A range of shear rates should also be used for the predicted flowrates. A typical range of shear rates for field conditions is between 10 and 1,000/s.

Rotational rheometer geometries are available as parallel plate, cone-plate, and concentric cylinder. Cone and plate geometry is recommended for waxy crude oils because shear stress and shear rates are uniform throughout the sample surface. These factors are critical for waxy crude oils whose rheological properties are sensitive to shear rates. It should be noted that the gap at the cone tip should be sufficiently large to avoid any effects on the wax microstructure. Slip between a sample and geometry surface can occur but this can be prevented by using roughened surfaces (Table 10.2).

The parameters used for a standard waxy crude oil rheology laboratory tests are tabulated in Table 10.2.

The laboratory test result's accuracy, repeatability, and reproducibility are highly dependent on the pretreatment procedure to redissolve waxes and completely remove thermal and stress history from sample. A typical procedure involves heating a sample 10°C–20°C above WAT while mixing in a closed container to minimize light end losses. The homogenized sample is stored in an oven at temperature above WAT for 24h. Subsampling into smaller volumes are also done when the sample temperature is higher than a specified value. It is also common to heat, mix the subsamples, and maintain the temperature higher than WAT for 1–2h before being tested.

TABLE 10.2

Waxy Crude Oil Rheological Measurement Parameters

Property	Major Factors
Viscosity	1. Temperature: initial and final 2. Cooling rate 3. Shear rate 4. Amount of gas dissolved in liquid 5. Amount of water—applicable for oil–water emulsion 6. Presence of crystal modifiers, for example, asphaltenes, resins, and flow improver chemicals 7. Water cut, water-in-oil emulsion
Gel yield stress	1. Cooling rate 2. Final temperature 3. Shear rate during cooling 4. Duration at gelling condition (aging) 5. Shear stress applied to break gel 6. Presence of crystal modifiers, for example, asphaltenes, resins, and flow improver chemicals 7. Amount of gas dissolved in liquid 8. Amount of water—applicable for oil–water emulsion gel

10.2.3.7 Wax Deposition

The basic wax deposition model is based on mass diffusion mechanism that assumes diffusion occurs in a laminar sublayer near to a wall. The dissolved concentration at bulk phase and sublayer is considered constant radially but varies as a function of bulk temperature, while concentration in the layer adjacent to the wall corresponds to the inner wall temperature. The temperature gradient in the laminar sublayer is calculated from a heat transfer equation. The model has the following form:

$$\frac{dM_w}{dt} = \rho_w D_w A_w \frac{dC}{dr} = \rho_w D_w A_w \frac{dC}{dT}\frac{dT}{dr}$$

where

$\frac{dMw}{dt}$ is the rate of wax deposited (kg/s),

ρ_w is the density of the solid wax (kg/m^3),

D_w is the diffusion coefficient of the wax in the oil phase (m^2/s),

A_w is the area of wax deposition (m^2),

$\frac{dc}{dr}$ is the wax concentration gradient (weight fraction/m) of wax concentration over pipe radial coordinate r (1/m),

C is concentration of dissolved waxes in the oil phase (weight fraction),

$\frac{dc}{dT}$ is concentration of dissolved waxes as function of temperature (weight fraction/°C) = total wax −solid wax concentration (WPC),

dT/dr is the radial temperature gradient of the wall (°C/m).

The steps in forming wax deposit are shown in Figure 10.10. Wax deposition models have been developed for single phase and multiphase flow. Zhenyu Huang et al.[7] have published a comprehensive volume on wax deposition. The assumptions and simplifications made in the deposition models mean that laboratory experiments conducted in similar conditions to the field are recommended.

Several types of devices are used to measure wax deposition rates, namely cold finger and flow loop. A flow loop schematic is shown in Figure 10.11.

Flow loop experiments are run based on three controllable parameters used to calibrate deposition model to experiments:

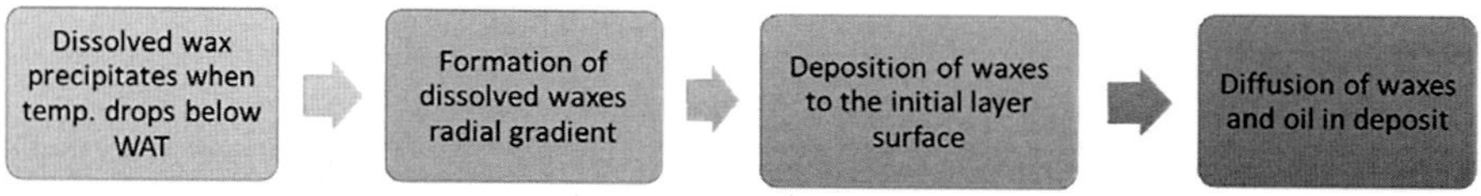

FIGURE 10.10
Wax deposition mechanism steps.

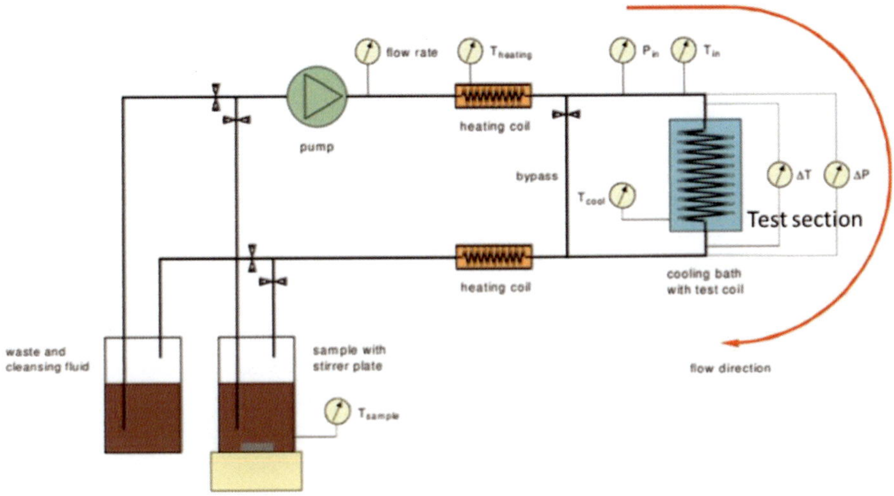

FIGURE 10.11
Flow loop schematic diagram (courtesy F5 Technologie GmbH).

- Shear rate—varying sample flowrates in a given test section inner diameter. For laminar flow, the shear rate is $\dot{\gamma}$ = (8 × velocity)/inner test section diameter.
- Temperature difference between oil sample at WAT and cold surface.
- Duration of tests is typically 6h or more. Sufficient sample volume should be used to avoid depletion of waxes in the test section.

The experimental variables used in a wax deposition laboratory tests program are shown in Table 10.3.

TABLE 10.3
Experimental Variables and Measured Properties for Wax Deposition Tests

Experimental Input Variables	Properties Measured or Calculated from Results
• Sample flowrates • Sample temperature held at WAT • Wall temperature varied by controlling coolant flowrate and temperature • Sample dynamic viscosity for a range of temperatures at various shear rates (rheometer) • Sample density at a range of temperatures (densitometer)	Pressure drop • Wax deposit thickness determined from test section pressure drop or other techniques • Deposit mass • Deposit mass flux • Deposit porosity (oil content) Temperature difference = sample temperature – wall temperature four temperature difference should be used

An example of a wax deposition flow loop tests is given in ref. 29. Figure 10.12 shows one of the results[8]. The three methods used measure wax deposit thickness were pressure drop, weight of test section, and a co-axial laser that measures the deposit thickness directly. All three methods gave a reasonably consistent wall thickness values.

To calibrate a wax deposition model, the model inputs are identical to sample properties and flow loop conditions. Calibration of the model to the laboratory results is done by adjusting parameters specific to the model but the diffusion coefficient is the main adjustable factor. The difficulty of conducting a flow loop test is the flowrate which has to be high enough to produce turbulent flow Reynolds number due to the small loop diameter, but this results in a shear rate which is not representative of field conditions. Most wax deposition tests are conducted to match field shear rates resulting in laminar flow or low turbulent flow Reynolds number (~10,000).

A second method to measure wax deposition rates is using a cold finger. A cold finger device consists of one or more stationary cylinder tubes into which a coolant flow in and out. The finger is inserted into a dead oil sample contained in a glass bottle which is stirred using a magnetic stirrer at the bottom. The shear rate and stress in a cold finger are not uniform unlike a flow loop, so the results have a much larger uncertainty for wax deposition model calibration. However, it is universally used to compare wax chemical inhibitor's performance due to ease of operation and small-sample volumes (~100 cm^3). It is also possible to run several tests simultaneously with some commercial units able to run 15–18 tests simultaneously. The quantitative results are acceptable for chemical testing as the purpose is to compare the

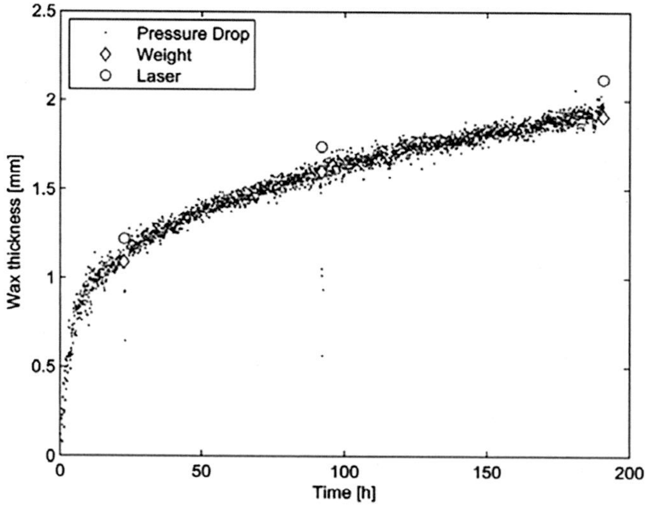

FIGURE 10.12
Wax thickness trend using three different methods (ref. 5.30).

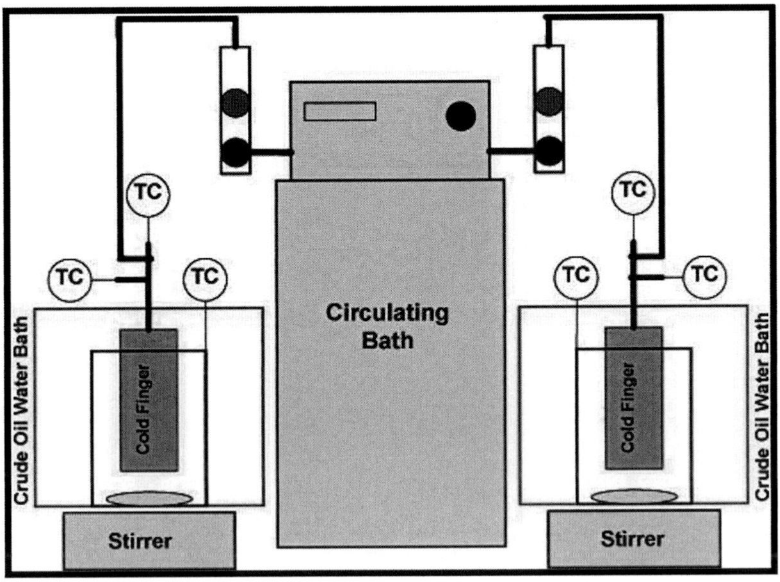

FIGURE 10.13
Schematic diagram of a cold finger.

relative performance of chemicals. A schematic of a cold finger device is shown in Figure 10.13.[9]

A standard program to test wax chemicals is to use a blank sample which has not been dosed with chemicals and samples dosed at varying concentrations. Results from cold finger tests include wax mass, wax deposition flux, photos, and a subjective evaluation of the deposit hardness. Figure 10.14 shows wax deposited from a cold finger test.

A third method to measure wax deposition is using a co-axial shear cell.[10,11] A sample is placed in between two cylinders with the outer cylinder is stationary, while the inner cylinder rotates. The main advantage of the cell is that the hydrodynamic conditions are identical to flow in a pipeline and achieve turbulent flow while using a relatively small-sample volume (~150 cm^3) compared a flow loop sample of 2–4 L. The Reynolds number of up to 600,000 was achieved in the cell as compared to the Reynolds number of up to 20,000 in a flow loop. It is also possible to test a live oil sample containing solution gas up to 103 MPa.

10.2.3.8 Rheology—Yield Stress

When liquid temperature falls to pour point temperature and below it, a solid gel that forms due to the formation of interlocking network of wax crystals yield stress (gel strength) is the force required to break down the wax structure in a gel. It can be used to determine the pumping or well pressure required to

FIGURE 10.14
Wax deposit from a cold finger test.

restart flow in a pipeline. The shear and thermal history of sample, temperature of the fluid before shutdown, cooling rates, and duration of the shutdown and the final fluid temperature of fluid can significantly affect yield stress.

Yield stress is measured using a rheometer or a flow loop. Laboratory data can be scaled up to full size pipelines using the following equation to calculate the restart pressure:

$$\Delta P = 4 * \Gamma y * (L/D)$$

where

ΔP = restart pressure which is the pressure difference between inlet and outlet pressure, Pa
Γy = gel yield stress, Pa,
L = length of gel, m
D = internal diameter of pipe, m.

The above equation is usually conservative because it assumes that the gel will break simultaneously throughout and methods[11–14] are available for a more accurate prediction of gel restart pressure. These methods are based on the concept of a yield front propagating starting with gel breaking at near the inlet, propagating downstream until of the gel has yielded to restart flow.

10.2.4 Prevention and Mitigation

Preventing and mitigating wax issues are concerned with deposition, viscosity, and gelling issues. Prevention is achieved by designing a system and operating outside of waxing tendency in a defined envelope throughout field life. In situations where prevention is not technically or economically feasible, wax is allowed to form but mitigation measures are implemented.

Wax deposition prevention and mitigation can be achieved by several methods shown in in Figure 10.15.

Thermal insulation. Insulation is used to reduce heat transfer between the production fluid andambient condition. It maintains fluid temperature above WAT and completely eliminates wax deposition. There is a distance limitation when heat loss is too great to maintain fluid temperature at the outlet above WAT. Pipeline insulation is classified acccording the degree of insulation based on overall heat transfer coefficient, *U*. Table 10.4 shows the classifications.

Heating. An alternative method for wax prevention is applying external heating. This can be achieved by hot water or oil circulation or some form of electrical heating. When using a hot fluid to maintain temperature in a system, the hot fluid is circulated along an outer jacket annulus on the pipe to keep the temperature of production fluid above WAT. The hot fluid temperature drops gradually.

Pipeline insulation and insulated-heated methods are classified acccording the degree of insulation based on overall heat transfer coefficient, *U*. Table 10.4 shows the types.

Heating bundle or hot water circulation refers to the pipeline configuration designed to warm production fluid using hot water. Generally, hot water circulation is a closed system, so that water is recirculated to the topside.

Wet insulation is the most commonly used pipeline insulation for subsea developments. A wide range of insulating materials which are exposed to seawater are available. The common materials used are Neoprene, Syntactic Epoxy, Polyurethane, and Polypropylene. Depending on the material, the maximum operating temperature is 140°C. For application in ultra-deepwater, some materials have been qualified to 3,000 m water depth.

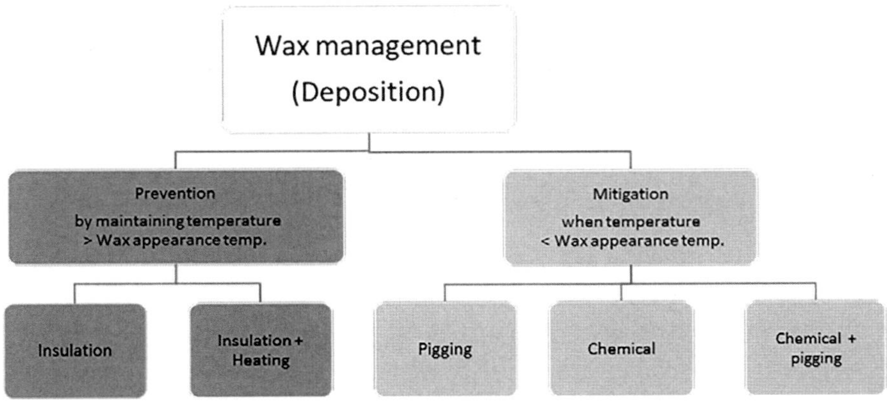

FIGURE 10.15
Wax deposition prevention and mitigation methods.

TABLE 10.4

Pipeline Insulation and Insulated-Heated Options

Type	*U*-Value (W/m^2K)	Remarks
Heating bundle with external hot fluid	3.0	
Wet insulation	2.0	Insulation in contact with seawater
Pipe in Pipe (PiP)	1.0	Insulation is protected by an external jacket. 0.5 W/m^2K using high-performance material
DEH—open loop with wet insulation	3.0	Nearly 30 installed systems Most tie-back distance = <20 km.
DEH—closed loop with PiP	1.0	A few installations with tie-back distance of 40 km. PiP; 0.5 W/m^2K using high-performance material
External heat traced	1.0	

Pipe-in-pipe achieves lower *U* value by using materials which have lower thermal conductivity but have to remain dry. An external jacket is used as a barrier between the insulation material and seawater. The common materials used are low-density polyurethane foam, aerogel, and Izoflex™.

Direct electric heating (DEH) passes an electric current through the pipeline wall using an electric cable piggybacked to the pipeline. The early systems used wet insulation but pipe-in-pipe (PiP) insulation with better insulation and lower electric power have been developed and used for longer tie-back distances (>20 km).

An alternative method to DEH is *external heat tracing* where heating bands or wires as a heating element are in contact with the pipeline. To protect the heating element and to provide better insulation, a PiP configuration is adopted. Fiberoptic cable to monitor temperature along the pipeline is also available.

In situations where prevention is not feasible, mitigation measures to control wax deposition are pigging, applying chemicals, and a combining pigging and chemicals.

Chemical inhibitor injection. Paraffin inhibitors are used to control the wax deposition rates. Typical paraffin inhibitors are hydrocarbon-soluble polymers or surfactants or a mixture of both. The chemicals are generally specific to the fluid although the practice by vendors is to test a family of similar chemicals on the fluid samples. Cold finger device is commonly used to compare the relative effectiveness of the chemicals. The compatibility with the injection system and other chemicals and effectiveness of the inhibitor must be verified by lab testing against the production fluid before injecting into the actual production system. Typical wax inhibitor is effectiveness is 50%–80% in terms of wax deposition rate and deposit amount.

Pigging. Pigging can be combined with chemical injection to remove the wax deposited on the inner wall of the pipe. For subsea system, to facilitate pigging, a flow line loop is required to launch and receive the pig at the topsides.

A subsea pig launcher can also be used but this system is rarely used in practice due to logistic and operational complexity in deep water fields.

During design, a general recommendation for pigging frequency is as follows:

- Wax thickness (2–6 mm), depending on the hardness of the wax deposit.
- Wax volume <50 bbl
- Pig pressure drop <3 bar

Many types of pigs can be used to remove wax deposit including bidirectional, foam, and bypass pigs; however, a bypass pig has been found from operating experience to be very effective to disperse wax ahead of the pig, reduce the risk of stuck pig, and extend pigging time interval.

10.2.4.1 Gelling/Pour Point Management

If the pour point temperature of the production fluid is higher than the minimum ambient temperature, the fluid is likely to solidify/gel after a shutdown. Managing the pour point of the fluid is very critical because once the fluid is allowed to solidify it can become very difficult to get fluid flowing again. In some instances, the pressure required to break a gelled fluid is higher than design pressure.

There are limited options to prevent gelling and due to the high consequence, the recommended method is to use chemicals called pour point depressant (PPD) to reduce pour point temperature below minimum ambient temperature. In cases where it is not possible to reduce pour point temperature below minimum ambient temperature, it might be possible to reduce gel yield stress to a point where it can be broken using pressure. Pressure can be applied to the gel using an oil such as diesel (preferably hot) or water with a pump. It should also be noted that live oil that contains dissolved gas has lower yield stress than dead oil. Shear and thermal conditions before and during shutdown also affect yield stress and should be evaluated thoroughly if flow restart by gel breaking is considered as a strategy.

10.3 Asphaltene

10.3.1 Chemistry

Asphaltenes are the heaviest, most aromatic, and most polar of the molecules in crude oil[15] and stabilized in crude oils by resins. Other components in crude oil have the opposite effect of resins, namely alkanes tends to reduce asphaltene solubility. Aromatic compounds such as toluene and benzene are good solvents of asphaltenes. The solvency effects of various types of components in crude oil on asphaltene and difficulty to define a distinct chemical structure led to asphaltene being defined as a solubility class. The solubility classes are saturates, aroma…

While chemical structures have not been defined precisely, elemental analysis showed that asphaltenes have carbon and hydrogen ratio of 1–1.2 ratio compared with 1–2 ratio for alkanes[16].

Sustained work over the last 20 years has produced a clearer picture of the asphaltenes chemical structure. The structure consists of aromatic rings, fused rings, aliphatic hydrocarbon chains, hetero atoms of nitrogen, sulfur and oxygen (NSO) atoms, and a small amount of metals such as vanadium and nickel. The chemical structure is thought to be two types[16]: Island and Archipelago with the island type more prevalent in crude oil. Figure 10.16 shows the two structures.

The molecular weight of asphaltene ranges from 500 to 1,000 g/gmol.

10.3.2 Phase and Physical Properties

The amount of asphaltene in a crude oil does not correlate to its stability at reservoir or high pressure or temperature. Crude oil with high asphaltene content can be stable while those with low asphaltene content are unstable. Two well-known examples are Maya crude which contain 17 wt% asphaltene is stable, while Hassi Messaoud crude has unstable asphaltene despite containing only 0.06 wt% asphaltene.

Asphaltene stability in reservoir oil depends on composition, pressure, and temperature. The changes of asphaltene stability are explained in a phase envelope as shown in Figure 10.17.

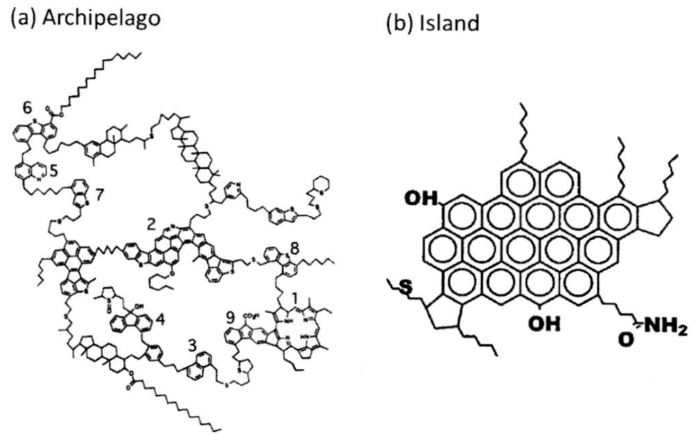

FIGURE 10.16
Asphaltene chemical structure. (Reprinted from Ref. 42 and 43.)

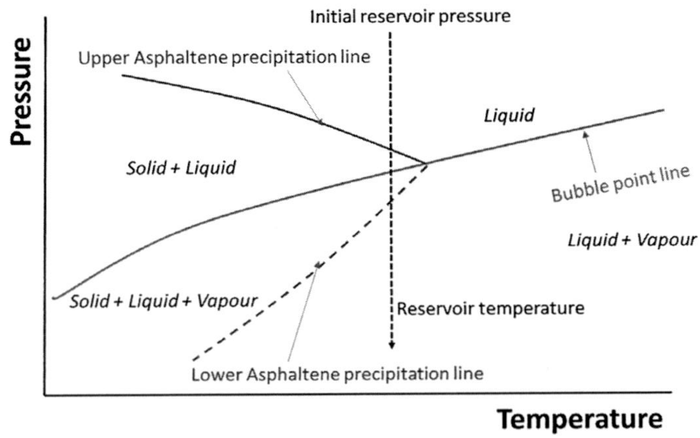

FIGURE 10.17
Asphaltene phase envelope.

At initial reservoir pressure, asphaltene is stable and dissolved in the liquid phase. When pressure reaches a point at the upper asphaltene line, asphaltene starts to precipitate. As pressure decreases further, more asphaltene precipitates. At the bubble point, solution gas is released from oil and reduces the amount of alkanes, which increases asphaltene stability. From the bubble-point pressure to the lower asphaltene line, less asphaltene precipitates. The minimum asphaltene solubility occurs at the bubble-point pressure with the maximum amount of solid precipitation.

The effects of asphaltene precipitation are as follows:

- In the reservoir, asphaltene solids can deposit onto the rock and change the fluid wettability from water wet to oil wet. The detrimental effect is to increase residual oil and reduce oil recovery.
- Asphaltene solids may aggregate into larger solids and plug the rock pore throats. The detrimental effect is to increase pressure drop and reduce well production capacity. The plugging is likely to occur near the wellbore.
- Deposition may occur in the tubing and reduce flow capacity.
- Asphaltenes solids suspended in the fluid can deposit in a separator where liquid velocity is low to aid gas–liquid separation. Figure 10.18 shows asphaltene solids accumulated in a separator at the Ula field[17].
- Asphaltene solids can lead to fouling in heat exchangers, heaters, and coolers and valves plugging.
- Solids can settle in tank, increasing liquid viscosity and reducing working capacity.
- Asphaltenes having polarity (electrically charged) can become emulsifying agent that stabilizes water-in-oil and oil-in-water emulsions.

There are several conditions that can de-stabilize asphaltenes as follows:

- For reservoir oils that are prone to precipitate asphaltene, pressure decrease due to production can lead to precipitation
- Mixing with natural gas for gas injection and gas lift
- Mixing with incompatible crude oils or condensate
- Mixing with CO_2

10.3.3 Laboratory Testing and Modeling

Asphaltene onset pressure (AOP). As shown in the phase diagram of a reservoir oil in Figure 10.18, there are two AOPs: upper AOP (UAOP) and lower AOP (lower AOP), which covers the upper and lower limits of solubility at a given temperature. Usually, only UAOP is measured as this is the point where asphaltenes start to precipitate when the pressure decreases from its initial value in the reservoir. The lower AOP is sometimes not measured or detectable because the changes of oil properties are more gradual than the upper AOP.

The sensitivity for particles detection is limited to 200 nm for the most sensitive method of solid detection system (SDS) that uses a laser light scattering with high-pressure microscope (HPM—1,000 nm limit) but measured AOP is still lower than the thermodynamic value. The methods to measure AOPs are summarized in Table 10.5.

The most common technique used in industrial application is light scattering. Figure 10.18 shows the result for a reservoir oil where the upper AOP is 7,393 psia and the bubble-point pressure is 7,195 psia.

Bubble-point pressure. Bubble-point pressure at a given temperature is where the solubility of the asphaltenes is at a minimum and when maximum asphaltenes precipitation will occur.

SARA (saturates–aromatic–resin = asphaltene). Saturates + aromatics + resins + asphaltenes

The data from a SARA analysis are dependent on the various solvents used in each procedure. The results depend on whether *n*-pentane or heptane has been used to precipitate asphaltene from a sample.

Empirical methods to evaluate asphaltene stability.

Measuring live oils AOPs is a time-consuming and costly process. First, special sampling techniques have to be used, namely samples must be taken at the bottom of a well (bottom hole sampling), asphaltene prevented from precipitation by using special sampling bottles and costly equipment used to measure onset pressure and relevant data. Therefore, methods have been researched and developed to infer asphaltene stability at high pressures and temperatures using dead oil samples.

Most of the methods rely on SARA data and refractive index measurements. The methods include De Boer et al.[18] Buckley et al.[19] Hong et al.,[20] and

TABLE 10.5

Experimental Methods to Measure AOP

Method	Description
Visual using high-pressure microscope	Sample is depressurized in a PVT cell while observing with high-pressure microscope. Normally used with light scattering (LS) technique *Disadvantages:* Cannot be used for dark oils
Gravimetric	Sample is depressurized in a PVT cell at several pressure steps. Samples are withdrawn from the bottom and flashed to atmospheric pressure. Analyze the mass of solid asphaltene in the sample *Advantages:* Can be used for dark oils *Disadvantages:* The method is time-consuming. Require large sample volumes = 600 cm^3 Accuracy depends on the careful selection of pressures and small pressure steps. Some oils may contain asphaltene that redissolves at lower pressures
Acoustic resonance technique	Sample is depressurized in a cell while measuring the resonance amplitude and frequency *Advantages:* Small-sample volume = 10 cm^3 Accuracy comparable to light scattering method. Short duration *Disadvantages:* Cannot detect lower AOP Sample cannot be mixed which could inaccurate detection due to signal scatter and noise
Filtration	At specified pressures, live oil in a PVT cell is pumped through a 0.5 micron filter. Subsamples are taken upstream and downstream to detect solids formation. Asphaltene precipitation is detected by a reduction in the absorbance value and/or an increase in the plugging tendency across the filter. *Advantages:* Detects UAOP and LAOP *Disadvantages:* Time-consuming Large sample volume Less accurate than LS to detect onset as initial precipitation is liquid like (ref. 6.4)
Viscosity	Conducted in a viscometer by detecting Inflexion point in the viscosity versus pressure data *Advantages:* Small volume *Disadvantages:* Detects onset when there is sufficient solids precipitated to increase viscosity. Poor sensitivity compared to other methods.

(*Continued*)

TABLE 10.5 (*Continued*)

Experimental Methods to Measure AOP

Method	Description
Light scattering (LS)	Uses a near-infrared (NIR) laser light (800 to 2,200 nm wavelength) to detect the drop of power at a detector opposite a transmitter. *Advantages:* Small volume, typically = 30 cm^3 Detects UAOP and LAOP *Disadvantages:* Detects onset when there is sufficient solids precipitated to increase viscosity. Poor sensitivity compared to other methods. Sensitive to depressurization rate Normally used with HPM to verify onset. Software can be used to measure particle size distribution.

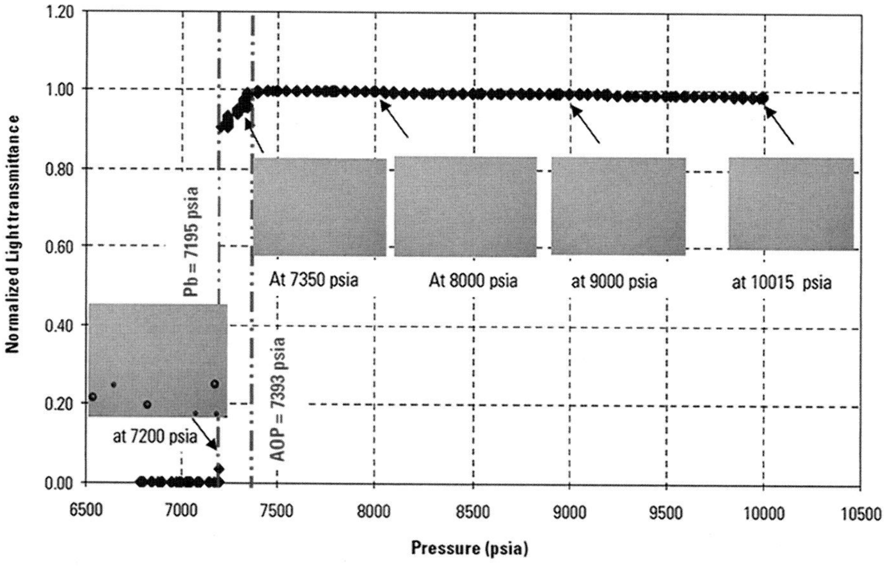

FIGURE 10.18
Example of an AOP for reservoir oil using light scattering technique.

colloidal instability index, CII[21,22] which are empirical based on correlating asphaltene instability from live oils tests or known fields with asphaltene to SARA and other measureable dead oil data. A similar empirical approach, the so-called titration method,[23] is to add a precipitant (*n*-pentane or *n*-heptane) to a sample until asphaltene precipitates. The mass ratio of precipitant to sample ratio is used a proxy for asphaltene in reservoir oils.

Asphaltene thermodynamic modeling uses compositional-based equation of state (EOS). There are several methods depending on theories. The theories on asphaltene precipitation are colloidal and solubility.

10.3.4 Prevention and Mitigation

Asphaltene forms due to a reduced pressure at different temperatures. In a production system comprising of reservoir, well, pipeline, and facilities, pressure drops cannot be prevented unlike for wax where temperature could be maintained to prevent wax deposition. Therefore, asphaltene solids are mitigated using pigging, chemicals (dispersant, solvent) or for wells, coiled tubing can be used to scrape solids and inject solvents.

Where asphaltenes form due to compositional changes from mixing with injected gas and incompatible oils, condensate, or diluents (e.g., Naptha), the ratio of the streams could be a feasible strategy to prevent or mitigate asphaltene formation.

10.4 Hydrate

Hydrates are crystalline solids that form between water and hydrate formers. Hydrate formers in oil and gas fields include methane, ethane, propane, iso-butane, *n*-butane, nitrogen, carbon dioxide, and hydrogen sulfide. *N*-Butane by itself cannot form a hydrate with water but it can enter a hydrate cage when methane is present. Once hydrates start forming, the particles can agglomerate to larger sizes. Hydrate can cause serious flow assurance problems to wells, pipelines, and facilities which could lead to a partial loss or complete interruption of gas and oil production[24].

The bonds that form between water and hydrate formers are hydrogen bonds due to the dipole forces between molecules. The strength of the bonds is such that water can freeze in hydrate form at relatively high temperature, for example, a mixture water and gas consisting of 60:40 mol% methane and ethane can form a hydrate at 18°C at 100 bar (Figure 10.19).

Hydrates can form the following three types of structures:

- Structure I consists of 46 water molecules that bond with methane, ethane, carbon dioxide, and hydrogen sulfide.
- Structure II consists of 136 water molecules that bond with nitrogen, propane, and iso-butane.
- Structure H consists of 34 water molecules that bond with hydrate formers such as cycloheptane but require methane to stabilize the structure. Structure H does not form in oil and gas fluids.

Gas hydrate is a non-stoichiometric clathrate because not all cavities of a hydrate unit cell are necessarily filled with guest molecules. The occupancy of hydrate cavities not only depends on the size of the guest molecules but also is related to the thermodynamic conditions (pressure and temperature). In oil and gas fields, structure II is more prevalent than structure I.

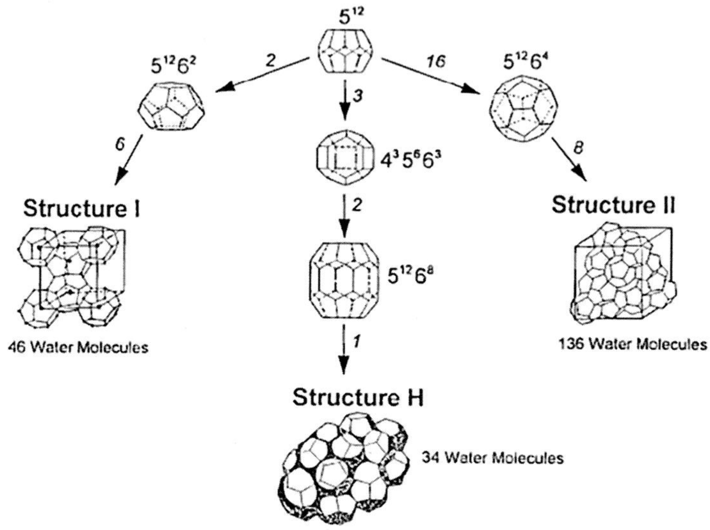

FIGURE 10.19
Hydrate structure.

10.4.1 Phase and Physical Properties

The phase behavior of hydrates is dependent on temperature and pressure, composition, the presence of salts, and chemical additives such as alcohol and glycols (Figure 10.20).

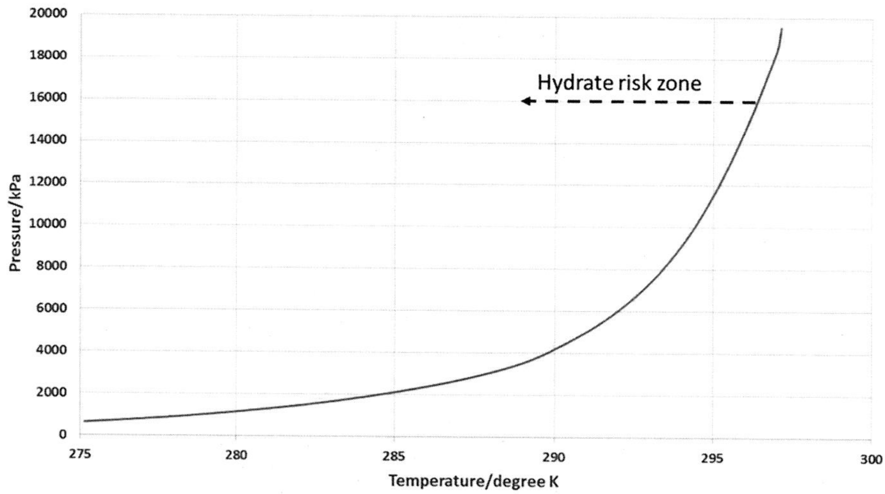

FIGURE 10.20
Hydrate phase envelope with pure water (salt free).

Prediction of the hydrate phase boundary is at thermodynamic equilibrium conditions when the last crystal has dissociated. For a given operating condition, the temperature difference between the dissociation point on the curve and the operating temperature is called subcooling. A certain degree of subcooling is needed to produce a stable hydrate nucleus and hydrate growth. Hydrate formation can occur anytime below hydrate dissociation curve but higher subcooling increases hydrate formation risk. Other than increasing risk, a higher subcooling increases the amount of water or hydrate former converted to hydrate due to the driving force being the subcooling temperature.

Most oil and gas fields are designed and operated outside the thermodynamic or dissociation curve except when specialized chemicals are used to control the formation of hydrate by kinetic or potential to agglomerate.

Salts dissolved in water acts as hydrate inhibitor which is beneficial when formation water breaks through as it can reduce the requirement of costlier inhibitors such as alcohols and glycols. Salts dissolve in water and dissociate into ions and some water molecules which are polar tend to bond to salt ions instead of the hydrate formers. Figure 10.21 shows the hydrate depression effect of 5 and 10 wt% NaCl in water. At 100 bar, hydrate dissociation temperature is reduced by 3 K for 5 wt% salt.

Monoethylene glycol (MEG) and methanol (MEOH) are two of the most widely used thermodynamic hydrate inhibitors (THIs) in the oil and gas industry. The effect of two inhibitors at 10 wt% concentration in water is shown in Figure 10.22.

Another class of hydrate inhibitors is low-dosage hydrate inhibitors (LDHI). There are two types: kinetic hydrate inhibitor (KHI) hinders hydrate nucleation and rate of growth while anti-agglomerant (AA) does not prevent

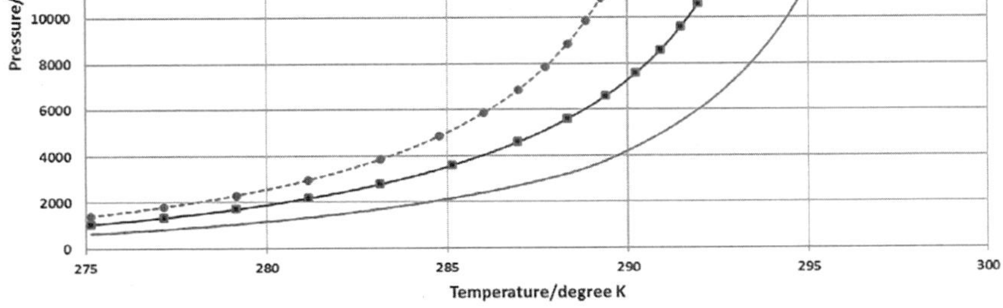

FIGURE 10.21
Hydrate phase envelope with pure water, 5 and 10 wt% NaCl in water.

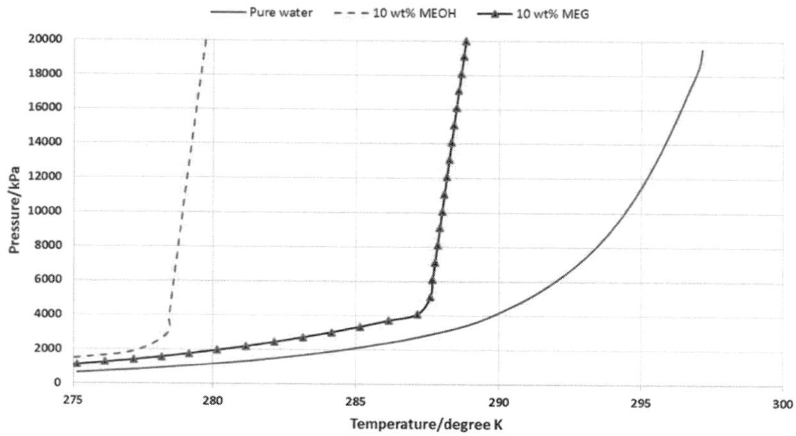

FIGURE 10.22
Hydrate phase envelope with pure water, 10 wt% MEOH and 10 wt% MEG in water.

hydrate formation but prevents the crystals from agglomerating and forming a blockage.

10.4.2 Laboratory Testing and Modeling

The hydrate phase is typically modeled using the solid solution of van der Waals and Platteeuw[25] as implemented by Parrish and Prausnitz.[26] The hydrate phase model is based on the thermodynamic equilibrium where temperatures, pressures, and fugacities are equal in all phases. Hydrate models have advanced to a stage where predictions are sufficiently accurate to be used for design without conducting experiments in most situations.

Classical EOSs such as Soave–Redlich–Kwong and Peng–Robinson are widely used in the industry. However, these EOSs are relatively simple because they do not take into account associative molecules (water, alcohols, and glycols) that interact through hydrogen bonding. Predictions for mixtures containing hydrogen bonding compounds can be inaccurate without readjustment of equations' parameters and even then there are limitations. Cubic-Plus-Association (CPA) model is a cubic EOS that combines the cubic SRK EOS (Soave, 1972) and association terms and has proven to be accurate for hydrate predictions.[27]

When hydrate dissociation boundaries need to be experimentally measured, there are established methods that can be used that could good accuracies of ±0.2°C.

10.4.2.1 Hydrate Formation Correlations

The authors have developed correlations for predicting the hydrate formation.[27–32] These correlations have their own limitations and are not precise but they can provide a reasonable value to plan the control and mitigation.

10.4.2.2 Makogan

$$\text{Log}\, P = \beta + 0.0497\ (t + kt^2) - 1$$

where p is pressure in MPa and t is temperature in C. Elgibaly and Elkamal developed correlation to calculated the β and k in the equation:

$$B = 2.681 - 3.811\gamma = 1.679\gamma^2$$

$$k = -0.006 + 0.011\gamma + 0.011\gamma^2$$

γ is gas gravity = MW/28.96

10.4.2.3 Kobayashi et al.

Kobayashi et al. developed an empirical equation that predicts the hydrate-forming temperatures at given pressures for systems including only hydrocarbons in limited range of temperatures, pressures, and gas specific gravities:

$$\begin{aligned} T = l / \Big[& A_1 + A_2(\text{In}\ \gamma_g) + A_3(\text{In}\ P) + A_4(\text{In}\ \gamma_g)^2 + A_5(\text{In}\ \gamma_g)(\text{In}\ P) \\ & + A_6(\text{In}\ P)^2 + A_7(\text{In}\ \gamma_g)^3 + A_8(\text{In}\ \gamma_g)^2(\text{In}\ P) \\ & + A_9(\text{In}\ \gamma_g)(\text{In}\ P)^2 + A_{10}(\text{In}\ P)^3 + A_{11}(\text{In}\ \gamma_g)^4 \\ & + A_{12}(\text{In}\ \gamma_g)^3(\text{In}\ P) + A_{13}(\text{In}\ \gamma_g)^2(\text{In}\ P)^2 \\ & + A_{14}(\text{In}\ \gamma_g)(\text{In}\ P)^3 + A_{15}(\text{In}\ P)^4 \Big] \end{aligned} \tag{1}$$

10.4.2.4 Berge Correlation

The equations developed by Berge are temperature explicit, i.e., temperature is calculated directly for a given pressure and specific gravity of the gas. The equations for predicting hydrate temperatures are as follows:

For $0.555 \le \gamma_g \le 0.58$:

$$\begin{aligned} T = & -96.03 + 25.37 \times \ln P \\ & - 0.64 \times (\ln P)^2 + \frac{\gamma_g - 0.555}{0.025} \times \left[80.61 \times P + 1.16 \times \frac{10^4}{P + 599.16} \right] \\ & - \left(-96.03 + 25.37 \times \ln P - 0.64 \times (\ln P)^2 \right) \end{aligned}$$

For $\leq \gamma_g \leq 1.0$

$$T = \frac{\left\{80.61 \times P - 2.1 \times 10^4 - 122 \times \frac{10^3}{\gamma_g - 0.535} - \left[1.23 \times 10^4 + 1.71 \times \frac{10^3}{\gamma_g - 0.509}\right]\right\}}{\left[P - \left(-260.42 - \frac{15.18}{\gamma_g - 0.535}\right)\right]} \quad (2)$$

Hydrate formation correlation using gas gravity and pressure

Towler and Mokhtab

$$T = 13.47 \ln(p) + 34.27 \ln(\gamma) - 1.675 \ln(p)\ln(\gamma) - 20.35$$

Motiee

$$T - 282.24469 + 78.99667 \log(P) - 5.352544\,9 \log(P)^2$$
$$+ 349473877\gamma - 150.8546\gamma^2 - 27.604065 \log(P)\gamma$$

where

T is hydrate formation temperature in °F,
P is pressure in psi
γ gas gravity

McCain

McCain developed an empirical relation for predicting the effect of brine on the temperature of hydrate formation. This equation is valid between 0.55 and 0.68 of gas gravity.

$$\Delta T = AS + BS^2 + CS^3$$

where

ΔT is depression in temperature, °F,
S is salinity of ionic liquid in weight percent, and coefficients A, B, and C are function of gas gravity. The values of these constants are calculated as follows:
$A = 2.20919 - 105746\gamma + 12.1601\gamma^2$
$B = -0.106056 + 0.722692\gamma - 0.85093\gamma^2$
$C = 0.00347221 - 0.0165564\gamma + 0.049764\gamma^2$

10.4.3 Prevention and Mitigation

Conventional gas hydrate flow assurance methods rely on avoiding hydrate formation thermodynamically, while new methods are based on delaying

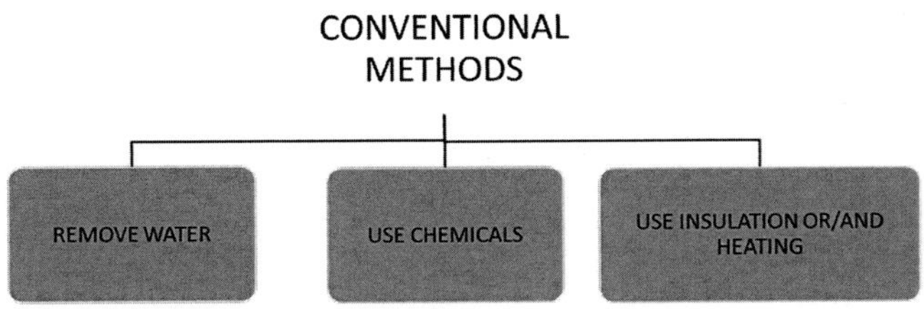

FIGURE 10.23
Conventional hydrate prevention methods.

hydrate formation and/or preventing hydrate plug formation/agglomeration by controlling solid hydrate particle sizes (Figure 10.23).

Current methods for avoiding gas hydrate problems are generally based on one or a combination of the following three techniques/methods:

Dehydration of natural gas. This is done normally by glycol (Tri-ethylene glycol) dehydration, solid adsorbents sieve, or refrigeration. The risk of hydrate formation is removed by separating water and drying the gas. If there is no water present, it is not possible for hydrate to form.

Chemical methods involve the injection of chemicals to prevent or delay hydrate formation. The most common chemical additives used to control hydrates in gas systems are the THIs, such as methanol and ethylene glycol. Alternatively LDHI such as KHI or AA can be used. KHI does not alter the thermodynamics of hydrate formation, but the polymers modify the kinetics of hydrate formation by prevention of nucleation and slowing crystals growth.

As Anti Agglomeratants do not prevent hydrate formation thermodynamically but prevent the crystals from agglomerating and forming a blockage. Laboratory tests need to be conducted to evaluate and select LDHI.

Insulation and/or heating methods can be passive systems that involve passive insulation (wet, PiP, etc.) or active systems that involve some form of heating the pipeline with a hot fluid or electrical heating. The temperature of production fluids is maintained in order to prevent the system entering the hydrate zone. Evaluation of a range of well flowing temperatures, flowrates, and operating need to be conducted to ensure hydrate-free conditions using steady state and transient thermal-hydraulic models.

10.5 Emulsion

Oil production is mostly associated with water during primary, secondary, and tertiary phases. Presence of two immiscible phases oil and water facilitate formation of emulsion during production and processing stages. The emulsion is formed in reservoir, wellbore, wellhead, and surface facilities. An emulsion is a mixture of two immiscible liquids in which droplets of one liquid become stably suspended in the other liquid because of either very slow coalescence or a barrier to coalescence. The phase which is present in the form of droplets is the dispersed phase (internal phase) and the phase in which droplets are suspended is called the continuous or external phase. An emulsion is thermodynamically unstable and kinetically stable. Thus, emulsion formation during primary, secondary, and tertiary phases of oil production is costly problem in terms of oil production loss and expensive chemicals used to prevent their formation and mitigation later. Emulsion has long been of great practical interest due to their widespread use in everyday life. It is common in the food, cosmetics, pulp and paper, pharmaceutical, and agricultural industries.[33–36]

Types of emulsion

The produced oilfield emulsions are classified in the following three categories:

- Water-in-oil emulsion
- Oil-in-water emulsion
- Multiple or complex emulsion

A schematic of these emulsions is shown in Figure 10.24.

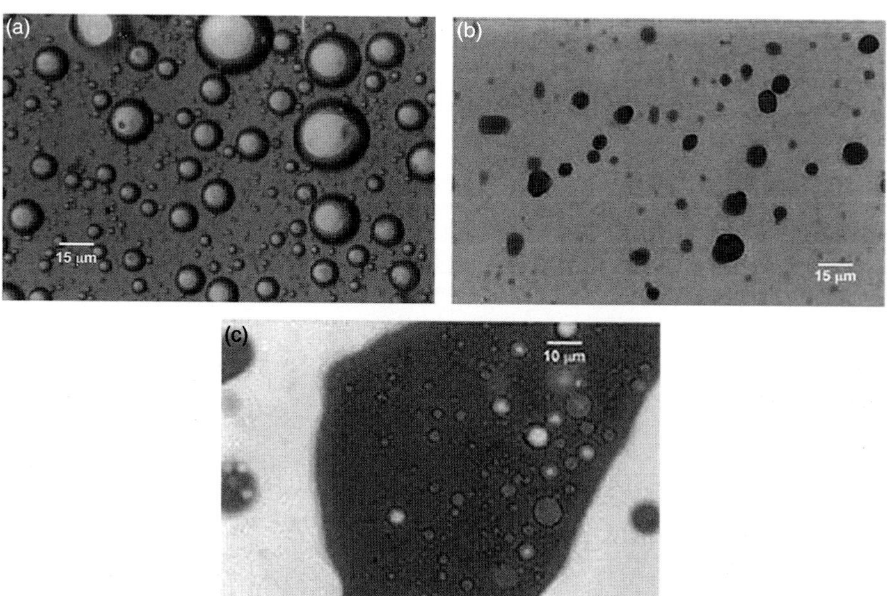

FIGURE 10.24
(a) Water-in-oil emulsion.[33] (b) Oil-in-water emulsion.[33] (c) Water-in-oil-in-water emulsion.[33]

The common colors of emulsions are dark reddish brown, gray, or blackish brown; however, any color can occur depending on the type of oil and water at a particular facility. Emulsion brightness is sometimes used to characterize an emulsion. Oilfield emulsions are characterized by several properties including:

- Appearance and color
- BS&W
- Droplet size
- Bulk and interfacial viscosities

10.5.1 Formation of Emulsion

Crude oil emulsion is formed in the presence of emulsifying agent when oil and water come into contact and mixing takes place. These two are critical parameters for formation of emulsion. There are numerous sources of mixing takes place during production, transportation, and processing of crude oil. Sources of mixing is defined as the amount of shear including:

- Flow in porous media
- Perforation, lifting (pumps and gaslift)
- Flow through tubing, flow lines, wellheads
- Valves, fittings, and chokes
- Surface equipment
- Release of gas bubbles due to phase changes

Mixing cannot be avoided as it is inherent part of the production of oil. When the mixing is higher the droplet size of water dispersed in oil is smaller and emulsion is tighter. Laboratory studies have shown that water droplets can vary in size from less than 1 μm to more than 1,000 μm. The second important ingredient for emulsion formation is presence of emulsifying agent. Type and amount of emulsifying agents define the tightness of the emulsion. Crude oil composition is complex and contains a large number of compounds. Crudes are lighter and heavier in nature and this has an impact on emulsion formation. Emulsifying agents are found mostly in heavier components. Emulsifying tendency varies based on the presence of heavier components. Crude with a small amount of emulsifier forms a less stable emulsion and separates relatively easily. Other crudes contain the right type and amount of emulsifier, which lead to very stable or tight emulsions.

Usually emulsions are not formed within the petroleum layer at the time of accumulation and occurrences. Emulsion formation begins during the movement of petroleum to the wellbore of producers and intensifies during further flow in tubing to wellhead and transport in pipes. Emulsion formation is prominent where mixing of oil and water is more pronounced. Degree of emulsion formation depends upon the production methodology and mechanism of fluid flow in reservoir. Therefore, it is defined by the character of the oil wells, time of its operation, and physical–chemical properties of the crude oil. In the early stage of production, oil rate is high and water production is less or almost absent. Water production increases with time after breakthrough or initiation of water injection. When two fluids are produced simultaneously, the formation of emulsion takes place. This leads to emulsion formation of movement of the oil–water mixture. In the case of production using pumps, shear forces are increased and formation of emulsion is increased.

Emulsion formation is common in all types of enhanced oil recovery (EOR) techniques, for example, chemical, gas based, and thermal methods. Formation of emulsion is sever in surfactant flooding as surfactants are partly soluble in oil and water. Because of this characteristics surfactant concentrates at the interface of oil and water where they form the interfacial film. This helps to lower the IFT and promotes the emulsification of the droplets. Naturally occurring emulsifiers in the crude oil include asphaltenes and resin, organic acid, and bases. Alkali is injected in reservoir along with surfactant and polymers. Alkali reacts with acidic component in crude oil and forms in situ surfactant which in turn forms emulsion. CO_2 is injected in oil reservoir in both miscible and immiscible mode and improves the oil recovery. CO_2 miscibility in the reservoirs triggers the precipitation of asphaltenes and which is one of the main source of formation of emulsion. Steam flooding helps to reduce the viscosity of crude oil and facilitate the oil production. Both oil and water production increases and emulsion is formed. Emulsion is also formed in the reservoir with the following activities and additives used:

- Drilling fluids
- Stimulation chemicals
- Corrosion inhibitors
- Scale inhibitors
- Wax
- Asphaltene control agents

10.5.2 Emulsion Stability

Viscosity is an important flow parameter that defines the flow rate in reservoir and wellbore. This parameter also defines the emulsion behavior. Interfacial films are responsible for emulsion stability. The following parameters affect the stability of emulsion:

- Heavy polar fractions in the crude oil
- Viscosity of the continuous phase
- Viscosity of the dispersed phase
- Volume fraction of the dispersed phase
- Temperature
- pH of brine and its composition
- Average droplet size and size distribution
- Presence of solids in addition to the dispersed phase liquid
- Shear rate
- Nature and concentration of the emulsifying agent

The effect of increased temperature is the sum of changes in several parameters. For instance, changes in the solubility of the crude oil surfactants or injected treating chemicals may occur as a result of increasing temperature. The density of the oil is reduced faster than the density of water as temperature increases, thereby accelerating the settling.

Fine solids are also produced along with oil and they act as mechanical stabilizers. These fine solids include clay particles, sand, silt, asphaltenes and waxes, corrosion products, shale particles, inorganic and organic scales that collect at the oil/water interface and drilling mud. These particles when smaller than emulsion droplets are collected at interface and are influenced by both oil and water. Thus, particle size, inter-particle interactions, and wettability of fine particles affect the stability of emulsion (Figures 10.25–10.27).

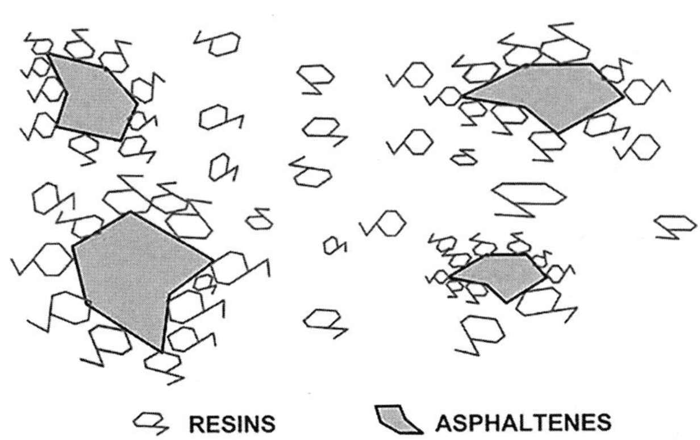

FIGURE 10.25
Asphaltene-resin micelles.[33]

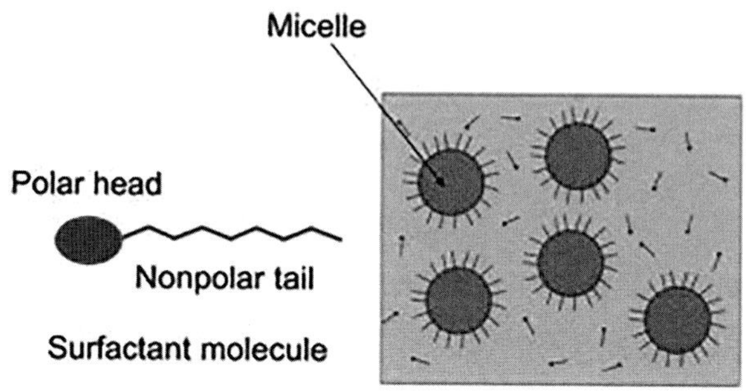

FIGURE 10.26
Surfactant molecule and formation of micelles.[33]

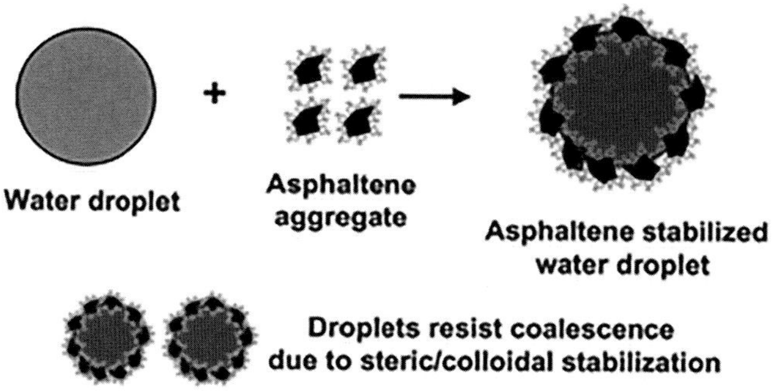

FIGURE 10.27
Emulsion stabilization with asphaltenes.[33]

10.5.3 Laboratory Studies

Extensive laboratory studies are carried to understand the following aspects of emulation. Oil and brine samples and injected fluids such as water, chemical, and gas are collected for the experiment:

- Emulsion tendency test
- Interfacial tension test
- Viscosity test
- Droplet size and its distribution
- Rheology of Emulsion
- Demulsification

The emulsion tendency test is conducted via bottle test method. In order to create the emulsion, dehydrated tubes are filled with fluid mixtures. These tubes are shacked at different rpm after homogenization and a tight

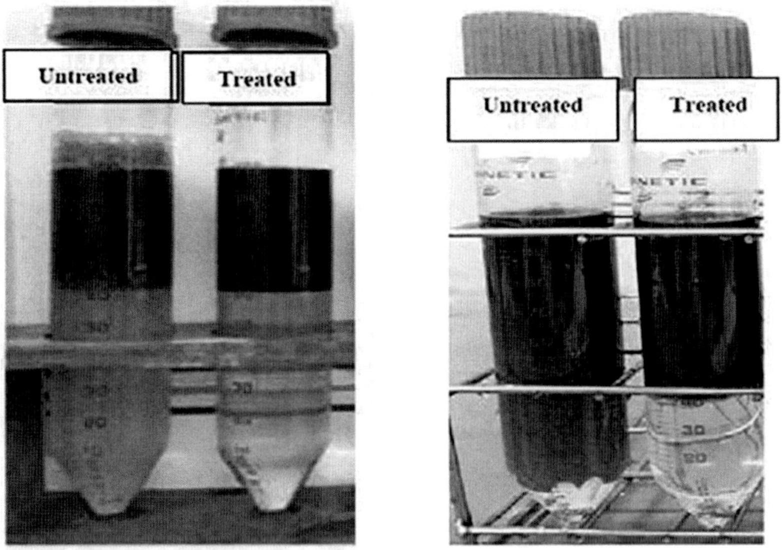

FIGURE 10.28
Treated and untreated emulsion of two crude samples.[34]

emulsion is created. The ratio of oil and water varies in the emulsion study. Generally, tests are conducted using water cut of 20%, 40%, 60%, and 80% (Figure 10.28). Injection of surfactant tends to produce stable emulsion due to ultra-low IFT between oil and water. Emulsion severity is studied at varying production scenario and worst case is considered for selection of demulsifier. Alkali injection generates dissolution of carbonate and silicate scaling tendency due to high pH.

Droplet size and its distribution is measured using microscopy and image analysis method. Light scattering, X-ray scattering, and neutron scattering also sued measure the droplet size. Typical droplet size varies from 0.1 μm to 100 μm. The smaller the size of droplets, the tighter the emulsion is. The longer resident time will be required in separator for its separation or removal. This will necessitate larger size of separator that will add cost to the project (Figure 10.29).

The viscosity of emulsion is always greater than the constituent compounds that formed the emulsion. This is due the fact that emulsion is non-Newtonian in nature. Temperature also affects the viscosity of emulsion. Viscosity decreases with increasing the temperature. It has been observed that viscosity of emulsion increases with an increase in water cut also. The water-in-oil emulsion is classified into three main groups: stable, meso-stable, and unstable. IFT is measured using spinning drop tensiometer. Interfacial can be understood as the energy or work required to increase the size of the interface between two adjacent phases that are immiscible.

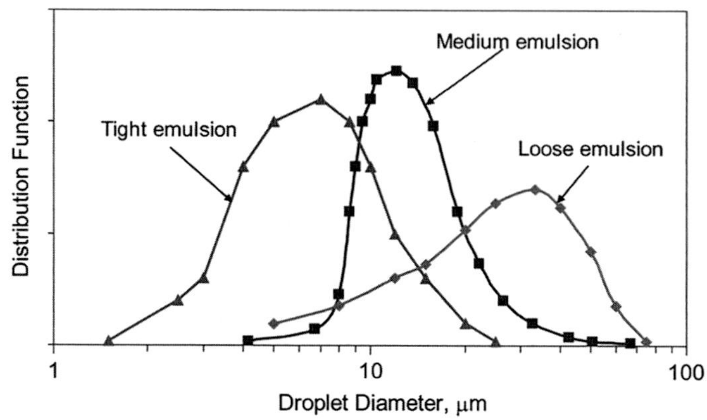

FIGURE 10.29
Nature of emulsion with the size of droplets.[33]

10.5.4 Demulsification

Demulsification is the process of breaking of the emulsion into water and oil phases. There are several methods that can be used to separate them.

Thermal methods. Viscosity of oil is reduced with increasing the temperature and destabilizes the interfacial film of emulsion. It is a cost-intensive process.

Mechanical methods, such as free water knock-out (FWKO) drums, desalter, etc., are also used to separate the emulsion.

Chemical methods. This is most widely used method. Demulsifiers are chemical that are injected in the flow stream to control the emulsion. Bottle tests are carried out to analyze the combination of essential parameters such as demulsifier dosage, resident time, heat, degree of agitation to synthesize the emulsion, and agitation effects after demulsifier injection in emulsion. Demulsifiers that give the best oil quality with the lowest BS&W, fastest water separation, and sharpest oil–water interface are selected for application.

10.5.4.1 Reservoir Souring, Corrosion, and Aquathermolysis

Water injection is conducted in reservoirs to maintain the reservoir pressure. When filed is located in offshore, sea water is the injection water source. Injection water is prepared compatible to the formation water and fluid. Microbes are present in the reservoir crude oil. Injection water also contains various sulfates. When injection water containing sulfate is injected into the reservoir bacteria present in the oil, these sulfate ions are converted into H_2S. These bacteria are called sulfate-reducing bacteria (SRB). Generation of H_2S turns the oil sour and the process is called reservoir souring. H_2S is very dangerous gas both to equipment and personnel. It is toxic. H_2S is soluble in water and creates acid which is highly corrosive. H_2S reacts with Fe and form iron sulfide (FeS) which acts as a protective layer over steel. In order control the corrosion and making operation safer due to H_2S, biocides are added to injection water. This adds cost to the operation.

Corrosion also takes due to presence of CO_2. Carbon dioxide is colorless gas. CO_2 itself is not corrosive. When it comes in contact of water, it forms the carbonic acid. This acid causes the corrosion. CO_2 is present in dissolved gases of crude oil. Sometimes thermal decomposition of carbonates also generates CO_2. CO_2 causes pitting, ringworm, and erosion types of corrosion. Corrosion by CO_2 increases as the partial pressure of CO_2 increases because there is a direct relation between carbonic acid in the condensed water and CO_2 partial pressure. API standard for CO_2 corrosion is as follows:

$P_{CO_2} < 7$ psi: no corrosion

7 psi $< P_{CO_2} <$ 30 psi moderate corrosion risk

$P_{CO_2} > 30$ psi severe risk

Corrosion due to CO_2 is maximum in the temperature range of 80°C–100°C.

Heavy oil and extra-heavy oil contain a significant mass fraction of sulfur. Sulfur varies 2%–8%. During steam flooding or in situ combustion, temperature in reservoir reaches very high 200°C–600°C. Aquathermolysis reactions occur which result in the generation of acid gas (H_2S and CO_2) which are toxic and/or highly corrosive.[37] As such, the determination of the level and extent of Aquathermolysis that will occur in the thermal operation (which is a very strong function of operating temperature and oil composition) is an essential screening and design parameter. Typically, the Aquathermolysis also results in partial upgrading of the heavy oil and bitumen (reduction in viscosity and sulfur content and increase in API gravity and volatiles fraction). The parameters that impact the Aquathermolysis are as follows:

- Oil type
- Presence/absence of water generally free water is required to promote significant Aquathermolysis reactions
- Presence/absence of core material—core material is not required to create Aquathermolysis, but degeneration of carbonates in rock at high temperatures can impact CO_2 generation levels and rates in some formations.
- Temperature level (typically 200°C, 250°C, 275°C, 300°C, 325°C, and 350°C (or cracking temperature)—Aquathermolysis rates are highly temperature dependent and are very slow at temperatures of less than 200°C for most oils.
- Exposure time (typically 3, 5, 15, and 30 days)—this suggests that most reactivity occurs in the first 5–10 days, but this is a function of temperature and the higher the temperature, the longer the period of reactivity.
- Partial pressure of inert or reactive gas (N_2, C_1, H_2, He, etc.)

References

1. William, R.T., Donald, S.B., and Donald, V.H. 1955. Properties of paraffin waxes. *Industrial & Engineering Chemistry* 47(6): 1219–1226. doi:10.1021/ie50546a049.
2. Musser, B.J., and Kilpatrick, P.K. 1998. Molecular characterization of wax isolated from a variety of crude oils. *Energy Fuels* 12(4): 715–725. doi:10.1021/ef970206u.
3. Japper-Jaafar, A., Bhaskoro, P.T., and Mior, S. 2016. A new perspective on the measurements of wax appearance temperature: Comparison between DSC, thermomicroscopy and rheometry and the cooling rate effects. *Journal of Petroleum Science and Engineering* 147: 672–681.
4. ASTM D-5853-17a. 2017. *Standard Test Method for Pour Point of Crude Oils.* American Society for Testing and Materials.
5. Venkatesan, R., Singh, P., and Fogler, H.S. 2002. Delineating the pour point and gelation temperature of waxy crude oils. *SPE Journal* 7(4): 349–352.
6. Zhao, Y., Paso, K., Kumar, L., Safieva, J., Sariman, M.Z.B., and Sjöblom, J. 2013. Controlled shear stress and controlled shear rate nonoscillatory rheological methodologies for gelation point determination. *Energy Fuels* 27: 2025–2032.
7. Huang, Z., Zheng, S., and Fogler, H.S. 2015. *Wax Deposition: Experimental Characterizations, Theoretical Modelling, and Field Practices.* Boca Raton, FL: CRC Press.
8. Hoffmann, R. and Amundsen, L. 2010. Single-phase wax deposition experiments. *Energy Fuels* 24: 1069–1080. doi:10.1021/ef900920x.
9. Jennings, D.W. and Weispfennig, K. 2005. Effects of shear and temperature on wax deposition: Coldfinger investigation with a gulf of Mexico crude oil. *Energy & Fuels* 19: 1376–1386.
10. Zougari, M., Jacobs, S., Ratulowski, J., Hammami, A., Broze, G., Flannery, M., Stankiewicz, A., and Karan, K. 2006. Novel organic solids deposition and control device for live-oils: Design and applications. *Energy & Fuels* 20:1656–1663.
11. Akbarzadeh, K. and Zougari, M. 2008. Introduction to a novel approach for modeling wax deposition in fluid flows. 1. *Industrial & Engineering Chemistry Research* 47:953–963.
12. Davidson, M.R., Nguyen, Q.D., Chang, C., and Rønningsen, H.P. 2004. A model for restart of a pipeline with compressible gelled waxy crude oil. *Journal of Non-Newtonian Fluid Mechanics* 123:269–280.
13. Vinay, G., Wachs, A., and Agassant, J.-F. 2005. Numerical simulation of non-isothermal viscoplastic waxy crude oil flows. *Journal of Non-Newtonian Fluid Mechanics* 128:144–162.
14. Zhao, Y., Kumar, L., Paso, K., Ali, H., Safieva, J., and Sjöblom, J. 2012. Gelation and breakage behaviour of model wax-oil systems: Rheological properties and model development. *Industrial & Engineering Chemistry Research* 51:8123–8133
15. Speight, J.G. 2010. *The Chemistry and Technology of Petroleum*, J. G. Speight, (4th ed.). Boca Raton, FL: CRC Press.
16. Vargas, F.M. and Tavakkoli, M. 2018. *Asphaltene Deposition: Fundamentals, Prediction, Prevention, and Remediation.* Boca Raton, FL: CRC Press
17. Thawer, R., Nicoll, D.C.A., Dick, G. 1990. Asphaltene deposition in production facilities. *SPE Production Engineering* 5(4): 475–480.

18. De Boer, R.B., Leerlooyer, K., Elgner, M.R.P., and Bergen, A.R.D. 1995. Screening of crude oils for asphaltene precipitation: Theory, practice and the selection of inhibitors. *SPE Production and Facilities* 10: 55–61.24987.
19. Buckley, J.S., Hirasaki, G.J., Liu, Y., Von Drasek, S., Wang, J.X., and Gill, B.S. 1998. Asphaltene precipitation and solvent properties of crude oils. *Petroleum Science and Technology* 16: 251–2810.
20. Hong, E. and Watkinson, P. 2004. A study of asphalene solubility and precipitation. *Fuel* 83:1881–1887.
21. Tianguang, F. and Buckley, J.S. 2002. Rapid and accurate SARA analysis of medium gravity crude oils. *Energy & Fuels* 16:1571–15710.
22. Jamaluddin, A.K.M., Creek, J., Kabir, C.S., McFadden, J.D., D'Cruz, D., Manakalathil, J., Joshi, N. and Ross, B. 2002. Laboratory techniques to measure thermodynamic asphaltene instability. *Journal of Canadian Petroleum Technology* 41(7): 44–52.
23. Andersen, S.I. 1999. Flocculation onset titration of petroleum asphaltenes. *Energy Fuels* 13(2), 315–322. doi:10.1021/ef980211d.
24. Dendy Sloan Jr., E. 2000.*Hydrate Engineering*. Richardson, TX: SPE.
25. van der Waals, J.H. and Platteeuw, J.C. 1959. Clathrate solutions. *Advances in Chemical Physics* 2, 1–57.
26. Parrish, W.R. and Prausnitz, J.M. 1972. Dissociation pressures of gas hydrates formed by gas mixtures. *Industrial & Engineering Chemistry Process Design and Development* 11(1):26–35.doi:10.1021/i260041a006.
27. dos Santos, L.C., Abunahman, S.S., Tavares, F.W., Ahón, V.R.R., and Kontogeorgis, G.M. 2015. Cubic plus association equation of state for flow assurance projects. *Industrial & Engineering Chemistry Research* 54(26): 6812–6824.doi:10.1021/acs.iecr.5b01410.
28. Kobayasi, R. et al. Petroleum Engineering Handbook Edited by Bradley, H.B. Richardson, TX: SPE.
29. McCain, W.D. 1990. *The properties of Petroleum Fluids*. Tulsa, OK: PenwellPublishing Co.
30. Makogan, Y.F. 1981. *Hydrate of Hydrocarbon*. Tulsa, OK: Penwell Publishing Co.
31. Motiee, M. 1991. Estimate possibility of hydrate. *Hydrological Processes* 70(7):98–99.
32. Towler, B.F. and Mokhtab, S. 2005. Quickly Estimation Hydrate Formation Temperature in Natural Gas, Hydrological Process.
33. Kokal, S. 2005. Crude oil emulsion: A state of the art review. SPE Production and Facilities.
34. Chandran, K. et al. Rheological behavior on treated Malaysia crude oil. In *4th International Conference on Applied Sciences.*
35. Kilpatrick, P.K. 2012. Water –in-crude oil emulsion stabilization: Review and unanswered questions. *Energy & Fuels* 26: 4017–4026.
36. Yee, H.V. et al. 2013. Managing chemical EOR (ASP) effects on formation damage and flow assurance in Malay basin. IPTC 16777.
37. Bennion, D.B. et al. Laboratory procedures for optimizing the recovery from high temperature thermal heavy oil and bitumen recovery operations. JCPT paper2007-206.

11

EOS and PVT Simulations

11.1 Introduction

Numerical finite-difference simulation has become a powerful tool to assist on field management guiding and ranking different development alternatives (including EOR). Fundamental to the predictive power of the numerical models is a proper representation of the hydrocarbon fluid behavior with changes of pressure and composition. While PVT laboratory and field observation tests may be used directly to estimate initial volumes and give insight on the flow behavior, use of EOS is the industry standard for projects that involve significant compositional changes and/or deal with complex fluid behavior. Use of PVT correlations calibrated to the laboratory and field observations is common on nonvolatile fluids undergoing primary and/or waterflooding.

Fluid characterization strategy, therefore, should be aligned with the current and expected development strategy, leaving room to incorporate different development options as the field matures. Two main numerical model types are popular for the finite-difference formulation, namely black oil and compositional. The former uses tables to represent the variation of PVT properties with pressure—and in some cases CO_2/N_2 injection—and the latter uses an EOS to account for pressure and compositional changes. A common misconception on the use of an EOS is the assumption of wide applicability to different processes, which may lead to misguided decisions; that is, an EOS is only valid to the ranges and processes it was tuned for, which is valid for most depletion development scenarios but not for any compositional processes. These processes do require the use of special laboratory calibration (see Chapter 4) to calibrate and validate the selected EOS results.

11.2 Black Oil and Compositional Models

Selection of model type should be a function of both hydrocarbon fluid and field development. Primary depletion on hydrocarbons up to 30–35 API may be accurately represented with a black oil formulation, while compositional processes (injection of solvent, CO_2, and some cases N_2, for example) do require the use of a compositional model to better represent the component interaction inside the reservoir and thus the compositional front. Highly volatile and retrograde reservoirs are more suited for a compositional modeling.

Much like those discussed in Chapter 6, selection of simulation model should be substantiated with use of 2D and 3D to represent the force balance expected in the reservoir (capillary, gravity, and viscous) as well as diffusive and dispersive effects.[1] Figure 11.1 shows a comparison between the results of the black oil and compositional models.

While it is understood that the compositional model, provided that EOS was tuned to specific process the field is undergoing, will provide a more rigorous estimation of the interaction between the liquid and/or gas components for each pressure and temperature changes in the reservoirs, results are not without uncertainty and should be treated thus. We can agree, however, that these results are closest to the expected behavior among all models and, therefore, for the purposes of defining the simulation approach model, may be used to rank the different models (black oil vs. compositional, multicomponent, reduced component, etc.).

Selection of EOS was discussed in Chapter 6 but it is important to highlight to the reader that each EOS exhibits a different behavior for different fluid types and processes. Traditionally, at the start of any calibration all available EOSs are tested against the observed data and are ranked on how close they are able to reproduce the laboratory behavior as well as the pressure-saturation locus; the top two ranking EOSs are then used for the detailed calibration (see Chapter 6) exercise, including testing of the EOS behavior under dynamic

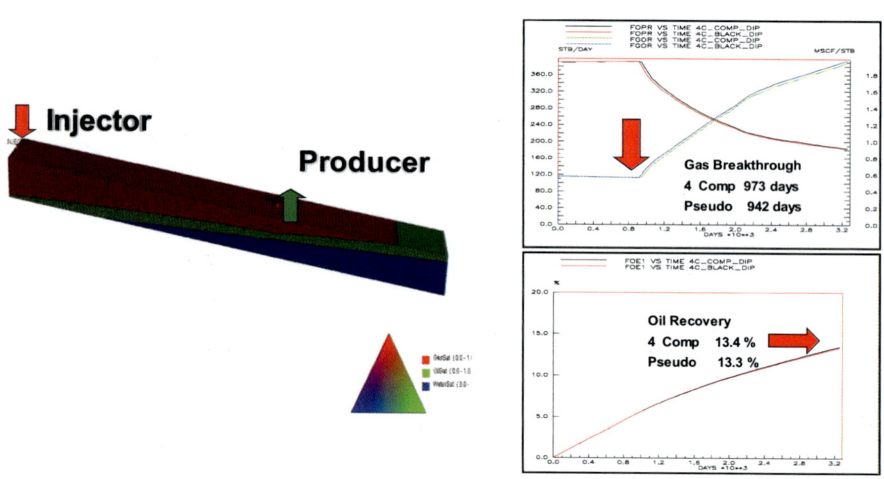

FIGURE 11.1
Black oil vs. compositional model comparison.

conditions. Should any convergence or mismatch issues arise during the process sensitivities on the top-ranking EOS may be performed to identify the better performing EOS to be carried out onto the full field. It is not uncommon that performance of EOS is different under dynamic conditions, particularly when composition, pressure, and temperature are near to the saturated locus, where small compositional changes may change the phase (liquid/vapor) of the hydrocarbon fluid while solving the finite-difference linear equations, making the selection of the most stable EOS extremely important for the consistency and robustness of the subsequent numerical model predictions.

There are several EOS strategies proposed in the literature (see also Chapter 6). However, it is important to understand the calibration strategy as it may impact on the predictive power of the compositional model. Several authors have investigated and proposed different calibration strategies. Coats and Smart[2] proposed splitting the plus fraction using Whitson's method, and then systematically regressed on the binary interaction coefficients (BIC) between the C1 and the heaviest component for saturation pressure calibration. Ω_a and Ω_b of the same hydrocarbon components to match the other PVT properties. Whitson and Brule[3] proposed that the Coats and Smart's method could be further improved if volume translation was added during the calibration stage. Similarly, using P_c and T_c in lieu of Ω_a and Ω_b yielded similar results as the original regression. Christensen[4], on the other hand, used a method to calibrate the SKR EOS using a three-step procedure involving regression on the C7+ molecular weight (within the defined uncertainty ranges), followed by volume translation parameters. Other authors have proposed variation and combination of these methods. However, in our opinion, selection of the calibration strategy is case dependent and should be approached on a step-by-step manner (as described in Chapter 6) so that the technical integrity of the EOS is preserved.

11.3 PVT Model Validation

Upon selecting, calibrating, and testing the EOS and/or black oil PVT model, the next stage is to validate the numerical model that predicts fluid properties with field data. Oil, water, and gas density should be compared with RFT/MDT/static pressure surveys to ensure gravity consistency (see Figure 11.2). Similarly, surface API gravity should be compared with the well observations. While it is expected that these parameters would have been part of the EOS/correlation calibration process, any deviations should be addressed back at that stage, often when they are exported compatibility or model set up issues arise. Given gravity importance's this validation step is a must for any black oil and compositional simulation model setup.

Similarly, solution gas–oil ratio or oil–gas ratio depending on the reservoir fluid type should be compared with the field observations, unless, of course

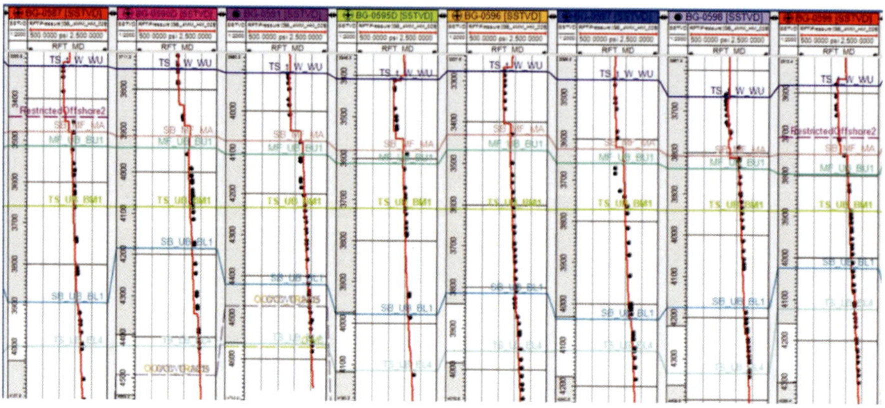

FIGURE 11.2
Example of RFT density validation.[5] The red solid line shows the numerical model results of pressure vs. depth. Dots correspond to RFT pressure observations.

in the rare case where production from gas cap and oil leg are commingled. Production before the reservoir reaches saturation pressure should follow the solution gas (or solution oil for retrograde/rich gas) until gas and/or oil comes out of solution and reaches high enough saturations to start flowing inside the reservoir. Figure 11.3 shows an example of condensate production and reservoir pressure. In this case used gas rate as control while pressure and condensate were the calibration variables. Considering that the reservoir exhibits no severe compartmentalization and sands were fairly continuous, condensate production becomes a function of PVT (EOS) at pressures higher than saturation pressure, and a combination of condensate saturation functions (multi-phase flow) and PVT (EOS) for pressures under the saturation locus. The calibration/validation of the former is straightforward depending on the fluid description, whereas the latter does require a pressure match to identify any potential inconsistencies on the generated PVT. Nevertheless, it is always a good practice, particularly for reservoirs using compositional modeling, to include EOS as part of the calibration exercise.

Oil viscosity should also be validated using field observations, including live and dead oil to ensure that mobilities in the numerical model are in line with the reservoir conditions.

Accuracy of the field and laboratory measurements should be incorporated onto the analysis to avoid over/underestimating key PVT properties; traditionally GOR/OGR are hardest to measure. Given the limited quantities produced prior to the crossing the saturation line, it is accepted that errors in the vicinity of ±5%–10% may occur.[6] Similarly, saturation pressure errors are lower with a range of ±5% same as fluid densities. Compositions, on the other hand, vary according to the equipment and fluid type but ranges of 10%–15% on the molecular weight predictions of the C7+ fractions.

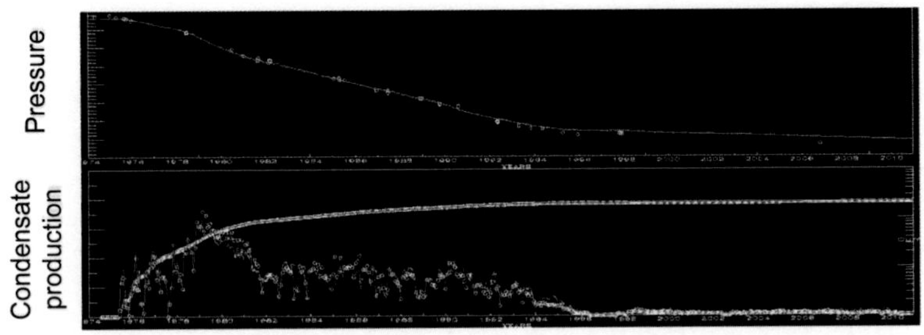

FIGURE 11.3
Example of condensate production.

Estimation of the heavy ends physical properties (C7+, C11+, C22+, etc.) to be used on any EOS calibration is an important step prior to the tuning stage, they are often estimated using correlations (based on the molecular weight and density of the plus fraction) and are, as discussed, considered to be a calibration parameter. In our opinion, splitting/groping and characterization of the heavy fractions do have a significant effect on the calibration process, particularly if the mole fractions are larger. There are several published approaches to estimate critical properties of the heavy fractions, using empirical correlations[7,8] based on PVT sample regressions and/or physical observations. Whitson et al.[3] recommended the following procedures for the C7+ characterization:

- Use either Twu[9] or Lee and Kesler[10] correlation to estimate P_c and T_c.
- Use Kesler and Lee[11] correlation for acentric factors.
- Use volume translation to match specific gravities.
- Specific gravities and boiling temperatures from Soreide[12] correlations.
- Finally, estimate BIC using Nagy and Shirkovskiy[13] correlation for the nonhydrocarbon and modified Chueh and Prausnitz[14] for the hydrocarbon pairs.

It should be noted that majority of commercial EOS tuning packages do include most of these correlations for the plus fraction characterization. It is a good practice to test a few combinations of these correlations to determine which better suits the fluid to be analyzed as well as the ones that provide better EOS stability.

11.3.1 Sample Contamination Validation

As described in Chapter 3, sample contamination is one of the key sources of uncertainty during fluid characterization. While there are several published techniques that deal with estimating the contamination level, there is no accepted methodology to quantify/define the viability of using slightly contaminated samples for PVT characterization. Considering the importance of fluid data on the development decisions it is, in our opinion, not desirable and thus different uncontaminated samples should be planned. It is important, however, to learn to recognize contamination levels and use this information as a criterion to select samples for laboratory and characterization analysis.

The issue of filtrate contamination on downhole samples is not new. Sampling technology has been advancing to mitigate contamination risks by introducing downhole fluid analyzers and increasing drawdown capacities of sampling chambers (see Chapter 3). Nevertheless, it is of course a necessity to test the samples for contamination prior to any analysis. Gozalpour et al.[15] proposed two main methods that are used to test for mud filtrate (particularly oil-based mud), assuming that the mole fraction of the heavy components ($C8^+$) of the uncontaminated samples and their corresponding molecular weights follow an exponential decay,[16] such that if a plot of log of mole fraction vs. molecular weight should follow a straight line (Figure 11.4).

The two methods subject to data availability are as follows: skimming and subtraction. The former relies on a single-stage flash of a contaminated single-phase reservoir sample and composition of gas and oil is measured. Live fluid (single phase) composition—is obtained from the combination of the flashed gas and liquid to the field measured gas–oil ratio (GOR) and fluid density. A plot of mole fraction with molecular weight (see Figure 11.4) should yield a straight line on the uncontaminated sample and thus compositions (mole fractions) of the contaminated components may be adjusted to follow the exponential decay line. This method is reported to yield reasonable results[17] for limited contamination and reservoirs, where the fluid follows the C8+ exponential decline.

The second method, subtraction, requires the mud filtrate's composition, which is then plotted on the same semi-log plot (see Figure 11.5). Following

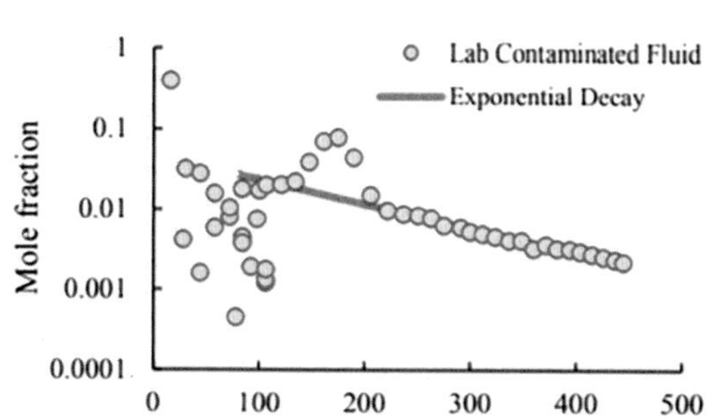

FIGURE 11.4
Mole fraction vs. molecular weight example.[17]

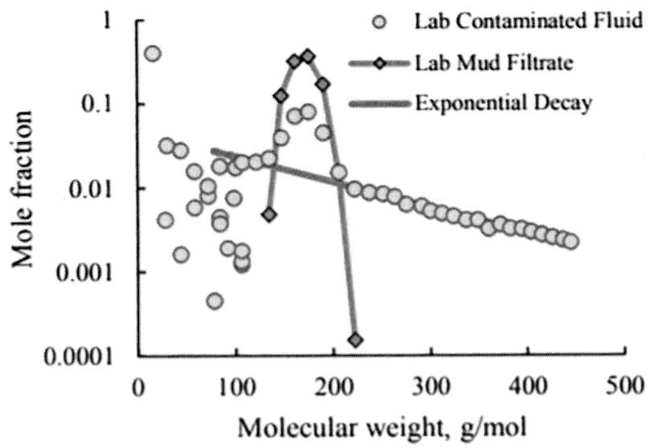

FIGURE 11.5
Mole fraction vs. molecular weight subtraction method.[17]

the same principles of the skimming method, a component material balance combined with a linear regression may be used to determine the composition of the uncontaminated sample. This method yields a very reliable estimation of the contamination level.

11.3.2 EOR Injection Considerations

When an EOR agent is injected onto the reservoir, particularly if dealing with light oils at miscible-to-near-miscible conditions, influence of dispersive and convective flow needs to be accounted for in the upscaling process. The former is related to molecular diffusion between the components of the reservoir fluid and the injectant, and the latter is dominated by reservoir heterogeneity. Dispersive effects are related to the contact area between the EOR agent and the reservoir fluid (which increases as the EOR agent moves thru the reservoir, thus making it scale dependent). Majority of commercial simulators operate under the assumption of steady-state condition (for component interactions) for a given cell and time step; that is, full contact exists between the components. This interaction may be reduced (and in fact most commercial simulators introduced features to do so) by preventing a portion of the reservoir oil to contact the EOR agent. The effect of mixing has been investigated by several authors,[18–21] identifying convection and dispersion as a key process to the performance of a near-miscible and miscible floods. The effects of grid size on miscible and near-miscible processes are illustrated in Figure 11.6, where as the finite-difference grid gets coarser, fingering and mixing effects are minimized—function of grid size and steady-state equilibrium within the cell, thus resulting in a different displacement profile.

As expected, injectant arrival to the producers, miscibility, and contacted oil are directly related to the model resolution. Therefore, should the size

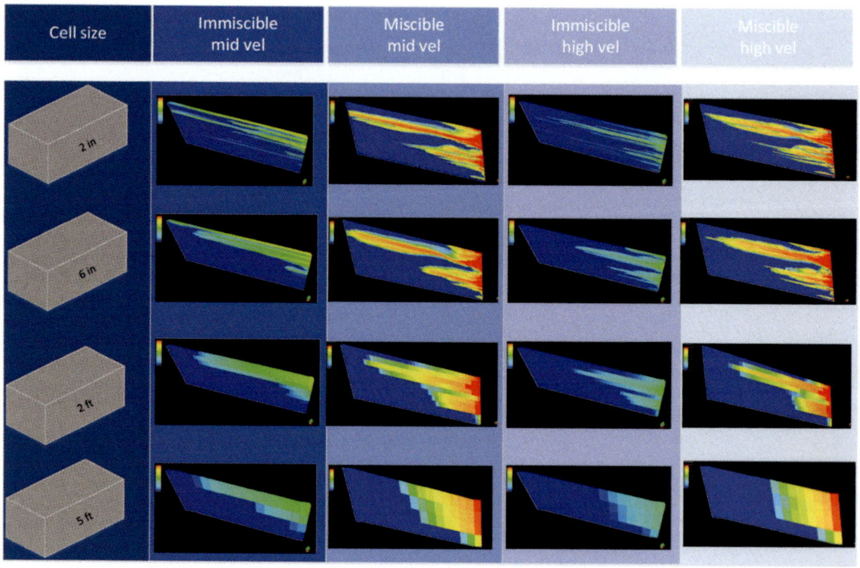

FIGURE 11.6
Effect of grid resolution on miscible and near-miscible flooding.

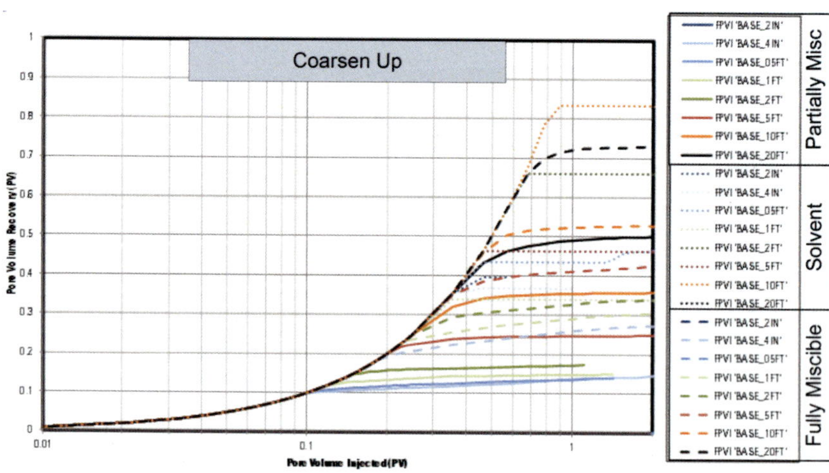

FIGURE 11.7
Effect of grid size and displacement process on a coarsening upward sequence.[1]

effect not be properly addressed benefits of the injection maybe highly overestimated[1]. Figure 11.7 shows an investigation on the effect of gridding at different levels of miscibility, clearly showing as the grid size increases so does the error on the recovery estimation.

As discussed previously, there are several avenues when it comes to upscale miscible and near-miscible flooding. Saturation functions may be used to account for the velocity and saturation differences (the larger the grid cell, the lower the total velocities and the fluid saturations), while fugacity coefficients may be used to account for the dispersive flow. Lastly, the amount of contacted oil—even for miscible floods—is a function of the pore throat/pore body structure, and of course the oil that is not contacted won't be mobilized. Therefore, it is expected that a portion of the oil saturation within the cells won't be contacted by the EOR agent and thus will remain trapped within the pore structure. Care must be taken when introducing this concept to the commercial simulators as it will impact the overall velocities and ultimately displacement efficiency of the agent. Its value, however, should be higher than zero (note that under the steady-state assumption for the finite-difference formulation, any oil present on the cells will contact the EOR agent unless otherwise specified).

11.4 PVT Correlations

While calibration of an EOS with field and laboratory data is preferable for fluid characterization, the process is highly dependent on the type, amount, and quality of the data available. Thus in some instances, where data are lacking, PVT correlations may be used to estimate the necessary PVT properties and thus continue with the investigation.

PVT correlations have been widely used in the industry, and several authors have described their utilization in detail with some adaptations to accommodate slightly different fluid types. The most popular correlations include Standing,[22,23] Lasater,[24] Beggs and Robinson,[25] Vasquez and Beggs,[26] Glaso,[27] Kartoatmodjo,[28] Egbogah and Jack,[29] Al-Marhoun,[30] Majeed et al.,[31] among others.

Standing used 22 mixtures of Californian crudes, while Lasater used his correlation on 137 samples from Canada, USA, and South America. Beggs and Robinson used 600 crude oil systems for their viscosity correlation. Vasquez and Beggs used 6,004 data points divided on high and low API gravities (>30 API and <30 API), whereas Glaso used 45 hydrocarbon samples. Al-Marhoun combined 69 PVT analyses from the Middle East and Kartoatmodjo used nearly 1,400 samples from South East Asia, California, and Alaska.

Selection of a given correlation should weight the fluid type (gravity, expected formation volume factors, viscosity, expected saturation pressures) and composition (known presence of impurities). Table 11.1 shows a summary of the ranges used to calibrate each correlation and may be used as a reference for correlation selection ranking.

TABLE 11.1

PVT Correlations Summary

	Standing	Lasater	Glaso	Kartoat-modjo	Vasquez-Beggs	Al-Marhoun	Rollins-McCain Creeger	Petrosky-Farhand	Labedi
Tank oil API	16.5–63.8	17.9–51.1	22.3–48.1	14.4–58.95	15.3–59.5	19.4 –44.6	18–53.5	16.3–45	32.2–48
Bubble point (psia)	130–7,000	48–5,780	165–7,142	0–6,040	15–6,055	130–3,573	–	1574–6,523	520–6,358
Temperature (F)	100–258	82–272	80–280	75–320	170 (mean)	74–240	–	114–288	128–306
Bo @ Psat (bbl/STB)	1.024–2.15	–	1.025–2.588	1.022–2.747	1.028–2.226	1.032–1.997	–	1.1178–1.6229	1.088–2.92
R_s (Scf/STB)	20–1,425	3–2,905	90–2,637	0–2,890	0–2,199	26–1,602	–	217–1,406	–
Separator gas gravity (air = 1)	–	–	–	0.4824–1.668	0.511–1.351	–	0.579–1.124	–	–
Total surface gas gravity (air = 1)	0.59–0.95	0.574–1.223	0.65–1.276	–	–	0.752–1.367	–	0.5781–0.8519	–
Separator pressure (psia)	265–465	15–605	415 (mean)	100	60–565	–	29.7–314.7	–	34.7–789.7
Separator temperature (F)	100 (mean)	34–106	125 (mean)	38–294	76–150	–	60–150	–	60–220
Reservoir pressure (psia)	–	–	–	10–6,000	141–9,515	20–3,573	–	1,700–1,0692	–
Stock tank GOR (Scf/STB)	–	–	–	–	–	–	4–220	–	–
Separator GOR (Scf/STB)	–	–	–	–	–	–	12–1,742	–	–

Source: De Ghetto et al.[32]

It is important to note that the input properties (such as GOR) can be expressed on the appropriate units and reflect the actual property that can be used to generate the correlation. GOR and any other volumetric estimation are a function of separator conditions for example. Therefore, it is important to understand each correlation's assumptions and provide data on the appropriate format.

Depending on the data availability, correlation estimations may be corrected to observed information; that is, PVT tables generated thru the correlations may be adjusted to better honor saturation pressures, viscosities, and gas–oil-ratios, thus compensating for correlation deviations as well as separator conditions. This process, however, should be approached with care to ensure that the predicted tables are still consistent and unphysical values are avoided. A comprehensive comparison of these correlations' behavior was discussed by De Ghetto et al.[32] where the predictive power of each correlation was tested using a diverse sample of 195 crude oils, pointing at strengths and weaknesses of each correlation.

Accuracy of the correlations, as discussed above, varies, thus when field data are limited, the range of plausible PVT properties may be wide as much as each correlation was developed to accommodate a wide range of hydrocarbon samples, the reality is that compositions and fluid behaviors are complex and hard to accommodate on these two-dimensional correlations. Furthermore, field measurements are not without inaccuracies compounding on the need to understand the implications on each PVT property prediction. It is advisable, therefore, to perform a simple uncertainty analysis to investigate the effect of each measurement on the prediction of key PVT properties (such as formation volume factors, solution gas, and live oil viscosity), which are likely to affect volume and field performance estimations. One such analysis is shown in Figure 11.8 where the corrected correlation inputs were varied following their expected uncertainty distribution and formation volume factors were calculated. This suggests that how different estimations can be (ranging from 1.25 to 1.61). It is of course expected that these ranges can be reduced thru data acquisition on the field as well as performance evaluation. This approach may be repeated to evaluate the top-ranking PVT correlations.

11.5 Impact of PVT Uncertainty on Volume and Recovery Estimations

PVT data are used in reservoir engineering calculations such as volumetric estimates, material balance, and numerical simulation. These PVT data have a direct impact on development decisions, thus requiring not only

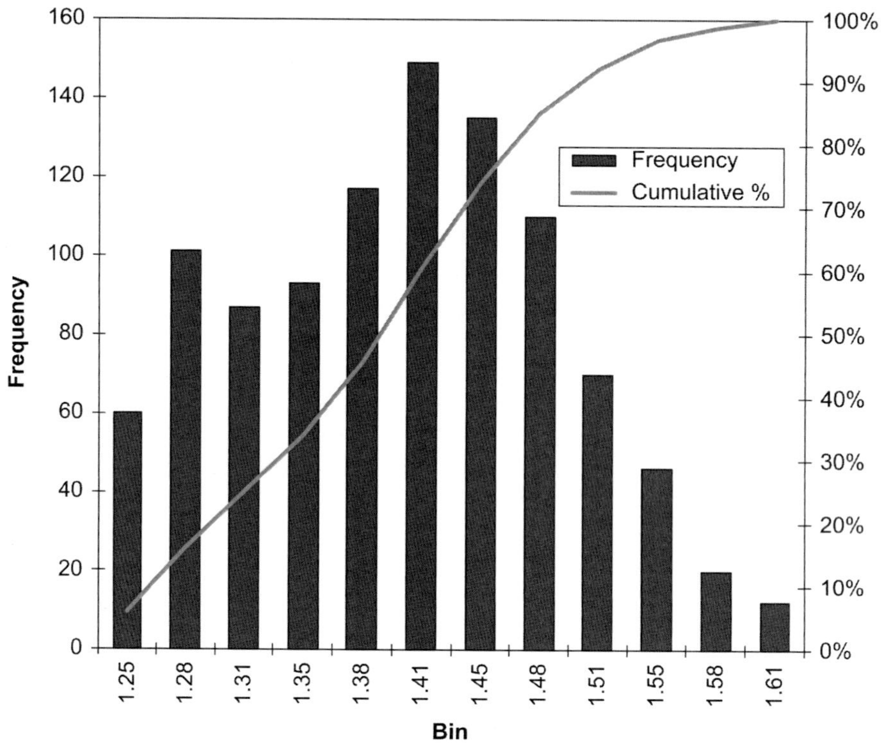

FIGURE 11.8
Formation volume factor for oil distribution.

consistency but also accuracy as the uncertainty in PVT parameters introduces uncertainty in calculated parameters.

Formation volume factors for oil and gas (B_o and B_g) impact parameters in the estimation of oil and gas in-place volumes. This is critical as these parameters vary with depth and if appropriate values are not used the volume estimates are either overestimated or underestimated. These aspects become critical in asset acquisitions. Similarly, oil compressibility is also used in estimating volume above the bubble-point pressure.

Oil in-place by the volumetric method is given by

$$N(t) = \frac{V_b \phi(p(t))\left(1 - S_w(t)\right)}{B_o(P(t))}$$

where

$N(t)$ = oil in-place at time t, STB
V_b = 7758 A h = bulk reservoir volume, bbl
7758 = bbl/acre-ft

A = area, acres
h = thickness, ft
$\varphi(p(t))$ = porosity at reservoir pressure p, fraction
$S_w(t)$ = water saturation at time t, fraction
$B_o(p(t))$ = oil formation volume factor at reservoir pressure p, bbl/STB
$p(t)$ = reservoir pressure at time t, psia

Gas in-place by the volumetric method is given by

$$G(t) = \frac{V_b \phi(p(t))(1 - S_w(t))}{B_g(p(t))}$$

where
$G(t)$ = gas in-place at time t, Scf
V_b = 43,560 A h = bulk reservoir volume, ft^3
43,560 = ft^3/acre-ft
A = area, acres
h = thickness, ft
$\varphi(p(t))$ = porosity at reservoir pressure p, fraction
$S_w(t)$ = water saturation at time t, fraction
$B_g(p(t))$ = gas formation volume factor at reservoir pressure p, ft^3/Scf
$p(t)$ = reservoir pressure at time t, psia

Material balance calculation

The general form of material balance equation is as follows:

$$N_p[B_o + B_g(R_p - R_s)] + W_p B_w - W_{inj} B_{winj} - G_{inj} B_{ginj}$$
$$= N\left\{[B_o - B_{oi} + B_g(R_{si} - R_s)] + \frac{B_{io}}{B_{gi}} m(B_g - B_{gi}) + B_{oi}(1 + m)\left(\frac{c_f + c_w S_w}{1 - S_w} \Delta p\right)\right\}$$
$$+ W_e B_w$$

$$G_p = G \cdot \left(1 - \frac{B_{gi}}{B_g}\right)$$

$$\text{Or } \frac{P}{z} = \left(\frac{P}{z}\right)_i \cdot \left(1 - \frac{G_p}{G}\right) \text{ psia}$$

These equations reveal that there are a number of PVT parameters that can cause errors in results if uncertainty is associated with them (see Figure 11.8).

Variation in PVT parameters must be incorporated in numerical simulation. Interfacial tensions (IFT) and changes in viscosity and their areal and vertical variations play an important role in EOR modeling.

PVT ranges should be included in the definition of most likely, high and low scenarios to adequately capture the effect of fluid uncertainties on the field performance.

PVT parameters, such as salinity and GOR variations, in reservoir management and surveillance also provide important technical information for understanding the field performance and planning of remedial actions.

References

1. Moreno, J., Flew, S., and Gurpinar, O. EOR: Challenges of translating fine scale displacements into full field models-part 2. SPE 165298-MS presented at the 2013 *SPE Enhanced Oil Recovery Conference, Kuala Lumpur Malaysia*, July 2–4.
2. Coats, K.H. and Smart, G.T. 1986. Application of a regression based EOS PVT program to laboratory data. *SPERE* 1(3): 277–299.
3. Whitson, C.H. and Brule, M.R. 2000. *Phase Behaviour*, vol. 20. Monograph SPE Henry L. Doherty Series. Richardson, TX: SPE.
4. Christensen, P.L. 1999. Regression to experimental PVT data. *Journal of Canadian Petroleum Technology* 30: 1–9.
5. Al-Samhan, A.M., Gurpinar, O.M., Ansarizadeh, M., Wang, J.T., Moreno, J.O., Sinha, S., Garmeh, G., Giddings, M.A., Rincon, A., Kovyazin, D., Samad Ali, S., Gomez, E., Bangale, M.R., Burman, K.R., Al-Houti, R.A., Filak, J.M., Al-Rashaid, M., Ma, E.D., and Bond, D.J. Challenges and solutions of history matching a giant reservoir model: An example from Greater Burgan field, Kuwait SPE-186065-MS, presented at the *2017 SPE Reservoir Characterization and Simulation Conference, Abu Dhabi*, May 8–10.
6. Trengrove, R.D., Hann, J.H., and Skates, J.R. The impact of PVT data quality on hydrocarbon recovery predictions SPE paper 22988 presented at the 1991 *SPE Asia-Pacific Conference Held in Perth, Australia*, November 4–7.
7. Sancet, G.F. Heavy fraction C7+ characterization for PR-EOS. SPE 113026-STU presented at the 2007 *SPE International Student Paper Contest at the SPE Annual Technical Conference and Exhibition, Anaheim, CA*, November 11–14.
8. Riazi, M.R. and Dauberte, T.E. 1987. Characterization parameters for petroleum fractions. *Industrial & Engineering Chemistry Research* 26(24): 755–759.
9. Twu, C.H. 1984. An internally consistent correlation for predicting the critical properties and molecular weight of petroleum and coal-tar liquids. *Fluid Phase Equilibria* 16: 137.
10. Lee, B.I. and Kesler, M.G. 1975. A generalized thermodynamic correlation based on three parameter corresponding states. *AIChE Journal* 21: 510.
11. Kesler, M.G. and Lee, B.I. 1976. Improve predictions of enthalpy of fractions. *Hydrocarbon Processing* 55: 153.

12. Soreide, I. 1989. Improved phase behaviour predictions pf petroleum reservoir fluids from a cubic equation of state. *Dr. Ing dissertation*, Norwegian Ins. Of technology, Trondheim, Norway.
13. Nagy, Z. and Shirkovskiy, A.I. Mathematical simulation of natural gas condensation process using the Peng-Robinson equation of state. Paper SPE 10982 presented at the 1982 *SPE Annual Technical Conference and Exhibition, New Orleans*, September 26–29.
14. Chuen, P.L. and Prausnitz, J.M. 1968. Calculation of high-pressure vapor-liquid equilibria. *Industrial & Engineering Chemistry* 60: 13.
15. Gozalpour, F., Danesh, A., Tehrani, D.H., Todd, A.C., and Thodidi, B. Predicting reservoir fluid phase and volumetric behaviour from samples contaminated with oil-based mud SPE 56747-MS presented at the 1999 *SPE Annual Technical Conference and Exhibition, Houston, TX*, October 3–6.
16. Katz, D.L. 1983. Overview of phase behaviour in oil and gas production. *Journal of Petroleum Technology*, 35: 1205–1214.
17. Zuo, J., Adriaan, G., Pfeiffer, T., Achourov, V., Chen, L., Mullins, O.C., Edmunson, S., and Partouche, A. Advances in quantification of miscible contamination in hydrocarbon and water samples from downhole to surface laboratories. SPWLA paper presented at the *58th Annual logging symposium, Oklahoma City, USA*, June 17–21, 2017.
18. Garmeh, G. and Johns, R. Upscaling of miscible floods in heterogeneous reservoirs considering reservoir mixing. Paper SPE 124000 presented at the 2009 *SPE Annual Technical Conference and Exhibition, New Orleans*, October 4–7.
19. Jerauld, G.R. A case study in scaleup for multi-contact miscible hydrocarbon gas injection. SPE 39626 presented in the 1998 *SPE/DOE Improved Oil Recovery Symposium, Tulsa*, April 19–22.
20. McGuire, P.L. and Stalkup, F.I. 1995. Performance analysis and optimization of the Prudhoe Bay miscible gas project. *SPERE* 10(2) 88–93.
21. Gharbi, R., Peters, E.J., and Elkamel, A. 1998. Scaling miscible fluid displacements in porous media. *Energy & Fuels* 12: 801–811.
22. Standing, M.B. 1962. Oil-system correlations. In *Petroleum Production Handbook*, Edited by T.C. Frick, Vol. 2, Chap. 19. Richardson, TX: SPE.
23. Standing, M.B. 1947. A pressure-volume-temperature correlation for mixtures of California oils and gases. *API Drilling and Production Practice*, 275–287.
24. Lasater, J.A. 1958. Bubble point pressure correlation. *Transaction AIME* 213: 379–381.
25. Beggs, H.D. and Robinson, J.R. 1975. Estimating the viscosity of crude oil systems. *JPT* 27: 1140–1141.
26. Vasquez, M.E. and Beggs H.D. 1977. Correlations for fluid physical property prediction. SPE 6719.
27. Glaso, O. 1980. Generalised pressure-volume-temperature correlations. *JPT* 32: 785–795.
28. Kartoatmodjo, T. 1990. New correlations for estimating hydrocarbon liquid properties (Thesis), The University of Tulsa, The Graduate School.
29. Egbogah, E.O. and Ng, J.T. 1990. An improved temperature-viscosity correlation for crude oil systems. *Journal of Petroleum Science and Engineering* 5: 197–200.
30. Al-Marhoun, M.A. 1988. PVT correlations for middle east crude oils. *JPT* 40: 650–666.

31. Majeed, G.H.A., Kattan, R.R., and Salman, N.H. 1990. New correlation for estimating the viscosity ofunder saturated crude oils. *Journal of Canadian Petroleum Technology* 29(3): 80–85.
32. De Ghetto, G., Paone, F., and Villa, M. Reliability analysis on PVT correlations. SPE 28904 presented at the 1994 *European Petroleum Conference, London*, October 25–27.

12

Empirical Relations for Estimating Fluid Properties

Standard conditions. Oil industry uses standard condition for pressure and temperature in order to compare volumes or pressures.

The standard conditions used are as follows:

$$P_{sc} = 1.013\ 25 \text{ bara } (14.696 \text{ psia})$$

$$T_{sc} = 15.65°\text{C } (60°\text{F})$$

The letter *a* refers to absolute pressure, i.e., above vacuum.

Pressures are often expressed in barg or psig, which means "gauge pressure," i.e., taking as reference zero the atmospheric pressure (1.013 bara or 14.7 psia)

$$100 \text{ barg} = 100 \text{ bars} + 1.013 \text{ bar} = 101.013 \text{ bara}$$
$$3000 \text{ psig} = 3000 \text{ psi} + 14.7 \text{ psi} = 3014.7 \text{ psia}$$

The reference conditions in SI system are defined as follows:

$$P_{sc} = 1.013 \text{ bara } (\text{or } 14.7 \text{ psia})$$
$$T_{sc} = 15°\text{C}$$

Fluid characterization is important for development, production, processing, and transport of oil and gas fluids. It plays an important role in estimating the products during refining. A reservoir fluid analysis is often called a PVT (pressure, volume, and temperature) measurement. The following parameters are measured as a function of pressure in the laboratory and are used in engineering calculation:

1. Bubble-point pressure, p_b
2. Oil formation volume factor, B_o
3. Gas formation volume factor, B_g
4. Solution gas–oil ratio, R_s
5. Oil viscosity, μ_o and gas viscosity, μ_g

6. Oil density, ρ_o
7. Gas gravity, γ_g
8. Oil compressibility, C_o

PVT data are used in following calculations:

- How much oil or gas is present?
- How much can be recovered?
- How fast it can be recovered?
- Reservoir management and strategies
- Surface facility design

In view of nonavailability of laboratory measure parameters, these parameters are generated using empirical correlations. Industry experts have developed a number of correlations for various fluid characterization parameters such as bubble-point pressure, formation volume factor, density, gas–oil ratio, oil and gas viscosity, Z-factor, and compressibility. They have used a large number of PVT analysis available worldwide. Some of the famous correlations for bubble-point pressure are mentioned below.

12.1 Bubble-Point Pressure Correlations

This is the first pressure where, at constant temperature, the oil system releases a first bubble of gas. This gas can affect the oil recovery.

12.1.1 MB Standing Correlation

$$p_b = 18.2\left(\left(\frac{R_s}{g_g}\right)^{0.83} \cdot \frac{10^{0.00091T}}{10^{0.0125\gamma_{API}}} - 1.4\right)$$

12.1.2 Glaso

$$\log(p_b) = 1.7669 + 1.7447 \cdot \log(x) - 0.30218 \cdot (\log(x)^2)$$

$$p_b = 10^{\log(pb)}$$

$$x = \left(\frac{R_s}{\gamma_g}\right)^{0.816} \frac{T^{0.172}}{\gamma_{API}^{0.989}}$$

12.1.3 Vasquez and Beggs

The correlation divides the data into two groups: one for oil gravity over 30°API and the other below 30°API

$$p_b = \left(\frac{R_s}{C_1 \gamma_g \cdot e\left(C_3 \left(\frac{\gamma_{\text{API}}}{T + 459.67} \right) \right)} \right)^{\frac{1}{C_2}}$$

12.1.4 Al-Marhoun

Al-Marhoun developed the correlation using laboratory measured data from Middle East oil and gas fields:

$$p_{bp} = -64.138910 + 0.702362 \times 10^{-2} \cdot x - 2.278475 \times 10^{-9} \cdot x^2$$

$$x = R_S^{0.722569} \cdot \gamma_g^{-1.879109} \cdot \gamma_o^{3.04659} \cdot T^{1.302347}$$

12.1.5 De Ghetto et al. for Heavy Oil and Extra-Heavy Oil

Heavy oil

$$p_h = \left(\frac{56.434 \cdot R_s}{\gamma_{g(p_{sp})} \times 10^{10.9267} \left(\frac{\gamma_{\text{API}}}{T + 459.67} \right)} \right)^{\frac{1}{1.2057}}$$

$$\gamma_{g(p_{sp})} = \gamma_g \left(1 + 0.5912 \cdot \gamma_{\text{API}} \cdot T_{sp} \cdot \log\left(\frac{p_{sp}}{114.7} \right) \times 10^{-4} \right)$$

$$p_b = \left(\frac{R_s}{\gamma_g} \right)^{\frac{1}{1.1128}} \cdot \frac{10.7025}{10^{(0.0169 \cdot \gamma_{\text{API}} - 0.00156T)}}$$

12.1.6 Hanafi et al.

$$p_b = 3.205 \cdot R_s + 157.27$$

12.1.7 Petrosky and Farshad

$$p_b = 122.727\left(\frac{R_S^{0.5774}}{\gamma_g^{0.8439}} \times 10^x - 12.34\right)$$

where

$$x = 4.561 \times 10^{-5} \cdot T^{1.3911} - 7.916 \times 10^{-4} \cdot \gamma_{\text{API}}^{1.541}$$

12.1.8 Velarde et al.

$$p_b = 1091.47\left(R_{sb}^{0.081465}\gamma_g^{-0.161488} \times 10^x - 0.740152\right)^{5.354891}$$

where

$$x = 0.013098 \cdot T^{0.282372} - 8.2 \times 10^{-6} \cdot \gamma_{\text{API}}^{2.176124}$$

12.1.9 Omar and Todd

Omar and Todd developed the correlation using data from Malaysian fields:

$$P_b = 18.2\left[\left(Rs / \gamma_g\right)^X * \text{EXP}\left\{0.00091\, T_f - 0.0125\, \gamma_{\text{API}}\right\} - 1.40\right]$$

where

$$X = A_1 + A_2 B_{ob} + A_3\gamma_g + A_4 B_{ob}^2 + A_5\gamma_g^2 + A_6\{1/(\gamma_g B_{ob})\}$$
$$A_1 = 1.4256,$$
$$A_2 = -0.2608,$$
$$A_3 = -0.4596$$
$$A_4 = 0.04481,$$
$$A_5 = 0.2360,$$
$$A_6 = -0.1077$$

12.1.10 Lasater Bubble-Point Pressure

This correlation uses specific gravity of oil and gas, reservoir temperature, GOR, and molecular weight of oil:

$$p_b = \frac{p_f(T + 459.67)}{\gamma_g}$$

where

$$x_g = \left[1 + \frac{\gamma_0}{7.521 x 10^{-6} R_S M_0}\right]^{-1}$$

$$p_f = e^{\left(\frac{(x_g - 0.15649)}{0.33705}\right)_{-0.59162}}$$

12.1.11 Dew-Point Correlations

Elsharkawy model and constants

$$P_d = A_0 + A_1\ T_f + A_2\ xH_2S + A_3\ xCO_2 + A_4\ xN_2 + A_5\ xC_1 + A_6\ xC_2 + A_7\ xC_3$$
$$+ A_8\ xC_4 + A_9\ xC_5 + A_{10}\ xC_6 + A_{11}\ xC_7 + A_{12}\ MW_7 + A_{13}\ \gamma_7$$
$$+ A_{14}\ (xC_7 + MW_7 +) + A_{15}(MW_7 + /\gamma_7 +) + A_{16}(xC_7 + MW_7 + /\gamma_7 +)$$
$$+ A_{17}\ \{xC_7 + /(xC_1 + xC_2)\} + A_{18}\{xC_7 + /(xC_2 + xC_3 + xC_4 + xC_5 + xC_6)\}\ldots$$

where

P_d = Dew-point pressure (DPP), psia
T_f = reservoir temperature (°F)
'x' = composition expressed as mole fraction
MWC_{7+} = molecular weight of the C_{7+}
γ_{7+} = specific gravity of the C_{7+} fraction.
The constants A_0 through A_{18} are;
$A_0 = 4268.85$,
$A_1 = 0.094056$,
$A_2 = -7157.87$,
$A_3 = -4540.58$,
$A_4 = -4663.55$,
$A_5 = -1357.56$,
$A_6 = -7776.10$,
$A_7 = -9967.99$,
$A_8 = -4257.10$,
$A_9 = -1417.10$,
$A_{10} = 691.5298$,
$A_{11} = 40660.36$,
$A_{12} = 205.26$,
$A_{13} = -7260.32$,
$A_{14} = -352.413$,
$A_{15} = -114.519$,
$A_{16} = 8.133$,

$A_{17} = 94.916$ and
$A_{18} = 238.252$

In the above empirical model, A_{17} and A_{18} account for the interaction between heavy fraction and light fraction, and heavy fraction and intermediate fraction, respectively.

Humoud and Al-Marhoun

$$Ln(P_d) = \beta_0 + \beta_1 Ln(T) + \beta_2 (Ln\ R_m) + \beta_3 Ln\ \left(P_{sp}\ T_{sp}\right) + \left(\beta_4 / T_{pr}\right) + \left(\beta_5 / P_{pr}\right)$$
$$+ \left(\beta 6 /_{\gamma c7+}\right) R_m = R_{sp} .\ \gamma_{gsp} / \gamma_{c7+}$$

where

T reservoir temperature in °F
T_{sp} primary separator temperature
P_{sp} primary separator pressure
P_{pr} and T_{pr} are pseudo-reduced pressure and pseudo-reduced temperature
γ_{c7+} specific gravity of C_{7+}
γ_{gsp} separator gas specific gravity
R_{sp} gas oil ratio Scf/bbl

And coefficients are

$B_0 = 43.777183$
$\beta_1 = -3.594131$
$\beta_2 = -0.247436$
$\beta_3 = -0.053527$
$\beta_4 = -4.291404$
$\beta_5 = -3.698703$
$\beta_6 = -4.590091$

In case fluid compositions are not available then using gas gravity, P_{pc} and T_{pc} can be calculated as follows:

$P_{pc} = 694.5 - 55.3\ \gamma_g$
$T_{pc} = 208.5 + 213.7\ \gamma_g$

12.2 Solution Gas Oil Ratio Correlations

R_s is a measure of the gas dissolved in the oil at any given condition. At initial condition, it is termed as R_{si}. Its units are standard cubic meters of gas per stock tank cubic meter of oil, sm^3/sm^3 or standard cubic feet per stock tank barrel (Scf/STB) or thousands of standard cubic feet per stock tank barrel oil

(Mcf/STB). The value of R_s is constant above the bubble point, and decreases with decreasing pressure below the bubble point as gas is released to become free gas. It can be zero at surface condition:

12.2.1 Standing

$$R_s\left(\left(\frac{p}{18.2}+1.4\right)\cdot\frac{10^{0.0125\gamma_{API}}}{10.00091\,T}\right)^{\frac{1}{0.83}}\cdot\gamma_g$$

12.2.2 Glaso

$$R_s=\left(\frac{x\cdot\gamma_{API}^{0.989}}{T^{0.172}}\right)^{\frac{1}{0.186}}\gamma_g$$

$$\log(x)=\frac{-b\pm\sqrt{b^2-4ac}}{2a}$$

where

$x = 10\log(x)$
$a = -0.30218$
$b = 1.7447$
$c = 1.7669 - \log(p)$

12.2.3 Vasquez and Beggs

$$R_s=C_1\gamma_g p^{C_2}\cdot e\left(C_3\left(\frac{\gamma_{API}}{T+459.67}\right)\right)$$

12.2.4 Al-Marhoun

$$R_s=\left(\frac{x}{\gamma_g^{-1.879109}\cdot\gamma_0^{3.04659}\cdot T^{1.302347}}\right)^{\frac{1}{0.722569}}$$

$$x=\frac{-b\pm\sqrt{b^2-4ac}}{2a}$$

where

$a = -2.278475\times10^{-9}$
$b = 7.02362\times10^{-3}$
$c = -64.13891 - p$

12.2.5 De Ghetto et al.

Heavy oil

$$R_S = \frac{\gamma_{g(p_{sp})} \cdot p^{1.2057}}{56.434} \times 10^{10.9267} \cdot \frac{\gamma_{API}}{T + 459.67}$$

Extra-heavy oil

$$R_S = \gamma_g \left(\frac{p}{10.7025} \times 10^{(0.0169 \cdot \gamma_{API} - 0.00156T)} \right)^{1.1128}$$

12.2.6 Hanafi et al.

$$R_s = -49.069 + 0.312p$$

$$R_s = 0 \text{ when } p \leq 157.28$$

12.2.7 Petrosky and Farshad

$$R_s = \left(\left(\frac{p}{112.727} + 12.34 \right) \cdot \gamma_g^{0.8439} \times 10^x \right)^{1.73184}$$

where

$$x = 7.916 \times 10^{-4} \gamma_{API}^{1.541} - 4.561 \times 10^{-5} \cdot T^{1.3911}$$

12.2.8 Velarde et al.

Saturated oil

$$R_s = \left(\frac{\left(\frac{p_b}{1091.47} \right)^{\frac{1}{5.354891}} + 0.740152}{\gamma_g^{0.161488} \times 10^x} \right)^{\frac{1}{0.081465}}$$

12.3 Formation Volume Factor

Oil formation volume factor (B_o)

Formation volume factor relates unit reservoir volume (*rb*) of oil to an unit surface volume (STB). The reservoir volume includes dissolved gas, while the surface volume is essentially dead oil and does not include the released gas. It is dimension less number. Initial formation volume factor B_{oi} increases up to the bubble-point pressure and then starts decreasing with further decrease in pressure. It can never be less than 1.0. Its inverse is called shrinkage and is equal to $1/B_o$.

Gas formation volume factor, B_g

Gas formation volume factor B_g is defined as the volume of gas under reservoir conditions divided by the volume of gas at surface under standard conditions.

Let *sc* represent standard conditions. Then, we have

$$B_g = \frac{V_{reservoir}}{V_{sc}} = \frac{(znRT/p)}{(z_{sc}nRT_{sc}/p_{sc})}$$

$$= 0.3699Tz/p \quad \text{m}^3/\text{m}^3 \quad \text{or} \quad 0.283Tz/p \quad \text{ft}^3/\text{ft}^3$$

where

$T_{sc} = 273°\text{K}$ or $460°\text{R}$

$p_{sc} = 101\,\text{kPa}$ or 14.6 psi

$Z_{sc} = 1.0$

To convert B_g from ft^3/ft^3 to bbl/Scf, divide the above B_g by 5.615 to give

$$B_g = 0.005034Tz/p \quad \text{bbl/Scf}$$

The best method to determine the formation volume factor of oil is its determination in the lab. Industry experts have developed correlation for its estimation in the absence of lab measured data. The most popular correlations developed using a large number of laboratory measured data from a number of fields from different parts of the world are described below.

12.3.1 Arp's Correlation

$$B_o = 1.05 + 0.0005\ R_s$$

This correlation was developed with limited data and provides a rough estimate of formation volume factor.

12.3.2 Standing Correlation

Saturated oil

$$B_o = 0.972 + 1.47 \times 10^{-4}\left(R_S\left(\frac{\gamma_g}{\gamma_o}\right)^{0.5} + 1.25 \cdot T\right)^{1.175}$$

Undersaturated oil

$$B_o = B_{ob} \cdot e\left(C_o\left(p_b - p\right)\right)$$

The oil compressibility used in this equation is obtained from the Vasquez and Beggs correlation.

12.3.3 Glaso

Saturated oil

$$\log\left(B_o - 1\right) = -6.58511 + 2.91329 \cdot \log(y) - 0.27683 \cdot \log(y)^2$$

$$y = R_s\left(\frac{\gamma_g}{\gamma_o}\right)^{0.526} + 0.968\,T$$

Undersaturated oil

$$B_o = B_{ob} \cdot e\left(C_o\left(p_b - p\right)\right)$$

12.3.4 Vasquez and Beggs

Saturated oil

$$B_o = 1 + A_1\,R_s + A_2\,(T - 60)\left(\frac{\gamma_{\mathrm{API}}}{\gamma_g}\right) + A_3\,R_s\,(T - 60)\left(\frac{\gamma_{\mathrm{API}}}{\gamma_g}\right)$$

Undersaturated oil

$$B_o = B_{ob} \cdot e\left(C_o\left(p_b - p\right)\right)$$

where
B_{ob} is the formation volume factor at the bubble-point pressure.

12.3.5 Al-Marhoun

Saturated oil

$$B_o = 0.574095 + 7.723532 \times 10^{-4} \cdot T + 2.454005 \times 10^{-3} \cdot y + 3.727676 \times 10^{-5} \cdot y^2$$

$$y = R_s^{0.501538} \cdot \gamma_g^{-0.145526} \cdot \gamma_o^{-5.220726}$$

Undersaturated oil

$$B_o = B_{ob} \cdot e\left(C_o\left(p_b - p\right)\right)$$

The oil compressibility used in this equation is obtained from the Vasquez and Beggs's correlation.

12.3.6 Petrosky–Farshad Correlation

$$B_o = 1.0113 + 7.2046\left(10^{-5}\right)\left[R_s^{0.3738}\left(\gamma_g^{0.2914}/\gamma_o^{0.6265}\right) + 0.24626(T-460)^{0.5371}\right]^{3.0936}$$

12.3.7 Omar and Todd's Correlation for Malaysian Crudes

$$B_{ob} = 0.972 + 0.000147\left[R_s\left(\gamma_g/\gamma_o\right)^{0.5} + 1.25\,T_f\right]^X$$

where

$$X = A_1 + A_2\left(\gamma_{\text{API}}/\gamma_g\right) + A_3\gamma_g$$
$$A_1 = 1.1663,$$
$$A_2 = -0.762\text{E}^{-3},$$
$$A_3 = -0.03990$$

12.3.8 Hanafi et al.

Saturated oil

$$B_o = 0.0006 \cdot R_s + 1.079$$

Undersaturated oil

$$B_o = B_{ob}e\left(c_o\left(p_b - p\right)\right)$$

12.3.9 De Ghetto et al.—Heavy Oil and Extra-Heavy Oil

Saturated oil

$$B_o = 1 + A_1 R_s + A_2(T-60)\left(\frac{\gamma_{\text{API}}}{\gamma_g}\right) + A_3 R_s (T-60)\left(\frac{\gamma_{\text{API}}}{\gamma_g}\right)$$

where

A_1, A_2, and A_3 are Vasquez and Beggs's constants for API ≤30°:
$A_1 = 4.677 \times 10^{-4}$

$A_2 = 1.751 \times 10^{-5}$
$A_3 = -1.811 \times 10^{-8}$

Undersaturated oil

$$B_o = B_{ob}\, e\left(c_o\left(p_b - p\right)\right)$$

Extra-heavy oil

$$B_o = 1 + A_1 R_s + A_2 (T-60)\left(\frac{\gamma_{\text{API}}}{\gamma_g}\right) + A_3 R_s (T-60)\left(\frac{\gamma_{\text{API}}}{\gamma_g}\right)$$

12.3.10 Petrosky and Farshad

Saturated oil

$$B_o = 1.0113 + 7.2046 \times 10^{-5} \cdot \left(R_s^{0.3738}\left(\frac{\gamma_g^{0.2914}}{\gamma_o^{0.6265}}\right) + 0.24626 \cdot T^{0.5371}\right)^{3.0936}$$

Undersaturated

$$B_o = B_{ob}\, e\left(c_o\left(p_b - p\right)\right)$$

Total oil formation volume factor (Bt)

12.4 Oil Compressibility Correlations

The oil compressibility, C_o, is defined as the rate of change of volume with pressure divided by volume at a constant temperature. It thus has units of inverse pressure. (For an ideal gas, C_g is just the inverse pressure.) The oil compressibility is not strongly dependent on pressure. Typical C_o values are between 3 and 5×10^{-6} Psi^{-1}.

12.4.1 Vasquez and Beggs

Saturated oil

$$c_o = \frac{-1433 + 5R_{sb} + 17.2T - 1180\gamma_g + 12.61\gamma_{\text{API}}}{10^5 p} + \frac{\dfrac{B_g}{B_o} \cdot \dfrac{dR_s}{dp}}{5.6145835}$$

Undersaturated

$$c_o = \frac{\left(-1433 + 5 \cdot R_{sb} + 17.2 \cdot T - 1180 \cdot \gamma_g + 12.61 \cdot \gamma_{\text{API}}\right)}{10^5 p}$$

Coefficient	$\gamma_o \leq 30°$ API	$\gamma_o > 30°$ API
A_1	4.677E-04	4.670E-04
A_2	1.751E-05	1.100E-05
A_3	−1.811E-08	1.337E-09
C_1	0.0362	0.0178
C_2	1.0937	1.1870
C_3	25.7240	23.9310

12.4.2 Hanafi et al.

Saturated oil

$$c_o = e^{\left(\frac{2.582}{\rho_o} - 0.99\right)} \times 10^{-6} + \frac{B_g}{5.6145835 \cdot B_o} \cdot \frac{dR_s}{dp}$$

Undersaturated oil

$$c_o = e\left(\left(\frac{2.582}{\rho_{ob}}\right) - 0.99\right) \times 10^{-6}$$

12.4.3 De Ghetto et al.

Saturated oil

$$c_o = \frac{-2841.8 + 2.9646 \cdot R_{sb} + 25.5439 \cdot T - 1230.5 \cdot \gamma_{g(p_{sp})} + 41.91 \cdot \gamma_{\text{API}}}{p \times 10^5}$$

$$+ \frac{B_g}{5.6145835 \cdot B_o} \cdot \frac{dR_s}{dp}$$

Undersaturated oil

$$c_o = \frac{-2841.8 + 2.9646 \cdot R_{sb} + 25.5439 \cdot T - 1230.5 \cdot \gamma_{g(p_{sp})} + 41.91 \cdot \gamma_{\text{API}}}{p \times 10^5}$$

Extra-heavy oil

Saturated oil

$$c_o = \frac{-889.6 + 3.1374 \cdot R_{sb} + 20 \cdot T - 627.3 \cdot \gamma_{g(p_{sb})} - 81.4476 \cdot \gamma_{\mathrm{API}}}{p \times 10^5}$$

$$+ \frac{B_g}{5.6145835} \cdot \frac{dR_s}{db}$$

Undersaturated oil

$$c_o = \frac{-889.6 + 3.1374 \cdot R_{sb} + 20 \cdot T - 627.3 \cdot \gamma_{g(p_{sb})} - 81.4476 \cdot \gamma_{\mathrm{API}}}{p \times 10^5}$$

12.4.4 Petrosky and Farshad

Saturated oil

$$c_o = 1.705 \times 10^{-7} \cdot R_{sb}^{0.69357} \cdot \gamma_g^{0.1885} \cdot \gamma_{\mathrm{API}}^{0.3272} \cdot T^{0.6729} \cdot p^{-0.5906} + \frac{B_g}{5.6145835 \cdot B_o} \cdot \frac{dR_s}{dp}$$

where dR_s/dp is from Vasquez and Beggs

Undersaturated oil

$$c_o = 1.705 \times 10^{-7} \cdot R_{sb}^{0.69357} \cdot \gamma_g^{0.1885} \cdot \gamma_{\mathrm{API}}^{0.3272} \cdot T^{0.6729} \cdot p^{-0.5906}$$

where

$$2.464 \times 10^{-5} < c_o < 3.507 \times 10^{-5}$$

Compressibility of gas (C_g)

Compressibility of gas C_g is different than gas deviation factor Z. Gas equation for real gases are expressed as $PV = nZRT$

$$V = nRT\,Z/p$$

Differentiating:

$$\frac{dV}{dP} = nRT \cdot \left(\frac{1}{p^2} \cdot \left(p \cdot \frac{dz}{dp} - z\right)\right)$$

Or

$$\frac{1}{V}\frac{dV}{dP} = \frac{1}{z} \cdot \frac{dz}{dp} - \frac{1}{p}$$

$$C_g = -\frac{1}{V}\frac{dV}{dP} = \frac{1}{p} - \frac{1}{z} \cdot \frac{dz}{dp}$$

for ideal gas: $z = 1.0$ (const.) i.e., $C_g = \dfrac{1}{p}$

12.5 Viscosity Correlations

Viscosity is the measure of resistance to flow. Oil viscosity increases when reservoir pressure decreases below the bubble-point pressure due to release of solution gas and above the bubble point, the liquid molecules are forced more closed together and oil viscosity increases as the pressure increases. Oil viscosity is minimum at the bubble-point pressure. The oil viscosity is dependent on temperature and drops exponentially with an increase in the temperature.

There are two common units of viscosity:

a. dynamic viscosity (μ)
b. kinematic viscosity (ν)

The dynamic viscosity, μ, has units of centipoise (cP) or Pascal-seconds (Pa-s). Multiply cP by 0.001 to obtain Pa-s. Poise is gm/cm/s. The kinematic viscosity, ν, has units of m^2/s. A Stoke is defined as cm^2/s, and thus kinematic viscosity is often expressed as centistokes (cs). (Water kinematic viscosity is about 0.01 cm^2/s or 1 cs.) To convert dynamic viscosity into kinematic viscosity, divide it by density, ρ, and thus we have

$$\nu = \mu / \rho$$

12.5.1 Beggs and Robinson

Range 16–58° API and 70°F–295°F. This correlation overstates the viscosity of the crude if used below temperature 100°F–150°F.

$$\mu_{od} = 10^x - 1$$

where

$$x = yT^{-1.163}$$

$$y = 10^z$$

$$z = 3.0324 - 0.02023\,G$$

12.5.2 De Ghetto et al.

Deadoil

$$\log(\mu_{od}+1) = 2.06492 - 0.0179\cdot\gamma_{\text{API}} - 0.70226\cdot\log(T)$$

Saturated oil

$$\mu_{os} = -0.6311 + 1.078\cdot x - 0.003653\cdot x^2$$

$$x = \left(0.2478 + 0.6144\cdot 10^{-0.000845 R_s}\right)\cdot \mu_{od}{}^{(0.4731+0.5158)}$$

$$y = 10^{-0.00081 R_s}$$

Undersaturated oil

$$\mu_o = 0.9886\cdot\mu_{os} + 0.002763(p - p_b)\left(-0.01153\cdot\mu_{os}{}^{1.7933} + 0.0316\cdot\mu_{os}{}^{1.5939}\right)$$

Extra-heavy oils (API < 10°)
Dead oil

$$\log(\mu_{od}+1) = 1.090296 - 0.012619\cdot\gamma_{\text{API}} - 0.61748\cdot\log(T)$$

Saturated oil

$$\mu_{os} = 2.3945 + 0.8927\cdot x + 0.001567\cdot x^2$$

$$x = \left(-0.0335 + 1.0785\cdot 10^{-0.000845 R_s}\right)\cdot \mu_{od}{}^{(0.5798+0.3432\,y)}$$

$$y = 10^{-0.00081 R_s}$$

Undersaturated oil

$$\mu_o = \mu_{os} - \left(1 - \frac{p}{p_b}\right)\left(\frac{10^{-2.19}\cdot\mu_{od}{}^{1.055}\cdot p_b{}^{0.3132}}{10^{0.0099\gamma_O}}\right)$$

12.5.3 Hanafi et al.

$$\mu_o = e\left(7.296\cdot\rho_o^3 - 3.095\right)$$

12.5.4 Khan et al. (Saudi Arabian Oil)

Oil viscosity (API < 10°)

$P = p_b$

$$\mu_{ob} = \frac{0.09\sqrt{\gamma_g}}{\sqrt[3]{R_s} \cdot \theta_\gamma^{4.5}(1-\gamma_o)^3}$$

where

$$\theta_r = \frac{T + 459.67}{459.67} = \text{relative temperature}$$

$p > p_b$

$$\mu_o = \mu_{ob}\, e\left(9.6\times10^{-5}(p - p_b)\right)$$

$p < p_b$

$$\mu_{os} = \mu_{ob}\left(\frac{p}{p_b}\right)^{-0.14} e\left(-2.5\times10^{-4}(p - p_b)\right)$$

12.5.5 Ng and Egbogah

Dead oil

$$\log = (\mu_{od} + 1) = -1.7095 - 0.0087917 \cdot T_{pp} + 2.7523 \cdot \gamma_o$$
$$+\left(-1.2943 + 0.0033214 \cdot T_{pp} + 0.9581957 \cdot \gamma_o\right) \cdot \log\left(T - T_{pp}\right)$$

$-50°C < T_{pp} < 15°C$

Saturated

$$\mu_{os} = x \cdot \mu_{od}^y$$
$$x = 10.715(R_s + 100)^{-0.515}$$
$$y = 5.44(R_s + 150) - 0.338$$

where μ_{od} is defined using the modified Beggs and Robinson's correlation.

$$\log(\mu_{od} + 1) = 1.8653 - 0.025086 \cdot \gamma_o - 0.5644 \cdot \log(T)$$

Undersaturated

$$\mu_o = \mu_{ob}\left(\frac{p}{p_b}\right)^x$$
$$x = 2.6p^{1.187}\exp\left(-11.513 - 8.98\times10^{-5}p\right)$$

12.5.6 Vasquez–Beggs correlation

$$\mu_o = \mu_{ob}\left(p/pb\right)^m$$

where

$$m = 2.6p^{1.187} \times 10^a, a = -3.9 \times 10^{-5} \times p - 5$$

12.5.7 Gas Viscosity, μ_g

Gas viscosities increase with pressure and temperature. Typical gas viscosities are 0.01–0.05 cP. For a given pressure gradient, gas will flow 100,000 times faster than a heavy oil with a viscosity of 1,000 cP. Gas viscosities are often measured with a vibrating wire viscometer

12.5.7.1 Lee–Gonzalez–Eakin

$$\mu_g = 10^{-4} K \exp\left[X\left(\rho_g/62.4\right)^Y\right]$$

$$\rho_g = \text{gas density at reservoir } T \;\&\; P$$

$$K = \left(9.4 + 0.02\,\text{Ma}\right)T^{1.5}/\left(209 + 19\,\text{Ma} + T\right)$$

$$X = 3.5 + 986/T \; + 0.01\text{ Ma}$$

$$Y = 2.4 - 0.2X$$

12.5.8 Water Viscosity by Meehan

$$\mu_w = \mu_{wD}\left[1 \; + 3.5 \times 10^{-2} \times p^2 \left(T - 40\right)\right]$$

$$\mu_{wD} = A \; + B/T$$

$$A = 4.518 \times 10^{-2} + 9.313 \times 10^{-7} \times Y - 3.93 \times 10^{-12} Y^2,$$

$$B = 70.634 + 9.576 \times 10^{-10} Y^2$$

$$Y = \text{water salinity}$$

12.5.9 Beal

$$\mu_{od} = \left(0.32 + \frac{1.8 \times 10^7}{{}^\circ API^{4.53}}\right)\left(\frac{360}{T + 200}\right)^a$$

where

$$a = \text{antilog}\left(0.43 + 833/^{\circ}\,\text{API}\right)$$

12.6 Density and Gravity

Density (ρ) is defined as ratio of unit mass over unit volume. The oil density increases with decrease in pressure. Oil density has units of gm/cm^3 or kg/m^3 and is commonly expressed in °API.

API specific gravity

Another common measure of oil specific gravity is defined by $\gamma_{API} = (141.5/\gamma_O) - 131.5$, with units in °API.

Gas specific gravity, γ_g

The ratio of density of any gas at standard conditions (14.7 psia and 60 °F) to the density of air at standard conditions is based on the ideal gas law ($PV = nRT$).

This is also defined as the ratio of the molecular weight of the gas to the molecular weight of air (28.97). Gas is measured in moles, which is the weight of the gas divided by its molecular weight. The molecular weight of a gas of density ρ_g relative to air is 28.97 ρ_g.

12.6.1 Hanafi et al.

Saturated oil

$$\rho_o\left(2.366 - \left(\frac{1.358}{B_o}\right)\right)^{-1}$$

Undersaturated oil

$$\rho_o = \rho_{ob}\, e\left(c_o\left(p - p_b\right)\right)$$

Oil density ρ_o

$$\rho_o = 64.4\gamma_o + 0.0136 R_s \gamma_g / B_o$$

At pressures above P_{sat}

$$\mu_{od} = \left(0.32 + \frac{1.8 \times 10^7}{^{\circ}\text{API}^{4.53}}\right)\left(\frac{360}{T + 200}\right)^a$$

where

$$a = \text{antilog}\left(0.43 + \frac{8.33}{°\text{API}}\right)$$

$$\rho_o = \rho_{ob} \exp\left[c_o \left(p - p_b\right)\right]$$

c_o can be derived from Petrosky and Farshad's correlation.

Gas compressibility factor Z:
Gas deviation factor can be estimated in the lab using the following principal gas law:

$PV = nRT$ ideal gas
$PV = nZRT$ Real gas

Z is a function of both P and T. Commonly $Z = Z(p)$ i.e., assume conditions of isothermal depletion. For measurement of Z, single-phase reservoir gas sample is collected and recombination of separator gas and the produced liquid or in the case of solution gas reservoirs, sampling the liberated gas.

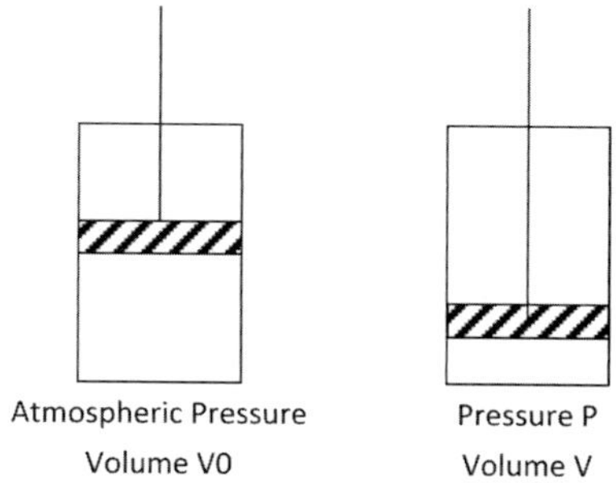

Thus, varying p and measuring V, $Z(p)$ at constant T can be obtained.
A graphical method has been suggested by Standing and Katz to calculate the Z-factor which uses the gas composition, their mole percentage, critical pressure (P_{pc}), and critical temperature (T_{pc}) data of individual components. Dimensionless parameters pseudo-reduced pressure (P_{pr}) and dimensionless pseudo-reduced temperature (T_{pr}) are calculated as follows:

$$P_{pc} = \sum_i n_i \cdot P_{c_i} : P_c = \text{critical pressure of } i$$

$$T_{pc} = \sum_i n_i \cdot T_c : T_c = \text{critical temperature of } i$$

$$P_{pr} = \frac{P}{P_{pc}} \text{ and } T_{pr} = \frac{T}{T_{pc}}, \text{at desired conditions } P \,\&\, T$$

where

P is pressure, psia

P_{pr} pseudo-reduced pressure, dimensionless

T is temperature of reservoir degree R

T_{pr} is pseudo-reduced temperature

P_{pc} and T_{pc} critical pressure and temperature and they can be defined as weighted sum of individual components

Generally T_{pr} is constant (isothermal depletion). Use Standing–Katz chart to retrieve Z as a function of P_{pr} at various T_{pr} (Figure 12.1).

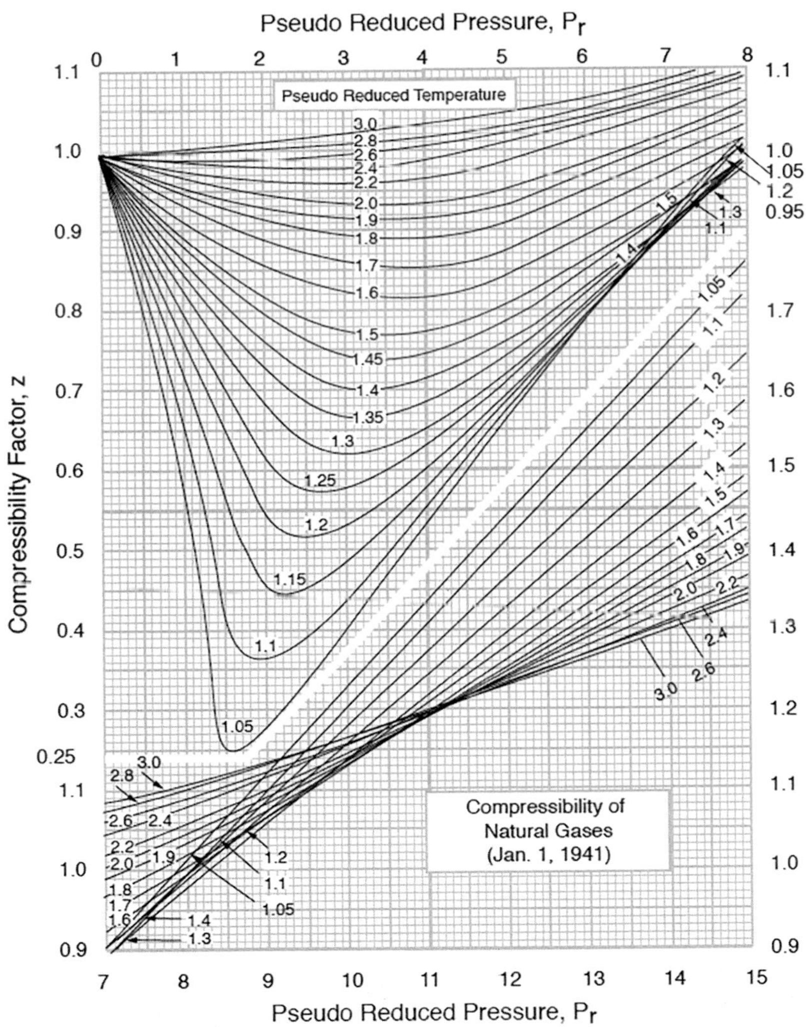

FIGURE 12.1
Compressibility factor of gas (Standing and Katz graphical method).

The pseudocritical properties of gases are generally computed using compositional data but can also be estimated from the gas gravity using the correlation presented in following equations

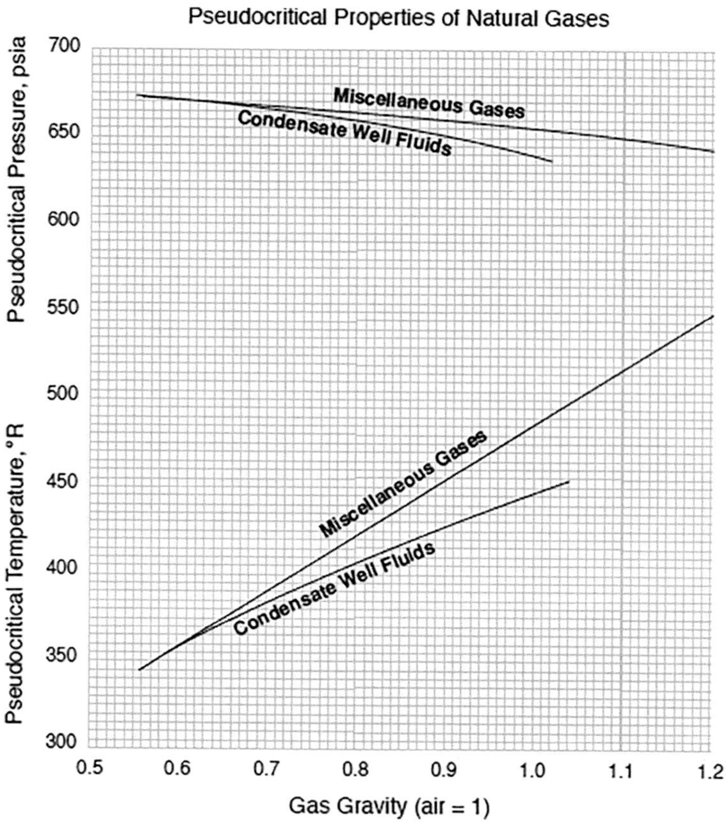

$$P_{pc} = 756.8 - 131.0\gamma_g - 3.6\gamma_g^2$$

$$T_{pc} = 169.2 + 349.5\gamma_g - 74.0\gamma_g^2$$

Correlation good for $0.57 \le \gamma_g \le 1.68$

All the above computations are valid for non-hydrocarbon impurities up to 5% in volume. Therefore, experimental determination of *Z*-factor becomes important.

Direct calculations of *Z*-factors were proposed by the following authors and are widely used:

- Hall Yarborough
- Dranchuk and Abu Kassem
- Dranchuk–Purvis and Robinson

Papay

$$Z = A = P_{pr}/T_{pr}\left[0.3648758 - 0.04188423\left(\frac{P_{pr}}{P_{pr}}\right)\right]$$

12.7 Minimum Miscibility Pressure

An ideal way of determining the minimum miscibility pressure (MMP) is using laboratory experiments. However, in absence of laboratory measurements number of correlations have been used to generate the MMP. These correlations use fluid characterization properties. Some of the main correlations are described as follows:

Firozabadi A and Aziz K

$$P_m = 9433 - 188\times 10^3\left((C_{c2} - C_{c5})/M_{c7+}T^{0.25}\right)$$
$$+ 1430\times 10^3\left((C_{c2} - C_{c5})/M_{c7+}T^{0.25}\right)^2$$

P_m = MMP
T temperature
$C_{c2} - C_{c5}$ concentration of intermediates in the reservoir including CO_2 and H_2S
M_{c7+} = molecular weight of C_{7+} of the reservoir oil

Connard

Connard suggested that MMP calculated by Peng–Robinson EOS agrees close to the vicinity of bubble-point pressure and thus leads to the following correlation:

$$P_m = 0.6909\, P_{m^*} + 0.3091 P_b$$

where
P_m = effective MMP
P_{m^*} = calculated miscibility pressure
P_b = bubble-point pressure

Glaso:

Glaso suggested the minimum miscibility correlation for crude oil with degree API more than 45 with N_2 as injectant:

$$P_m = 725 - 3.85x - 9.22y + 0.63z$$

and

$$P_m = 8788,8 - 9.17x - 1.17y - 1.25z \text{ when API is more than 45.}$$

where

P_m = MMP
X = concentration on C1 in reservoir
Y = concentration of C2–C5 in the reservoir
Z = reservoir temperature

Hudgins et al.

$$P_m = 5568\ e^{-R1} + 3641\ e^{-R2}$$
$$R1 = 792.06(C2 - C5)/M_{c7+}T^{0.25}$$
$$R2 = 2.158 \times 10^6 C_1^{5.6}\ 12\ 792/M_{c7+}T^{0.25}$$

12.8 Uncertainty in PVT Data

Certain degree of uncertainty exists in laboratory measurement. Also deviation between laboratory results and their matching in EOS matching is observed. A general trend of deviation is as follows:

Black oil PVT

Bubble point	±2%
GOR	±4%
Bo	±1%
Viscosity	±17%
Composition	±8%

Gas condensate PVT

Dew point	±5%
Liquid dropout	±12%
Z	±2%
CGR	± 6%
Bg	±2%
Compositions	±6%

Formation water PVT

Bw	±2%

Recovery factor estimation:

API has developed correlations for estimating the recovery efficiency using large number filed data for water-drive and depletion-drive reservoirs. These correlations are handy in the estimation of recovery of filed with limited data and few assumptions like abandonment pressure. (Arps et al.)

Water-drive reservoir

$$RE(\%\text{OOIP}) = 54.898 * \left[\frac{\phi(1-S_{wi})}{B_{oi}}\right]^{0.0422} * \left[\frac{K\mu_{wi}}{\mu_{oi}}\right]^{0.0770} * (S_{wi})^{-0.1903} * \left(\frac{P_i}{P_a}\right)^{-0.2159}$$

With $r^2 = 0.92$ & standard error of estimate = ±32% @ 95% confidence level.

Solution gas-drive reservoirs

$$RE(\%\text{OOIP}) = 41.815 * \left[\frac{\phi(1-S_{wi})}{B_{ob}}\right]^{0.1611} * \left[\frac{K}{\mu_{ob}}\right]^{0.0979} * (S_{wi})^{-0.3722} * \left(\frac{P_i}{P_a}\right)^{-0.1741}$$

With $r^2 = 0.87$ & standard error of estimate = ±40% @ 95% confidence level.

Water drive in carbonate reservoirs

$$RE(\%\text{OOIP}) = 52.70 * \left[\frac{\phi(1-S_{wi})}{B_{oi}}\right]^{0.4068} * \left[\frac{K\mu_{oi}}{\mu_{wi}}\right]^{0.0337} * (S_{wi})^{-0.04096}$$

with $r^2 = 0.58$ & standard error of estimate = ±20% @ 95% confidence level.

Russian technical literature used statistical data to establish an equation to estimate the recovery efficiency.

$$RE = 0.333 - 0.0089(\mu_o/\mu_w) + 0.121\ \log(k) + 0.0013t + 0.0038h$$
$$+ 0.149kp - 0.085Q^* + 0.173S_o - 0.00053S$$

where

μ_o and μ_w are viscosities of oil and water in cP
K is average permeability of rock in Darcy
h is net pay thickness in m
K_p net pay and gross thickness ratio
Q ratio of oil in place above bottom water to the oil in place
S_o oil saturation in %
S drainage area of the well ha, (Total area divided by number of wells)

Surgucsev et al.

Surgucsev et al. developed RE correlation for fractured porous vuggy carbonate reservoirs produced by water flooding:

$$RE = 1/\Phi\ S_o\left\{0.44\Phi - 0.0101\log(\mu_o/\mu_w) - 0.236SS^*/h\right\}$$

where

Φ is porosity
S and S^*– the average and effective drainage area of the well in ha
S = oil saturated area of the reservoir/number of wells
S^* = exploited area/number of wells
h is thickness
μ_o and μ_w viscosity of oil and water

Gas recovery factor

Gas recovery can be estimated in volumetric gas reservoirs using formation volume factor of gas at initial and abandonment pressure:

$$Gp/G = \left(1 - B_{gi}/B_{ga}\right)$$

Interfacial tension

Firoozabady and Ramey

Gas–oil

$$\sigma_{od} = \left(1.17013 - 1.694 \times 10^{-3} T\right)\left(38.085 - 0.259\gamma_{\text{API}}\right)$$

Oil–water

$$\sigma_{hw} = \left[\frac{1.58(\rho_w - \rho_h) + 1.76}{T_\gamma^{03125}}\right]^4$$

where

T_r is reduced temperature
Pseudo critical temperature of dead oil

$$T_{co} = 24.2787 K_w^{1.76544} \gamma_o^{2.12504}$$

Pseudo critical temperature of gas

$$T_{cg} = 169.2 + 349.5\gamma_{ghc} - 74.0\gamma_{ghc}^2$$

Pseudo critical temperature of live oil

$$T_{cm} = x_o T_{co} + x_g T_{cg}$$

Hydrate formation correlations

The authors have developed correlations for predicting the hydrate formation. These correlations have their own limitations and are not precise but they can provide a reasonable value to plan the control and mitigation.

Makogan

$$\log P = \beta + 0.0497\ \left(t + kt^2\right) - 1$$

where P is pressure in MPa and t is temperature in °C. Elgibaly and Elkamal developed correlation to calculated the β and k as follows:

$$\begin{aligned} &\mathrm{B} = 2.681 - 3.811 \\ &\gamma = 1.679\gamma^2 \\ &k = -0.006 + 0.011\gamma + 0.011\gamma^2 \\ &\gamma \text{ is gas gravity} = MW/28.96 \end{aligned}$$

Kobayashi et al.

Kobayashi et al. developed an empirical equation that predicts the hydrate-forming temperatures at given pressures for systems including only hydrocarbons in limited range of temperatures, pressures, and gas specific gravities:

$$\begin{aligned} T = l/\Big[& A_1 + A_2\left(\ln\gamma_g\right) + A_3\left(\ln P\right) + A_4\left(\ln\gamma_g\right)^2 + A_5\left(\ln\gamma_g\right)\left(\ln P\right) \\ & + A_6\left(\ln P\right)^2 + A_7\left(\ln\gamma_g\right)^3 + A_8\left(\ln\gamma_g\right)^2\left(\ln P\right) \\ & + A_9\left(\ln\gamma_g\right)\left(\ln P\right)^2 + A_{10}\left(\ln P\right)^3 + A_{11}\left(\ln\gamma\right)^4 \\ & + A_{12}\left(\ln\gamma_g\right)^3\left(\ln P\right) + A_{13}\left(\ln\gamma_g\right)^2\left(\ln P\right)^2 \\ & + A_{14}\left(\ln\gamma_g\right)\left(\ln P\right)^3 + A_{15}\left(\ln P\right)^4\Big] \end{aligned} \tag{12.1}$$

Berge correlation

The equations developed by Berge are temperature explicit, i.e., temperature is calculated directly for a given pressure and specific gravity of the gas. The equations for predicting hydrate temperatures are as follows:

$$\begin{aligned} &\text{For } 0.555 \le \gamma_g \le 0.58: \\ &T = -96.03 + 25.37 \times \ln P - 0.64 \times \left(\ln P\right)^2 \\ &\qquad + \frac{\gamma_g - 0.555}{0.025} \times \left[80.61 \times P + 1.16 \times \frac{10^4}{(P + 599.16)}\right] \\ &\qquad - \left(-96.03 + 25.37 \times \ln P - 0.64 \times \left(\ln P\right)^2\right) \end{aligned}$$

For $0.58 \le \gamma_g \le 1.0$

$$T = \frac{\left\{80.61 \times P - 2.1 \times 10^4 - 122 \times \frac{10^3}{\gamma_g - 0.535} - \left[1.23 \times 10^4 + 1.71 \times \frac{10^3}{\gamma_g - 0.509}\right]\right\}}{\left[P - \left(-260.42 - \frac{15.18}{\gamma_g - 0.535}\right)\right]} \tag{12.2}$$

Towler and Mokhtab developed hydrate formation temperature as a function of gas gravity and pressure:

$$T = 13.47 \ln(p) + 34.27 \ln(\gamma) - 1.675 \ln(p) \ln(\gamma) - 20.35$$

References

1. Arps, J.J. et al. 1967. A statistical analysis of recovery efficiency API bull d14 Oct 1967a review of bubblepoint pressure and oil formation volume factor correlations. *SPE Reservoir Evaluation & Engineering* 4(2): 146–160. SPE-71302-PA.
2. Abdul-Majeed, G.H. and Abu Al-Soof, N.B. 2000. Estimation of gas–oil surface tension. *Journal of Petroleum Science and Engineering* 27(3–4): 197–200.
3. Ahmed, T. 1989. *Hydrocarbon Phase Behavior.* Houston, TX: Gulf Publishing Co.
4. Al-Marhoun, M.A. 1992. New correlations for formation volume factors of oil and gas mixtures. *Journal of Canadian Petroleum Technology* 31(3): 22. PETSOC-92–03–02.
5. Beal, C. 1970. The viscosity of air, water, natural gas, crude oil and its associated gases at oil field temperatures and pressures, No. 3, 114–127. Richardson, TX: Reprint Series (Oil and Gas Property Evaluation and Reserve Estimates), SPE.
6. Beggs, H.D. and Robinson, J.R. 1975. Estimating the viscosity of crude oil systems. *Journal of Petroleum Technology* 27(9): 1140–1141. SPE-5434-PA.
7. Carr, N.L., Kobayashi, R., and Burrows, D.B. 1954. Viscosity of hydrocarbon gases under pressure. *Journal of Petroleum Technology* 6(10): 47–55. SPE–297–G.
8. Chew, J. and Connally Jr., C.A. 1959. A viscosity correlation for gas-saturated crude oils. *Transactions of the American Institute of Mining, Metallurgical, and Petroleum Engineers* 216: 23.
9. Connard, P.G. 1987. Discussion and correlation of nitrogen and lean gas miscibility pressure, *SPERE*, 258–259.
10. Dindoruk, B. and Christman, P.G. 2001. PVT properties and viscosity correlations for gulf of Mexico oils. *Presented at the SPE Annual Technical Conference and Exhibition, New Orleans*, 30 September-3 October. SPE-71633-MS.
11. Firoozabadi, A. and Aziz, K. 1986. Analysis and correlation of nitrogen and lean gas miscibility pressure. *SPERE* 1: 572–582.

12. Glasø, Ø. 1980. Generalized pressure-volume-temperature correlations. *Journal of Petroleum Technology* 32(5): 785–795. SPE-8016-PA.
13. Gomzikov, V.K. Development of petroleum reservoirs by Josef Papay Akademiai Budapest.
14. Hudgins, D.A.et al. Nitrogen miscible displacement of light crude oil: Alaboratory study, April 17–20 Tulsa.
15. Katz, D.L. and Lee, R.L. 1990. *Natural Gas Engineering—Production and Storage.* New York: McGraw-Hill.
16. Kay, W.B. 1936. Density of hydrocarbon gases at high temperature and pressure. *Industrial & Engineering Chemistry* 28, 1014–1019.
17. Kobayasi R et al: *Petroleum Engineering Handbook,* Edited by H.B. Bradley. Richardson, TX: SPE.
18. Lasater, J.A. 1958. Bubble point pressure correlations. *Journal of Petroleum Technology* 10(5): 65–67. SPE-957-G. Al-Shammasi, A.A. 2001.
19. Lee, A.L., Gonzalez, M.H., and Eakin, B.E. 1966. The viscosity of natural gases. *Journal of Petroleum Technology* 18(8): 997–1000. SPE-1340-PA.
20. Makogan, Y.F. 1981. *Hydrate of Hydrocarbon*. Tulsa, OK: Pennwell.
21. McCain Jr., W.D. 1990. *The Properties of Petroleum Fluids.* Tulsa, OK: PennWell.
22. Pedersen, K.S., Fredenslund, A., and Thomassen, P. 1989. *Properties of Oils and Natural Gases.* Houston, TX: Gulf Publishing Co.
23. Petrosky Jr., G.E. 1990. PVT correlations for Gulf of Mexico crude oils. *MS thesis,* University of Southwestern Louisiana, Lafayette, Louisiana.
24. Petrosky Jr., G.E. and Farshad, F. 1998. Pressure-volume-temperature correlations for Gulf of Mexico crude oils. *SPE Reservoir Evaluation & Engineering* 1(5): 416–420. SPE-51395-PA.
25. Piper, L.D., McCain Jr., W.D., and Corredor, J.H. Compressibility factors for naturally occurring petroleum gases. Paper SPE 26668 presented at the 1993*SPE Annual Technical Conference and Exhibition, Houston,* October 3–6.
26. Standing, M.B. 1981. *Volumetric and Phase Behavior of Oil Field Hydrocarbon Systems* (9thEdition). Richardson, TX: Society of Petroleum Engineers of AIME.
27. Stewart, W.F., Burkhardt, S.F., and Voo, D. Prediction of pseudocritical parameters for mixtures. Presented at the 1959 *AIChE Meeting, Kansas City, MO,* May 18.
28. Sutton, R.P. Compressibility factors for high-molecular-weight reservoir gases. Paper SPE 14265 presented at the 1985 *SPE Annual Technical Conference and Exhibition, Las Vegas, NV,* September 22–25.
29. Towler, B.F. and Mokhtab, S. 2005. Quickly estimation hydrate formation temperature in natural gas, hydrological process.
30. Vazquez, M. and Beggs, H.D. 1980. Correlations for fluid physical property prediction. *Journal of Petroleum Technology* 32(6): 968–970. SPE-6719-PA.
31. Vazquez, M.E. 1976. Correlations for fluid physical property prediction. *MS thesis,* University of Tulsa, Tulsa, Oklahoma.
32. Velarde, J., Blasingame, T.A., and McCainJr., W.D. 1997. Correlation of black oil properties at pressures below bubble point pressure—a new approach. Presented at the *Annual Technical Meeting of CIM, Calgary, AB,* June 8–11. PETSOC–97–93.
33. Whitson, C.H. and Brulé, M.R. 2000. *Phase Behavior,* vol. 20. Henry L. Doherty Monograph Series. Richardson, TX: SPE.

INDEX

D

R

S